T0180929

Advances in Intelligent Systems and Computing

Volume 759

Series editor

Janusz Kacprzyk, Polish Academy of Sciences, Warsaw, Poland
e-mail: kacprzyk@ibspan.waw.pl

The series "Advances in Intelligent Systems and Computing" contains publications on theory, applications, and design methods of Intelligent Systems and Intelligent Computing. Virtually all disciplines such as engineering, natural sciences, computer and information science, ICT, economics, business, e-commerce, environment, healthcare, life science are covered. The list of topics spans all the areas of modern intelligent systems and computing such as: computational intelligence, soft computing including neural networks, fuzzy systems, evolutionary computing and the fusion of these paradigms, social intelligence, ambient intelligence, computational neuroscience, artificial life, virtual worlds and society, cognitive science and systems, Perception and Vision, DNA and immune based systems, self-organizing and adaptive systems, e-Learning and teaching, human-centered and human-centric computing, recommender systems, intelligent control, robotics and mechatronics including human-machine teaming, knowledge-based paradigms, learning paradigms, machine ethics, intelligent data analysis, knowledge management, intelligent agents, intelligent decision making and support, intelligent network security, trust management, interactive entertainment, Web intelligence and multimedia.

The publications within "Advances in Intelligent Systems and Computing" are primarily proceedings of important conferences, symposia and congresses. They cover significant recent developments in the field, both of a foundational and applicable character. An important characteristic feature of the series is the short publication time and world-wide distribution. This permits a rapid and broad dissemination of research results.

More information about this series at http://www.springer.com/series/11156

Sanjiv K. Bhatia · Shailesh Tiwari
Krishn K. Mishra · Munesh C. Trivedi
Editors

Advances in Computer Communication and Computational Sciences

Proceedings of IC4S 2017, Volume 1

 Springer

Editors
Sanjiv K. Bhatia
Department of Computer Science
University of Missouri
Columbia, MO
USA

Shailesh Tiwari
Department of Computer Science
and Engineering
ABES Engineering College
Ghaziabad, Uttar Pradesh
India

Krishn K. Mishra
Department of Computer Science
and Engineering
Motilal Nehru National Institute
of Technology Allahabad
Allahabad, Uttar Pradesh
India

Munesh C. Trivedi
Department of Computer Science
and Engineering
ABES Engineering College
Ghaziabad, Uttar Pradesh
India

ISSN 2194-5357   ISSN 2194-5365   (electronic)
Advances in Intelligent Systems and Computing
ISBN 978-981-13-0340-1   ISBN 978-981-13-0341-8   (eBook)
https://doi.org/10.1007/978-981-13-0341-8

Library of Congress Control Number: 2018940399

This Springer imprint is published by the registered company Springer Nature Singapore Pte Ltd.
The registered company address is: 152 Beach Road, #21-01/04 Gateway East, Singapore 189721, Singapore

Preface

IC4S is a major multidisciplinary conference organized with the objective of bringing together researchers, developers, and practitioners from academia and industry working in all areas of computer and computational sciences. It is organized specifically to help computer industry to derive the advances in next-generation computer and communication technology. Researchers invited to speak will present the latest developments and technical solutions.

Technological developments all over the world are dependent upon globalization of various research activities. Exchange of information and innovative ideas is necessary to accelerate the development of technology. Keeping this ideology in preference, the International Conference on Computer, Communication and Computational Sciences (IC4S 2017) has been organized at Swissôtel Resort Phuket Patong Beach, Thailand, during October 11–12, 2017.

This is the second time the International Conference on Computer, Communication and Computational Sciences has been organized with a foreseen objective of enhancing the research activities at a large scale. Technical Program Committee and Advisory Board of IC4S include eminent academicians, researchers, and practitioners from abroad as well as from all over the nation.

In this book, selected manuscripts have been subdivided into various tracks—Intelligent Hardware and Software Design, Advanced Communications, Intelligent Computing Techniques, Web and Informatics, and Intelligent Image Processing. All sincere efforts have been made to make it an immense source of knowledge for all and include 85 manuscripts. The selected manuscripts have gone through a rigorous review process and are revised by authors after incorporating the suggestions of the reviewers. These manuscripts have been presented at IC4S 2017 in six different technical sessions. A gift voucher worth EUR 150 (one hundred and fifty euros) for session-wise Best Paper Presentation of paper has been awarded by Springer at IC4S 2017 through which authors can select and buy an e-book from Springer Link.

IC4S 2017 received around 425 submissions from around 650 authors of 15 different countries such as USA, Iceland, China, Saudi Arabia, South Africa, Taiwan, Malaysia. Each submission has gone through the plagiarism check. On the basis of plagiarism report, each submission was rigorously reviewed by at least two reviewers with an average of 2.3 per reviewer. Even some submissions have more than two reviews. On the basis of these reviews, 85 high-quality papers were selected for publication in this proceedings volume, with an acceptance rate of 20%.

We are thankful to the keynote speakers—Prof. Phalguni Gupta, IIT Kanpur, India; Prof. Jong-Myon Kim, University of Ulsan, Ulsan, Republic of Korea; Dr. Nitin Singh, MNNIT Allahabad, India; Mr. Aninda Bose, Senior Editor, Hard Sciences, Springer Nature, India; and Dr. Brajesh Kumar Singh, RBS College, Agra, India, to enlighten the participants with their knowledge and insights. We are also thankful to the delegates and the authors for their participation and their interest in IC4S 2017 as a platform to share their ideas and innovation. We are also thankful to Prof. Dr. Janusz Kacprzyk, Series Editor, AISC, Springer, for providing continuous guidance and support. Also, we extend our heartfelt gratitude to the reviewers and Technical Program Committee members for showing their concern and efforts in the review process. We are indeed thankful to everyone directly or indirectly associated with the conference organizing team, leading it toward the success.

Although utmost care has been taken in compilation and editing, a few errors may still occur. We request the participants to bear with such errors and lapses (if any). We wish you all the best.

Columbia, USA   Sanjiv K. Bhatia
Ghaziabad, India   Shailesh Tiwari
Allahabad, India   Krishn K. Mishra
Ghaziabad, India   Munesh C. Trivedi

Contents

About the Editors

Dr. Sanjiv K. Bhatia received his Ph.D. in Computer Science from the University of Nebraska, Lincoln, in 1991. He presently works as Professor and Graduate Director (Computer Science) at the University of Missouri, St. Louis, USA. His primary areas of research include image databases, digital image processing, and computer vision. He has published over 40 articles in those areas. He has also consulted extensively with industry for commercial and military applications of computer vision. He is an expert in system programming and has worked on real-time and embedded applications. He serves on the organizing committee of a number of conferences and on the editorial board of international journals. He has taught a broad range of courses in computer science and was the recipient of Chancellor's Award for Excellence in Teaching in 2015. He is a senior member of ACM.

Dr. Shailesh Tiwari is currently working as a professor in Computer Science and Engineering Department, ABES Engineering College, Ghaziabad, India. He is also administratively heading the department. He is an alumnus of Motilal Nehru National Institute of Technology Allahabad, India. He has more than 15 years of experience in teaching, research, and academic administration. His primary areas of research are software testing, implementation of optimization algorithms, and machine learning techniques in software engineering. He has also published more than 40 publications in international journals and in proceedings of international conferences of repute. He has served as a program committee member of several conferences and edited Scopus and E-SCI-indexed journals. He has also organized several international conferences under the banner of IEEE and Springer. He is a senior member of IEEE, a member of IEEE Computer Society, and an executive committee member of IEEE UP Section. He is a member of reviewer and editorial board of several international journals and conferences.

Dr. Krishn K. Mishra is currently working as a visiting faculty, Mathematics and Computer Science Department, University of Missouri, St. Louis, USA. He is an alumnus of Motilal Nehru National Institute of Technology Allahabad, India, which is also his base working institute. His primary areas of research include evolutionary algorithms, optimization techniques, and design and analysis of algorithms. He has also published more than 50 publications in international journals and in proceedings of international conferences of repute. He is serving as a program committee member of several conferences and also editing few Scopus and SCI-indexed journals. He has 15 years of teaching and research experience during which he made all his efforts to bridge the gaps between teaching and research.

Dr. Munesh C. Trivedi is currently working as a professor in Computer Science and Engineering Department, ABES Engineering College, Ghaziabad, India. He has rich experience in teaching the undergraduate and postgraduate classes. He has published 20 textbooks and 81 research papers in international journals and proceedings of international conferences. He has organized several international conferences technically sponsored by IEEE, ACM, and Springer. He has also worked as a member of organizing committee in several IEEE international conferences in India and abroad. He is on the review panel of IEEE Computer Society, *International Journal of Network Security, Pattern Recognition Letters and Computer & Education* (Elsevier's journal). He is also a member of editorial board for *International Journal of Computer Application, Journal of Modeling and Simulation in Design and Manufacturing* (JMSDM) and *International Journal of Emerging Trends and Technology in Computer Science & Engineering*. He has been appointed member of board of studies as well as in syllabus committee of different private Indian universities and member of organizing committee for various national and international seminars/workshops. He is an executive committee member of IEEE UP Section, IEEE Computer Society Chapter, IEEE India Council, and also IEEE Asia Pacific Region 10. He is an active member of IEEE Computer Society, International Association of Computer Science and Information Technology, Computer Society of India, and International Association of Engineers and a life member of ISTE.

Part I
Intelligent Computing Techniques

Multi-view Ensemble Learning Using Rough Set Based Feature Ranking for Opinion Spam Detection

Mayank Saini, Sharad Verma and Aditi Sharan

Abstract Product reviews and blogs play a vital role in giving an insight to end user for making purchasing decision. Studies show a direct link between product reviews/rating and revenue of product. So, review hosting sites are often targeted to promote or demote products by writing fake reviews. These fictitious opinions which are written to sound authentic known as deceptive opinion spam. To build an automatic classifier for opinion spam detection, feature engineering plays an important role. Deceptive cues are needed to be transformed into features. We have extracted various psychological, linguistic, and other textual features from text reviews. We have used mMulti-view Ensemble Learning (MEL) to build the classifier. Rough Set Based Optimal Feature Set Partitioning (RS-OFSP) algorithm is proposed to construct views for MEL. Proposed algorithm shows promising results when compared to random feature set partitioning (Bryll Pattern Recognit 36(6):1291–1302, 2003) [1] and optimal feature set partitioning (Kumar and Minz Knowl Inf Syst, 2016) [2].

Keywords Multi-view ensemble learning · Opinion spamming
Opinion mining · Feature set partitioning · Rough set

1 Introduction

A fake review can make a huge potential impact on consumer behavior and their purchasing decision. That is why this field is getting attention throughout academia and businesses. Manually, it is difficult to detect deceptive opinion spam from reviews which are generally unstructured in nature. Features play a vital role in building auto-

M. Saini (✉) · S. Verma · A. Sharan
Jawaharlal Nehru University, New Delhi, India
e-mail: mayank82_scs@jnu.ac.in; mayanksaini1986@gmail.com

S. Verma
e-mail: sharadlnx@gmail.com

A. Sharan
e-mail: aditisharan@jnu.ac.in

© Springer Nature Singapore Pte Ltd. 2019
S. K. Bhatia et al. (eds.), *Advances in Computer Communication and Computational Sciences*, Advances in Intelligent Systems and Computing 759,
https://doi.org/10.1007/978-981-13-0341-8_1

matic deceptive spam review classifier. We need to link between daily words uses to deceptive behavior to build new features. From feature perspective opinion spamming problem has two issues. First to identify and construct new features, second to use them effectively. We have extracted various aspects of review such as linguistic, psychological, quantitative textual features, etc., by incorporating corresponding feature sets.

Multiple sources and different perspectives of features form different views of datasets. Classic machine learning algorithms fails to exploits these views to enhance classification accuracy. On the other hand, MEL uses consensus and complementary information for better classification results. Moreover, when a natural split of feature set does not exist, a split can be manufactured to enhance the performance. We have proposed rough set based Optimal Feature Set Partitioning (RS-OFSP) to generate multiple views from original feature set. Previously proposed, OFSP algorithm is sensitive to the order in which the feature is arrived. Due to random ordering of features, it does not deliver a consistent output. To overcome this issue, RS-OFSP first ranked the feature in order of their relevance using rough set theory. A feature might be strongly relevant, weakly relevant, or nonrelevant based on its discriminatory or prediction power. Rough sets define strong and weak relevance for discrete features and discrete targets.

The rest of the paper is organized as follows. The second section focuses on various works related to opinion spamming in consideration with different approaches. Section 3 explains the theoretical framework and proposed algorithm of the paper. In the penultimate section, i.e., Sect. 4, experimental details and result analysis have been described. The last section comprises of the conclusion as well as the future work.

2 Related Work

Opinion spam detection techniques mainly rely on three information sources to extract the features: review text, reviewer characteristics, and product information. Review text is a foremost source for information as other information is not available in most of the related datasets. In this area, the key challenges are a lack of proper deceptive review dataset and no access to spammers' identity to the analysts.

Initially, opinion spam problem has been treated as duplicate review identification problem [3]. However, this assumption is not appropriate. One of the finest works in the field of deceptive opinion spam identification has been done by integrating psychology and computational linguistics by Ott et al. [4]. The author claimed that best performance was achieved by using psychological features along with unigrams by using linear support vector machine. We have also used this same crowd sourced dataset to perform our experiments.

Another approach Co-training for Spam review identification (CoSpa) [5] has been proposed to detect deceptive spam reviews. Lexical terms were fetched from reviews and used as first view. For second view, Probabilistic Context-Free Grammars rules were constructed from review text. Support Vector Machine is used as the base learner. This co-training based approach used unlabeled dataset to improve classification accuracy. The results shows improvement in identification of deceptive spam reviews even though two views were not independent.

Opinion spamming can be done individually or may involve a group [6]. If it involves a group then it can be even make more impact due to its size. A significant impact on sentiment can be made on a target product. Their work was based on the assumption that a reviewers group is working together to demote or promote a product. The author has used frequent pattern mining to find a potential spammer group and used several behavioral models derived from the collusion phenomenon among fake reviews and relation models.

Previous attempts for spam/spammer detection used reviewer's behaviors [7], text similarity and linguistics features [4, 8], review helpfulness, rating patterns, graph-based method [9], and temporal activity-based method [7].

Other interesting findings include rating behaviors [10], network-based analysis of reviews [10], topic models based approach [11], review burstiness and time series analysis [12, 13], and reviewers with multiple user id's or accounts [14]. Some other recently proposed studies include machine learning based approaches [8, 15, 16] and hybrid approaches [17].

3 Feature Identification and Construction

This section has explained the features and their theoretical significance for building deceptive opinion classifier. This work is based on the observation that these features help us to distinguish between deceptive and truthful reviews.

3.1 Bigrams/Unigrams

Unigrams and bigrams have been used to get the context of the review. Some generic preprocessing like removing stop words, extra white spaces have been done before generating Document-Term Matrix.

3.2 Parts of Speech

A different genre of text has a difference in POS distribution. To utilize this fact, we have used common parts of speech such as noun, personal pronouns, impersonal pronoun, comparative and superlative adjectives, adverbs, articles, etc., to differentiate between the deceptive and truthful reviews.

3.3 Quantitative Feature

A review with more factual numbers and fewer emotion words has higher chances of being truthful. Word count, sentence count, numbers (thousand, third), and quantifiers (many, few, much) are calculated and used as a quantitative text feature.

3.4 Psychological and Linguistic Features

Word play is deceptive and so is the human being. Review Language plays a major role in identifying the hidden intentions. Apart from syntactic, semantic, and statistical features, we need to establish a link between use of regular words and deceptive behavior to catch spammers. Many studies have been done by the linguists and psychologists to find the association between the nonverbal clues to deceptive behavior. To fetch these clues, we have used Linguistic Inquiry and Word Count (LIWC) [18]. It gives a quantitative value for each review in psychologically meaningful categories.

3.5 Readability

Readability can be defined as: "the degree to which a given class of people find certain reading matter compelling and comprehensible [19]." It is observed that a review that has been written by average US citizen contains simple, familiar words and usually, fewer jargons compare to one written by professionally hired spammer.

Various readability metrics have been suggested to identify how readable text is. To be specific, we computed Automated Readability Index (ARI), Coleman Liau Index (CLI), Chall Grade (CG), SMOG, Flesch-Kincaid Grade Level (FKGL), and Linsear (LIN).

3.6 Lexical Diversity

Lexical diversity is another text characteristic that can be used to distinguish between deceptive and truthful opinions. For a highly lexically diverse text, the word choice of the writer needs to be different and diversified with much less repetition of the vocabulary. Moreover, previous researches have shown that lexical diversity is comparatively higher in writing than in speaking [20, 21]. We have computed Type-Token Ratio, Guiraud's Root TTR, Dugast's Uber Index, Maas' indices, Measure of Textual Lexical Diversity, Moving-Average Type-Token Ratio, Carroll's Corrected TTR, Mean Segmental Type-Token Ratio, Summer's index, and Moving-Average Measure of Textual Lexical Diversity.

4 Related Concepts, Notations, and Proposed Method

A dataset might have different views based on its properties or multiple sources of collection. Contrast to co-training, MEL is a supervised learning algorithm and can have k number of views. Conventional machine learning algorithms combine all views to adapt learning. In contrast to single-view learning, MEL exploits redundant views based on consensus and complementary principles.

MEL is a two steps process. (a) View generation (b) View combination. View generation is corresponding to feature set partitioning. It basically divides the original features set into non-empty disjoint set of features. View validation is needed to ensure MEL effectiveness. Various linear and nonlinear ensemble methods exist to combine multiple views. We have used performance weighting ensemble method to combine output of view classifiers.

4.1 Feature Ranking Based on Degree of Dependency

Rough set theory is a mathematical approach to quantify imprecision, vagueness and uncertainty in data analysis. Rough sets provide the concept of core and reduct to determine the most important features/attributes from the information system. Reduct is the minimal subset of conditional attributes which maintain the quality of classification. The intersection of all reduct is called core. However, reduct and core do not provide the relevancy of each feature.

In order to find the relevancy of features, we need to discover the dependencies between conditional attributes and decision attribute. If M and N are the sets of attributes then in rough set theory, dependency is defined in the following way:

For M, $N \subset A$, it is said that N depends on M in a degree k ($0 \leq k \leq 1$), denoted $M \rightarrow N$, if

$$k = \gamma_M(N) = \frac{|POS_M(N)|}{|U|} \tag{1}$$

We measured the significance of an attribute by calculating the change in dependency after removing the attribute from the set of considered conditional attribute.

Table 1 shows RS-OFSP algorithm for feature ranking and optimal view generation. The first part depicts the procedure for ranking the features based on the concept of rough set. Once the ranked features are obtained, the second part takes it as features set and finds an optimal number of partitions or views.

Table 1 RS-OFSP Algorithm for view construction Algorithm

Input.
$M = X * D$ // A dataset containing n conditional attributes (X) defined as $X = \{x_1, x_2, x_3, x_4 \ldots x_n\}$ and 1 decision attribute (D)
k   // Number of Views
f_i // i^{th} classifier
Output.
X_{opt} // Set of optimally partitioned Views

Begin

  // Initialization
  for i = 1 to k
     xi = Φ;   // Creating two NULL views
     Ai = 0;   // initialize accuracies
  end

  // Feature Ranking
  dep = γC(D)   // degree of dependency of D on C
  for i = 1 to n
     R_i = dep - γ(C - x_i)(D) // relevancy of feature fi
  end
  X-sort = sort(R)   // sort the features according to its relevancy
  for attr = 1 to n
     for i = 1 to k
        X_{temp} = X_i ∪ { X-sort attr}   // adding attribute to i^{th} views
        A_{itemp} = fi (X_{temp});   // Evaluation of feature subset X_{temp}
        A_{idiff} = A_{itemp} - A_i ;
     end
     If max(A_{diff}) > 0 then
        t = argmax$_i$(A_{diff}) // t^{th} view which has A_{tdiff} > 0
        X_t = X_t ∪ { X-sort attr };
        At = A_{ttemp} ;
     end
  end
  return X_{opt} = { X1 , X2, X3........., X_k}

End:

5   Experiments and Results for MEL

This section has been divided into two subsections. The first subsection has given the description of datasets and experimental details. And the second subsection has shown results along with detail analysis.

5.1 Databases and Experimental Setup

In this work, we have used two datasets. One is a publicly available gold standard corpus of deceptive opinion spam given by Myle Ott [7]. Another dataset we have crawled from Yelp. Yelp is a review hosting commercial site which publicly filters the fake reviews. Yelp's filtering algorithm has evolved over the time to filter deceptive and fake reviews. Yelp's filter has also been claimed to be highly accurate by various studies. We have crawled 2600 truthful and 2600 deceptive/filtered reviews and both of these consist of 1300 positive and 1300 negative reviews. We have treated five-star rated review as a positive review and one star as a negative review. We have collected these reviews from hundred Chicago hotels. To maintain the class balance, we have selected the same number of filtered and non-filtered reviews from each hotel.

MEL has used Random Feature Set Partitioning (RFSP), Optimal Feature Set Partitioning (OFSP) and our proposed Rough Set Based Optimal Feature Partitioning (RS-OFSP) for view construction. For RFSP, the features were partitioned into equally in all views. Support Vector Machine (SVM), Naïve Bayes (NB), and K-Nearest Neighbor (K-NN) classifiers have been used for view accuracy. Performance weighting technique is used to ensemble the views. 10-fold cross validation has been performed for measuring all general accuracies. Experiments were carried out on Matlab-2015a.

5.2 Experimental Results and Analysis

Deception is a complicated psychological behavior which is related to cognitive processes and mental activity. From features point of view, psycholinguistic features along with unigrams and bigrams played an important role to get the context of the reviews. Readability, lexical diversity and subjectivity worked as complementary to improve the accuracy further. In readability measure for ott dataset, a significant difference in ARI (two-tailed t-test $p = 0.0194$), CLI (two-tailed t-test $p = 0.0202$), CG (two-tailed t-test $p = 0.0122$), SMOG (two-tailed t-test $p = 0.0264$), FKGL (two-tailed t-test $p = 0.0353$), and LIN (two-tailed t-test $p = 0.03$) for truthful and deceptive reviews. And for yelp dataset difference in ARI (two-tailed t-test $p = 0.0045$), CLI (two-tailed t-test $p = 0.03$), CG (two-tailed t-test $p = 0.02$), SMOG (two-tailed t-test $p = 0.01$), FKGL (two-tailed t-test $p = 0.01$), and LIN (two-tailed t-test $p = 0.03$) for truthful and deceptive reviews is quite significant too. For lexical diversity and subjectivity are also, a clear difference is seen between truthful and deceptive reviews.

Classification accuracy is used to measure the effectiveness of feature partitioning algorithms. Learning algorithms k-NN, NB, and SVM are used to estimate view accuracies in all methods. Figure 1 and Fig. 2 shows the comparison in classification accuracy of RFSP, OFSP with RS-OFSP methods using k-NN, NB and SVM on Ott and Yelp datasets, respectively. It shows the accuracy of ensemble classifiers improving with the increase in the number of partitions.

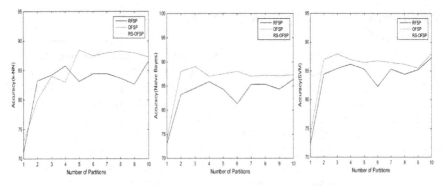

Fig. 1 Comparison of MEL classification accuracy using RFSP, OFSP with RS-OFSP methods on Ott's dataset

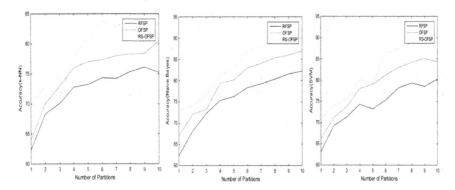

Fig. 2 Comparison of MEL classification accuracy using RFSP, OFSP with RS-OFSP methods on Yelp dataset

Analysis of MEL classification accuracy using RFSP, OFSP, and RS-OFSP is as follows. Using k-NN for view accuracy, MEL has performed better using RS-OFSP compared to RFSP for all values of k in both the datasets. But when compared to OFSP, RS-OFSP performed better for all values k except k = 5 and 6 in Ott dataset and k = 3 for Yelp dataset. Using NB for view accuracy, MEL has performed better using RS-OFSP compared to RFSP and OFSP for all values of k except k = 2 in Ott dataset and all values of k in Yelp dataset. Using SVM for view accuracy, MEL has performed better using RS-OFSP compared to RFSP and OFSP for all values of k except k = 2 in Ott dataset and all values of k in Yelp dataset.

Figures 3 and 4 show the comparison in classification accuracy using boxplot analysis of RFSP, OFSP with RS-OFSP methods on Ott dataset. For most of the value of k (number of the partition), RS-OFSP performance is better than the other two in both OTT and YELP dataset. We noticed more variance in classification accuracies in all the methods for Yelp dataset compare to Ott dataset. So clearly, RS-OFSP has an edge over others is a better method.

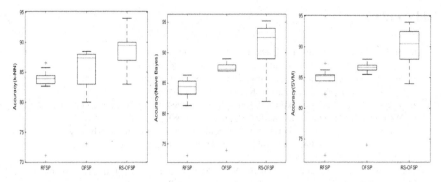

Fig. 3 Boxplot comparison of MEL classification accuracies using RFSP, OFSP with RS-OFSP methods on Ott's dataset

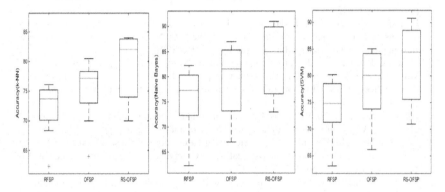

Fig. 4 Boxplot comparison of MEL classification accuracies using RFSP, OFSP with RS-OFSP methods on Yelp dataset

6 Concluding Remarks

This work has explored the use of multi-view ensemble learning algorithms more specifically multi-view ensemble learning for automated deceptive review detection. The effectiveness of MVL algorithms depends upon view construction and ensemble techniques. We proposed RS-OFSP for view construction and compared with RFSP and OFSP. MEL algorithm performed much better than baseline classifier using the same set of feature sets. This research also suggested that SVM classifier may be considered for classification using MEL in opinion spam detection, where RS-OFSP method is utilized as feature set partitioning. Overall, we find that, by exploring the consistency and complementary properties of different views, multi-view learning is rendered more effective, more promising, and has better generalization ability than single-view learning.

References

1. Bryll, R., Gutierrez-Osuna, R., Quek, F.: Attribute bagging: improving accuracy of classifier ensembles by using random feature subsets. Pattern Recognit. **36**(6), 1291–1302 (2003)
2. Kumar, V., Minz, S.: Multi-view ensemble learning: an optimal feature set partitioning for high-dimensional data classification. Knowl. Inf. Syst. (2016)
3. Jindal, N., Liu, B.: Opinion spam and analysis. In: Proceedings of the International Conference on Web Search Web Data Min. WSDM 08, p. 219 (2008)
4. Ott, M., Choi, Y., Cardie, C., Hancock, J.T.: Finding deceptive opinion spam by any stretch of the imagination. In: Proceedings of the 49th Annual Meeting of the Association for Computational Linguistics, pp. 309–319 (2011)
5. Zhang, W., Bu, C., Yoshida, T., Zhang, S.: CoSpa: a co-training approach for spam review identification with support vector machine. Information **7**(1), 12 (2016)
6. Mukherjee, A., Liu, B., Glance, N.: Spotting fake reviewer groups in consumer reviews. In: Proceeding WWW'12. Proceedings of the 21st International Conference on World Wide Web, pp. 191–200 (2012)
7. Lim, E.-P., Nguyen, V.-A., Jindal, N., Liu, B., Lauw, H.W.: Detecting product review spammers using rating behaviors. In: Proceedings of the 19th ACM International Conference on Information and Knowledge Management, pp. 939–948 (2010)
8. Hancock, J.T.: Negative deceptive opinion spam. In: Naacl, no. June, pp. 497–501 (2013)
9. Wang, G., Xie, S., Liu, B., Yu, P.S.: Review graph based online store review spammer detection. In: Proceedings—IEEE International Conference on Data Mining, ICDM, pp. 1242–1247 (2011)
10. Akoglu, L., Chandy, R., Faloutsos, C.: Opinion fraud detection in online reviews by network effects. In: Association for the Advancement of Artificial Intelligence, pp. 2–11 (2013)
11. Li, J., Cardie, C., Li, S.: TopicSpam: a topic-model-based approach for spam detection. In: ACL 2013—51st Annual Meeting of the Association for Computational Linguistics. Proceedings of the Conference, vol. 2, pp. 217–221 (2013)
12. Fei, G., Mukherjee, A., Liu, B., Hsu, M., Castellanos, M., Ghosh, R.: Exploiting burstiness in reviews for review spammer detection. In: Proceedings of the Seventh International AAAI Conference on Weblogs and Social Media, pp. 175–184 (2013)
13. Ye, J., Kumar, S., Akoglu, L.: Temporal opinion spam detection by multivariate indicative signals. In: ICWSM16, pp. 743–746 (2016)
14. Qian, T., Liu, B.: Identifying multiple userids of the same author, no. October, pp. 1124–1135 (2013)
15. Weichselbraun, A., Gindl, S., Scharl, A.: Enriching semantic knowledge bases for opinion mining in big data applications. Knowl.-Based Syst. **69**(1), 78–85 (2014)
16. Hernández Fusilier, D., Montes-y-Gómez, M., Rosso, P., Guzmán Cabrera, R.: Detecting positive and negative deceptive opinions using PU-learning. Inf. Process. Manag. **51**(4), 433–443 (2015)
17. Hu, N., Liu, L., Sambamurthy, V.: Fraud detection in online consumer reviews. Decis. Support Syst. **50**(3), 614–626 (2011)
18. Pennebaker, J.W., Boyd, R.L., Jordan, K., Blackburn, K.: The Development and Psychometric Properties of LIWC2015. Austin, Texas (2015)
19. McLaughlin, G.H.: SMOG grading: A new readability formula. J. Read. **12**(8), 639–646 (1969)
20. Crowhurst, M.: Spoken and written language compared. In: Language and Learning Across the Curriculum, p. 109. Oxford University Press (1994)
21. Johansson, V.: Lexical diversity and lexical density in speech and writing: a develop- mental perspective. Work. Pap. **53**, 61–79 (2008)

Analysis of Online News Popularity and Bank Marketing Using ARSkNN

Arjun Chauhan, Ashish Kumar, Sumit Srivastava
and Roheet Bhatnagar

Abstract Data mining is a process of evaluating practice of examining large preexisting databases in order to generate new information. The amount of data has been growing at an enormous rate ever since the development of computers and information technology. Many methods and algorithms have been developed in the last half-century to evaluate data and extract useful information to help develop faster. Due to the wide variety of algorithms and different approaches to evaluate data, several algorithms are compared. The performance of any algorithm on a particular dataset cannot be predicted without evaluating it with the same constraints and parameters. The following paper is a comparison between the trivial kNN algorithm and the newly proposed ARSkNN algorithm on classifying two datasets and subsequently evaluating their performance on average accuracy percentage and average runtime parameters.

Keywords Data mining · Classification · Nearest neighbors · ARSkNN

A. Chauhan
Department of ECE, Manipal Institute of Technology, Manipal University,
Manipal, India
e-mail: arjunchau@gmail.com

A. Kumar · R. Bhatnagar
Department of CSE, Manipal University Jaipur, Jaipur, India
e-mail: kumar.ashish@jaipur.manipal.edu

R. Bhatnagar
e-mail: roheet.bhatnagar@jaipur.manipal.edu

S. Srivastava (✉)
Department of IT, Manipal University Jaipur, Jaipur, India
e-mail: sumit.srivastava@jaipur.manipal.edu

© Springer Nature Singapore Pte Ltd. 2019
S. K. Bhatia et al. (eds.), *Advances in Computer Communication and Computational
Sciences*, Advances in Intelligent Systems and Computing 759,
https://doi.org/10.1007/978-981-13-0341-8_2

13

1 Introduction

We live in a world which is ever changing. Technology has developed a lot in the past few years and it has helped man evolve. Today in this world of digitalization, the flow of data is enormous. Not only is it ever changing but is also growing at an exponential rate. As recorded in 2012, about 2.5 exabytes of data are generated every day, and that number doubles every 40 months. More data cross [16] the Internet every second than were stored in the entire internet just 20 years ago. To keep up with such flow of data are in the form of images, videos, datasets, files, text, animations, etc. it is hard for humans to use it completely to benefit his own self.

In such scenarios, Big Data Analytics plays an important key role as it operates on large datasets. The very foundation of data mining comes from the analyses and computation of large volumes of data to obtain key information and harness it, efficiently and effectively [16]. The concept of data mining and machine learning comes into field. This idea was developed in the 1900s to help manage data regarding online data transactional processing. Many mathematicians and statisticians, around the end of the twentieth century, were working together to develop algorithms to help solve this problem. Many varied kinds of algorithms were developed but most of them were not promoted till recently only because of faster processing speeds and powerful machines. Due to the ever-growing data in this digital era, machine learning algorithms are redesigned to adapt to new information and help track it. Some algorithms like kNN, SVM, and gradient descent are the more commonly used algorithms [9]. However, the performance of these algorithms depends greatly on the size of the dataset they are being tested on. Machine learning algorithms can be broadly divided into four types: Regression, classification, clustering and rule extraction. Classification is a technique in which a model or a classifier is constructed to predict a particular labeled class [6]. It is a part of supervised learning. KNN and ARSkNN are used in classification and is part of supervised learning framework.

The kNN classifier is an accurate and efficient method to classify data but due to its core functionality being distance calculation among each training instances and testing instance, it is time consuming and process heavy. In order to overcome this, another algorithm was developed called the ARSkNN [12] which does not calculate distance to estimate similarity between training and testing instances. This algorithm is based on similarity measure known as Massim rather than the distance measure [11]. The approach not only increases accuracy of classification but also takes lesser time to classify the information. The most commonly used kNN definition for measuring distances is the Euclidean method. There are 76 similarity measures [2] apart from this which improve performance in some cases. All of them are directly or indirectly depend on the distance calculation. The choice of individual measure depends on characteristics of a dataset like if it is binary or not and if data are symmetric.

In the following paper, we shall compare and evaluate kNN and ARSkNN on the basis of accuracy and time required by them to classify. The datasets used for the demonstration are the bank marketing dataset and the Online News Popularity dataset from the UCI repository. This paper is divided into seven sections. The first section is the introduction. The second section contains the literature review which is followed by the experimental setup in next third section. The fourth section contains information about the datasets used followed by empirical evaluation, discussion and conclusion in fifth, sixth, and seventh sections, respectively.

2   Literature Review

Classification is a problem in machine learning which deals with categorizing unknown data into set of categories which are formed during training of the classifier. It is an example of pattern recognition and finds reason in the set of training data provided to it [9]. The choice of algorithm for classification depends on the size, quality, and nature of the data.

Figure 1 is a graphical representation of a classification problem. The red circles, blue triangles, and green squares are indicative of separate classes present in the dataset. The new data is of unknown class and needs to be classified. To do so many algorithms are used. In the following subsections, we have discussed two algorithms that are used for classification and their base principle of operation.

2.1   k-Nearest Neighbors

The kNN is a straightforward and efficient classification model in unsupervised learning [8]. It conservatively uses the distance function to train and predict in a given

Fig. 1 Example of classification

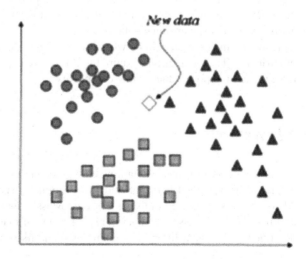

dataset. It is a straightforward lazy learner as it stores data, deletes any constructed query, and is dependent on the information stored by the training samples in the memory [1]. It simply memorizes all examples in the training set and then compares the test sample to them. For this reason, kNN is also called memory-based learning or instance-based learning. It is intensively used in the fields of statistical pattern recognition, data mining, image processing, and many others.

The working principle [10] of the algorithm is that if the number of k most similar instances nearest to the testing query Q belong to a particular class then it is concluded that the point belongs to the same class. The distance of the query Q from the k-nearest neighbors, i.e., the geometrically closest points is calculated (usually based on Euclidian distance) and is stored in the memory. Then, if all k points do not belong to the same class, a voting is passed and the class having maximum frequency of votes is assigned to be the class of the query Q. Different similarity measures can be used to optimize performance. The choice of similarity measure depends on the type of dataset and can only be found out by testing individual measures one at a time. The performance can also be increased by nearly 10% if outliers in a dataset are known [18]. The outliers are noise to the classifier and deviate it from predicting the outcome precisely.

In n-dimensional space, every point in the training dataset is represented as $P = (P_1, P_2, P_3, \ldots, P_n)$ and the queried instance is represented as $Q = (Q_1, Q_2, Q_3, \ldots, Q_n)$.

The distance from the query point is most commonly taken as the Euclidian distance can be given by

$$d_{Euc} = \sqrt{\sum_{i=1}^{n} |P_i - Q_i|^2} \tag{1}$$

The distribution of data in the dataset is detrimental for the time it takes the algorithm to run. The value of parameter of k also plays a very important role in determining the number of computational steps. It is also observed that kNN can be applied to higher dimensional data.

However, the problem will be when it is applied to dataset having a large amount of attributes. The calculations required will increase exponentially and would need large processing power due to time complexity.

2.2 ARSkNN

The ARSkNN is a supervised learning classification algorithm developed by Ashish Kumar, Prof. Roheet Bhatnagar and Prof. Sumit Srivastava and hence the name ARSkNN. It is an algorithm that is based on Massim and takes density distribution as a parameter to classify an query [11]. The idea of mass estimation was initially described in the paper titled Mass Estimation and its Applications which was proposed by Kai Ming Ting, Fei Tony Liu and James Tan Swee Chuan of Monash University, Australia and Guang-Tong Zhou, of Shandong University, China [19].

Massim, as explained, is simply dependent on the count of the number of occurrences (instances). It operates on the basis of the region of the point rather than the distance. The entire dataset is divided into several parts based on the number of splits. The mass of a particular point is calculated based on the density distribution. This split can be linear or multidimensional. A linear split would be effective for binary classification whereas multidimensional split can be used for multiclass classification.

The most important aspect of the ARSkNN which gives it an edge over kNN is the time complexity of ARSkNN [12] in modeling stage which is $\mathcal{O}(ntlogd)$ and in class assignment stage being $\mathcal{O}(kn)$ where, n is the number of total instances in dataset, d are the number of randomly selected instances from dataset, t is the number of random regions to be used to define mass similarity, and k is the number of nearest neighbors.

For the modeling stage, the time complexity can be reduced from $\mathcal{O}(ntlogd)$ to $\mathcal{O}((n+d)t)$ by using indexing technique.

3 Experimental Setup

The classification on two datasets by kNN and ARSkNN was executed on WEKA [5] Experimenter on Intel core i7 processor at 2.4 GHz 8 GB RAM and 10-fold cross-validation on version 3.7.10. The kNN classifier was used from lazy folder as IBK. The ARSkNN was executed after adding the .jar file into the classifier folder.

4 Dataset Information

We have analyzed the traditional kNN and ARSkNN upon two different datasets. The two datasets were obtained from the UCI Repository. Both of them are described in the following subsections and a brief summarization has been done in Table 1.

Table 1 Characteristics of datasets

Dataset name [Ref]	# Instances	# Attributes	# Classes	Domain
Bank marketing [14]	45211	17	2	Success in selling bank scheme
Online news popularity [4]	39797	61	2	Popularity of news on internet

4.1 Bank Marketing Dataset

The bank marketing dataset [13, 14] is used to predict whether or not a particular customer will subscribe to a term deposit or not based on 17 features. This is a binary classification problem as the outcome of this would either be a yes wherein the customer agrees to deposit money or a no in which case he does not. Such kind of classification can be helpful to not only banks but also other sectors like transportation industry, Online sales industry and Manufacturing industries to make their sales and offers more effective. The collected data can help predict whether an individual will agree or not to a particular scheme. Based on inputs taken from a group of people regarding a particular instance, the feedback received from them along with other features like the age, sex, occupational status, profession, marital status, etc., will help categorize people into different groups. This will help companies strategize their offers and optimize their target groups thus making the entire procedure more efficient.

4.2 Online News Popularity Dataset

The Online News Popularity dataset [4] contains information regarding several news instances that occurred in the recent past and is classified on the basis of whether it was popular or not. This information was collected from the American internet media company, BuzzFeed. This dataset is a binary dataset and hence data is classified into two classes. In this case dataset, it is specifically divided into the category whether the news is popular or not based on the number of shares it gets. The dataset is segmented into several attributes like number of words in title, length of article, number of links, number of images, etc. Such predictions can be useful and effective in delivering strong and immediate news information to the masses and help them become more aware about key issues. It can also be used by journalists to make their news more effective and gain popularity.

5 Empirical Evaluation

Above mentioned datasets are evaluated on two parameters. Average accuracy percentage and the average run time of the entire testing and training phase.

The accuracy of a classification is evaluated by calculating the number of correctly recognized class examples (true positives), the number of correctly recognized examples that do not belong to the class (true negatives), and examples that either were incorrectly assigned to the class (false positives) or that were not recognized as class examples (false negatives). The average accuracy percentage is taken over 10-fold cross-validation. It is a measure of the ability of the classifier. Classifier predicts the class label correctly and the accuracy of the classifier refers to how well a given classifier can guess the value of predicted attribute for a new data [3, 7, 15, 17].

The average runtime of the classifier is defined as the time required by the classifier to train and test the dataset [3, 7, 15, 17].

The following results are calculated for values of k being 1, 3, 5, and 10 and for number of sTrees being formed 10, 50 and 100 in ARSkNN. The highest accuracy percentage and worst average runtime are highlighted in Table 2 and Table 3 as bold texts respectively. As the data in Table 2 indicates, for $k = 1$ the ARSkNN outperforms the kNN algorithm in terms of accuracy. The accuracy is increased by up to 2.19% for $k = 1$. For $k = 3$, an improvement is observed in the kNN classifier but still the ARSkNN is better. For $k = 5$, the accuracy of kNN increases more and is almost equal to ARSkNN for a value of 100 for sTrees. At $k = 10$ the kNN is marginally better than ARSkNN.

For the dataset Online News Popularity, it is observed from Table 3 that not only does the ARSkNN perform better than kNN but also has a steady increase in the accuracy as the value of k increases. For a value of $k = 1$, the ARSkNN is better than the kNN classifier by a great margin beating it by 5.64%. Likewise for $k = 3$, the margin increases further. This general trend is observed for values of $k = 5$ and $k = 10$.

For average runtime parameter upon Bank Marketing dataset (Table 4), the time taken by the kNN classifier is very large compared to ARSkNN classifier. It is also observed that the time required increases at a fast rate as k increases for kNN whereas the time required reduces for ARSkNN for 10 sTrees. For $k = 1$, kNN required 19.03 s whereas ARSkNN only required 0.12 s for 10 sTrees. For $k = 3$, the time of kNN increases further to 19.99 s but stays constant for 10 sTrees. For value of k being 5 and 10, the time of kNN continuous to increase whereas for 10 sTrees, the time reduces further.

Table 2 Average accuracy (in %) for bank marketing

	$k = 1$	$k = 3$	$k = 5$	$k = 10$
IBK	87.15	88.65	88.97	**89.13**
ARSkNN with 10 sTrees	88.44	88.93	88.91	88.87
ARSkNN with 50 sTrees	89.21	89.07	88.97	88.88
ARSkNN with 100 sTrees	**89.34**	**89.13**	**89.03**	88.93

Table 3 Average accuracy (in %) for online news popularity

	$k = 1$	$k = 3$	$k = 5$	$k = 10$
IBK	57.35	59.32	60.33	61.31
ARSkNN with 10 sTrees	58.92	61.48	62.36	63.00
ARSkNN with 50 sTrees	**62.99**	**64.14**	**64.29**	**64.40**
ARSkNN with 100 sTrees	N.A.	N.A.	N.A.	N.A.

Table 4 Average runtime (in s) for bank marketing

	$k = 1$	$k = 3$	$k = 5$	$k = 10$
IBK	**19.03**	**19.99**	**20.96**	**21.71**
ARSkNN with 10 sTrees	0.12	0.12	0.11	0.11
ARSkNN with 50 sTrees	0.49	0.54	0.57	0.54
ARSkNN with 100 sTrees	1.09	1.09	1.18	1.32

Table 5 Average runtime (in s) for online news popularity

	$k = 1$	$k = 3$	$k = 5$	$k = 10$
IBK	**37.07**	**39.19**	**41.73**	**43.86**
ARSkNN with 10 sTrees	0.18	0.20	0.19	0.18
ARSkNN with 50 sTrees	0.90	0.90	0.90	0.93
ARSkNN with 100 sTrees	N.A.	N.A.	N.A.	N.A.

The classifier of ARSkNN with 100 sTrees could not be modeled due to lack of processing power and memory allotted to WEKA (Table 5).

For average runtime parameter upon Online News Popularity dataset, ARSkNN outperforms kNN by a clear margin. The time required by kNN is almost 206 times the amount of time required by ARSkNN. For $k = 1$, kNN classifier takes 37.07 s whereas ARSkNN takes only 0.18 s for 10 sTrees. The increment in the time for kNN for a value of $k = 5$ is much greater than the increment in ARSkNN classifier. It is also evident from the results that, the time required to classify data using kNN increases as the value of k increases, whereas using ARSkNN, the time remains almost the same or reduces as the value of k increases.

6 Discussion

The similarity between ARSkNN and kNN is that both of them are dependent on the training samples for detecting the output. However, the computational time for ARSkNN is significantly low because it does not calculate individual distances from each point every time rather it divides the entire plane into parts and creates similarity trees (sTrees) to classify data. In this paper, only Euclidian distance is taken as similarity measure for comparison. Further experimentations can be done by researchers using different similarity measures to compare and evaluate performance with ARSkNN. It is also observed that Massim is a more processor friendly operation than calculating individual distances. More research can be done to optimize the kNN algorithm using better means and approaches like ARSkNN did by using Massim.

7   Conclusion

As we have observed by these results, ARSkNN is a much more efficient and computational friendly algorithm. Not only does this result in lower computational time but also has higher accuracy when it comes to classification of data. The simplicity of this method is a very strong factor that supports the usage of this irrespective of the size of the dataset. Even though kNN is used quite frequently, its processing time and accuracy drop when it comes to large datasets.

Different similarity measures could be used to tweak the performance of the kNN classifier but the very core operation of the kNN is what makes it slow and memory inefficient. It has undoubtedly proven its functionality and is one of the best except when it comes to large datasets, wherein it becomes an enormous drawback. In this world of ever-growing data, it is hence a requirement to use more efficient and processor friendly algorithms which not only are accurate but also are less time complex.

References

1. Aha, D.W.: Feature weighting for lazy learning algorithms. In: Feature Extraction, Construction and Selection, pp. 13–32. Springer (1998)
2. Choi, S.S., Cha, S.H., Tappert, C.C.: A survey of binary similarity and distance measures. J. Syst. Cybern. Inform. **8**(1), 43–48 (2010)
3. Demšar, J.: Statistical comparisons of classifiers over multiple data sets. J. Mach. Learn. Res. 7(Jan), 1–30 (2006)
4. Fernandes, K., Vinagre, P., Cortez, P.: A proactive intelligent decision support system for predicting the popularity of online news. In: Portuguese Conference on Artificial Intelligence, pp. 535–546. Springer (2015)
5. Hall, M., Frank, E., Holmes, G., Pfahringer, B., Reutemann, P., Witten, I.H.: The weka data mining software: an update. ACM SIGKDD Explor. Newslett. **11**(1), 10–18 (2009)
6. Han, J., Pei, J., Kamber, M.: Data Mining: Concepts and Techniques. Elsevier (2011)
7. Hsu, C.W., Lin, C.J.: A comparison of methods for multiclass support vector machines. IEEE Trans. Neural Netw. **13**(2), 415–425 (2002)
8. Jiang, L., Cai, Z., Wang, D., Zhang, H.: Bayesian citation-knn with distance weighting. Int. J. Mach. Learn. Cybern. **5**(2), 193–199 (2014)
9. Kotsiantis, S.B., Zaharakis, I., Pintelas, P.: Supervised machine learning: A review of classification techniques (2007)
10. Kuang, Q., Zhao, L.: A practical GPU based kNN algorithm. In: International Symposium on Computer Science and Computational Technology (ISCSCT), pp. 151–155 (2009)
11. Kumar, A., Bhatnagar, R., Srivastava, S.: A critical review of mass estimation & its application in data mining techniques. In: 2014 IEEE International Advance Computing Conference (IACC), pp. 452–456. IEEE (2014)
12. Kumar, A., Bhatnagar, R., Srivastava, S.: Arsknn-a k-NN classifier using mass based similarity measure. Procedia Comput. Sci. **46**, 457–462 (2015)
13. Moro, S., Cortez, P., Rita, P.: A data-driven approach to predict the success of bank telemarketing. Decis. Support Syst. **62**, 22–31 (2014)
14. Moro, S., Laureano, R., Cortez, P.: Using data mining for bank direct marketing: an application of the crisp-dm methodology. In: Proceedings of European Simulation and Modelling Conference-ESM'2011, pp. 117–121. Eurosis (2011)

15. Othman, M.F., Yau, T.M.S.: Comparison of different classification techniques using weka for breast cancer. In: 3rd Kuala Lumpur International Conference on Biomedical Engineering 2006, pp. 520–523. Springer (2007)
16. Russom, P., et al.: Big data analytics. TDWI best practices report, fourth quarter, vol. 19, 40 (2011)
17. Salzberg, S.L.: On comparing classifiers: Pitfalls to avoid and a recommended approach. Data Min. Knowl. Discov. **1**(3), 317–328 (1997)
18. Shin, K., Abraham, A., Han, S.Y.: Improving kNN text categorization by removing outliers from training set. In: CICLing, pp. 563–566. Springer (2006)
19. Ting, K.M., Zhou, G.T., Liu, F.T., Tan, J.S.C.: Mass estimation and its applications. In: Proceedings of the 16th ACM SIGKDD International Conference on Knowledge Discovery and Data Mining, pp. 989–998. ACM (2010)

Fuzz Testing in Stack-Based Buffer Overflow

Manisha Bhardwaj and Seema Bawa

Abstract Due to rapid deployment of information technology, the threats on information assets are getting more serious. These threats are originated from software vulnerabilities. The vulnerabilities bring about attacks. If attacks are launched before the public exposure of the targeted vulnerability, they are called zero-day attacks. These attacks damage system and economy seriously. One such attack is buffer overflow attack which is a threat to the software system and application for decades. Since buffer overflow vulnerabilities are present in software, attackers can exploit thus obtains unauthorized access to system. As these unauthorized accesses are becoming more prevalent, there is need for software testing to avoid zero-day attacks. One such testing is fuzz testing, locates vulnerabilities in software and find deeper bugs. The Stack-based American Fuzzy Lop (SAFAL) model has been proposed. This model works for software to exploit vulnerabilities. The model begins the process of fuzzing by applying various modifications to the input file. The binaries are compiled using the AFL wrappers. Input test case file is provided to the model to execute the test cases. The target program resulted in various crashes and hangs that discovered stack buffer overflow vulnerabilities. A list of crashes, hangs, and queues is found in output directory. The model displays real-time statistics of the fuzzing process. The SAFAL model improves the quality of software as the hidden bugs are found. The effectiveness and efficiency of SAFAL model are hence established.

Keywords Fuzz testing · Stack buffer overflow · Symbolic execution · AFL

M. Bhardwaj (✉) · S. Bawa
Thapar University, Patiala, India
e-mail: manishabofficial@gmail.com

© Springer Nature Singapore Pte Ltd. 2019   23
S. K. Bhatia et al. (eds.), *Advances in Computer Communication and Computational Sciences*, Advances in Intelligent Systems and Computing 759,
https://doi.org/10.1007/978-981-13-0341-8_3

1 Introduction

Manually doing exploitation is one of the difficult and time-consuming processes as it requires low-level knowledge of computer systems and also analyzes the control and data flow of program. If the program is complex, analysis work will become difficult.

In software testing field, many techniques aim to find bugs in a program and generate test case to trigger those bugs. However, those test cases are meaningless and manual exploit generation is a difficult and time-consuming process because it not only requires low-level knowledge of computer systems, such as operating system internals and assembly language, but also analyzes the control and data flow of program execution by hand. If program under test is large or uses complex algorithms, the analysis work will be extremely difficult.

In software testing field, there are unknown vulnerabilities in program and by the generation of test cases, these vulnerabilities have been found that led to the less failure of the system in undesirable circumstances. For this reason, the work done here shows the exploitation of stack-based buffer overflow in software and fuzz testing to find the vulnerabilities which are a threat to the system.

Fuzz testing is a technique which tests an application in a way that breaks the program. So in traditional software testing, a software developer writes an application and is supposed to perform some functions, for example, a calculator application. Testing of calculator application, the requirement is to actually do the math correctly. If the application is working properly, then that is usually enough to pass the traditional tests.

The difference with fuzzing technique is that the fuzz tester provides invalid input to an application and causes the application to misbehave. Thus, in the example for the application of the calculator, mathematical operations are performed on numbers. When a fuzz tester does mathematical operations on alphabets or simply purges on the keyboard, then application resists with the malformed input.

Fuzz testing is a low effort sort of testing. In fuzz testing, the tester inputs the malformed data into the application and undergoes different test cases and sooner or later it might cause a crash. As the crashes are met, the software vendor spots the programming error and these errors are fixed. Hence, it improves the quality of software.

Buffer overflow in stack (stack buffer overrun) occurs when the program is written to a memory address present on stack when there is program's call. Buffer in a stack is fixed length. Buffer overflow in stack occurs when size of the buffer is fixed and more data written as per the defined size of the buffer located on stack. The data adjacent to the stack results in corruption and with the overflow causes the target program to crash or to operate incorrectly. Since there are active function calls on stack, buffer in stack causes overflow which derails execution.

Fig. 1 Buffer overflow in stack

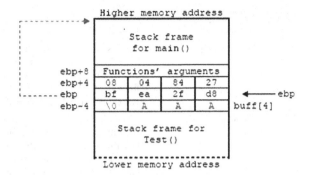

Stack buffer overflow are caused deliberately by an attack known as stack smashing. If the affected program is running with some privileges, or data are accepted from untrusted network (For example, a webserver) then there are more chances that the bug detected is a vulnerability which will affect the security of the system. The buffer in stack when overflows with data provided by the malicious user corrupts the stack by injection of executable code into the program and hence takes control over the process. Stack buffer overflow is one of the oldest vulnerabilities which allows the untrusted users to gain access into the system (Fig. 1).

2 Background Details and Related Work

For finding bugs in a software, fuzz testing or fuzzing is one of the testing techniques [1]. Since the bugs are security related so fuzz testing on these bugs are against interface of programs. Fuzzing does not ascertain completeness or correctness of program instead it focuses on identification of anomalies in an application by applying some protocols and attack heuristics [2].

Fuzzing process is commonly used by companies and open source projects so that the quality of the software can be improved by the analysts who discover the vulnerabilities and reports bugs [2]. In comparison with other software testing techniques, reverse engineering, and source code audits, fuzzing is found to be more effective and cost efficient. Comparison done with traditional software testing techniques or even source code audits and reverse engineering, fuzzing is found to be more effective and cost efficient [3].

API in-memory fuzz testing technique uses dynamic symbolic execution for locating routines and instruction belonging to target binary executable [4]. Input data is parsed and processed. In this method during testing stage, loops are constructed using binary instrumentation. The model proposed mutates the taint memory values present in each loop and construction of each loop for routines is performed by binary instrumentation. According to experiments performed, the designed model effectively detects vulnerabilities such as buffer overflows. Comparison done with

Table 1 Example of fuzzers

S.No.	Fuzzer	Description
1	Taming compiler fuzzer	Aggressive random testing tools which explores compile bugs. More than 1700 bugs were found using a single test case. Using ad hoc methods, undesirable test cases are filtered out. When large number of test cases was triggered, test cases ordered to priority [6]
2	SNOOZE	SNOOZE is a network protocol fuzzer tool. Identifies security flaws in network protocol. This fuzzer allows the tester to operate on stateful operation of a protocol and generate messages for each state and focus on vulnerability classes. SIP protocol was implemented as initial prototype to test programs. This tool exposed real-world bugs in the programs analyzed [7]
3	VUzzer	VUzzer is an application fuzzer that does not require any prior knowledge of application or input format. Static and dynamic analysis on applications is performed so there is maximum coverage and deeper paths can be explored. This fuzzing strategy evaluated three datasets: DARPA Grand Challenge binaries (CGC), real-world applications, and LAVA dataset. On all of these datasets, VUzzer yields significantly better results than other fuzzers [8]

Table 2 Processing time

Run time	Last new path	Last unique path	Last unique hangs
17 min 32 s	3 min 9 s	11 min 28 s	12 min 22 s

Table 3 Overall results

Cycles done	Total paths	Unique crashes	Unique hangs
0	58	6	1

traditional fuzzing tools indicates this proposed model eliminates the interruption of executed path and 95% execution speed is enhanced [4].

New model proposed for the fuzzer which detects heap-based buffer overflow based on concolic execution. The model proposed executes the concrete data of target program and the constraints of each path obtained are executed symbolically [5]. The path constraints obtained generate test data that traverses new execution path in target program. Heap buffer overflow was calculated for each executed path. The constraints obtained determine which input data have caused overflow in the executed path. This model again repeats the cycle by generating new test data which is obtained by combination of path and vulnerability constraints, which traverses path and again specifies vulnerabilities in the path [5]. The model tested different groups of programs and the results obtained show vulnerabilities for programs accurately (Tables 1, 2, 3, 4, 5, 6, 7, and 8).

Table 4 Map coverage

Map density	Count coverage
0.54%/0.83%	1.27 bits/tuple

Table 5 Findings in depth

Favored paths	New edges on	Total crashes	Total tmouts
1 (1.72%)	30 (51.72%)	69 (6 unique)	59 (10 unique)

Table 6 Path geometry

Levels	Pending	Pend fav	Own finds	Imported	Stability
2	58	1	57	n/a	100.00%

Table 7 Crashes

ID	Signature	Source file	Operation	Position
000000	11	000000	Flip1	12
000001	11	000000	Flip1	12
000002	06	000000	Flip1	27
000003	06	000000	Flip1	27
000004	06	000000	Flip1	118
000005	06	000000	Flip1	365
000006	06	000002	Flip1	365
000007	06	000003	Flip1	12

MiBO model was created detecting vulnerabilities due to buffer overflow caused by malicious inputs. For identification of MiBO vulnerabilities, white-box testing technique was used to analyze all execution paths. Black-box testing technique triggered MiBO vulnerabilities by providing different inputs, but limited coverage was achieved [9]. When the vulnerabilities were identified, i.e., there was a "hit", thus crash happened by the test input.This paper presented the model which identifies non-crash MiBO vulnerabilities using white-box testing technique, thus dynamically discovers likely memory layouts to help the fuzzing process. This is very challenging since memory addresses and layouts keep changing with the running of software. In different executions with different inputs, the layouts may also change. To address these challenges, we selectively analyze memory operations to identify memory layouts. If a buffer border identified from the memory layout is exceeded, an error will be reported. The fuzzing results will be compared with the layout for future input generation, which greatly increases the opportunity to expose MiBO vulnerabilities. This paper implemented a prototype called ArtFuzz and performed several evaluations. ArtFuzz discovered 23 real MiBO vulnerabilities (including 8 zero-day MiBO vulnerabilities) in nine applications.

Table 8 Hangs

ID	Source file	Operation	Position	Value
000000	000000	Flip1	290	–
000001	000000	Filp1	5	–
000002	000000	Filp1	92	–
000003	000000	Filp1	335	–
000004	000000	Arith8	195	–
000005	000000	Arith8	72	−9
000006	000002	Arith8	106	−18
000007	000003	Arith8	107	−22
000008	000003	Arith8	108	−3
000009	000003	Arith8	108	−9
000010	000003	Arith8	119	−14
000011	000003	Arith8	119	−25
000012	000003	Arith8	121	−29
000013	000003	Arith8	122	17
000014	000003	Arith8	124	−19
000015	000003	Arith8	132	13
000016	000003	Arith8	141	−6
000017	000003	Arith8	146	3

This paper presented a new technique, symbolic execution technique which analyses the vulnerabilities in code [10]. As business is becoming more dependent on software and computer programs so there is need for elimination of security bugs. This paper represents smart fuzzing which uses data from symbolic execution engine which automatically generates input which causes the crash. Symbolic execution engine used generates the data automatically that are against vulnerability issues. This paper also represents a method for user in clarifying error reports because some verification tools fail to verify the program and hence provides the wrong result, so users manually verify the program which becomes time-consuming, and error-prone tasks. This technique computes small queries to user. This proved advantage in program inspection and debugging.

New methods are proposed for detecting vulnerabilities is smart fuzzing. This technique uses concolic execution which detects buffer overflow vulnerability [11]. For each execution path, buffer overflow constraints and path are symbolically calculated. As the calculation of the vulnerability constraints have been done they are stored in stack memory. The vulnerability constraint generated determines the length of input data up to which the overflow in the stack buffer occurs. Estimation of length and address of variables in stack has been considered by the structure of stack and accessing the local variables in binary codes.

Buffer overflow constraints have been calculated on the basis of estimated length and addresses. To find the overflow in stack buffer, the vulnerability constraints considers which part of input data involves in overflow and thus helped in fuzzing. The results of this estimation has calculated stack overflow constraints.

A model called AURORA which is a robust kernel-based solution, to security problem which controls buffer overflow attacks. The model utilizes either addresses of buffer with stores input strings or signature which detects and blocks buffer overflow attacks in kernel. The model creates the signatures based on the buffer overflow attacks so there is no need to create new signatures for instances of new attacks [1]. When the process is under buffer overflow attack, the model allows the execution or termination without the cost of repeated crashes or idleness of process, thus the model is robust to control buffer overflow attacks. There is no need for modification of source code and furthermore model is compatible with application and operating system. This model has shown less than 1% overhead and least false positives and hence accurately blocks various buffer overflow attacks.

3 Proposed Framework

Fuzzing is powerful software testing technique for identification of security issued in the software. Bugs such as remote code execution, privilege escalation are found which are critical for software. American Fuzzy Lop (AFL) is a simple and rock-solid instrumentation-guided genetic algorithm.

These steps are performed on a software pdf crack using AFL fuzzing technique which finds the vulnerability in the software. High level of code coverage is there. This strategy causes afl-fuzz to reach a high level of code coverage for its testing. The fuzzer starts detecting when a new path is being triggered and more input files are created and identification of bugs started.

3.1 Binary Instrumentation

With the availability of source code, the instrumentation is being carried out by the tool which functions as a substitute for gcc or clang in any third-party code compilation process. By using afl-fuzz, most programs produce the fastest results.

If the source code is not available or if the source program recompilation of the destination program cannot/does not want to be done, AFL supports QEMU mode. This is done when the target program could not be built with afl-gcc.

3.2 Choosing Initial Test Cases

For the initial test cases, there is requirement of one or more input files that contain data which is expected by target application for the fuzzer to operate correctly. If the large corpus of data is there for screening, then afl-cmin utility is used. This afl-utility identifies the subset of files. Different code paths originate in target binary for these files.

3.3 Fuzzing Binaries

The process of fuzzing is performed by the model afl-fuzz utility. Fuzzing binaries with the test cases requires read-only directory, to store the findings requires a separate place, and the path to which binary can be tested.

/afl-fuzz—I input—o outout/path/to/program @@

Programs which receive input from file uses "@@", mark location in the command line of target and to where to place the input file name.

3.4 Parallelized Fuzzing

In afl-fuzz, every instance is generally associated with one core, i.e., on a multi-core system, for full utilization of software, parallelization is necessary. This parallel fuzzing mode provides interfacing AFL to others fuzzers and engines such as symbolic and concolic execution engine.

3.5 Interpreting Output

For interpreting the output, the fuzzer completes one queue, which may take some time. In the output directory, three subdirectories are created:

I. queue—In queue subdirectory, there are test cases for which every path is to be executed and the file is provided by the user. To shrink the size of the file, afl-cmin tool can be used. This generally offers equivalent edge coverage for subset of files.

II. crashes—In the tested program some of the test cases receives fatal signal such as SIGSEGV, SIGILL. So these crashes are stored in crash directory.

III. hangs—There are test cases that cause the test cases of the program to time out. Before the test cases hung, there is the default time limit which is larger of 1 s and value of −t parameter. To change the value, AFL_HANG_TMOUT setting is to be done.

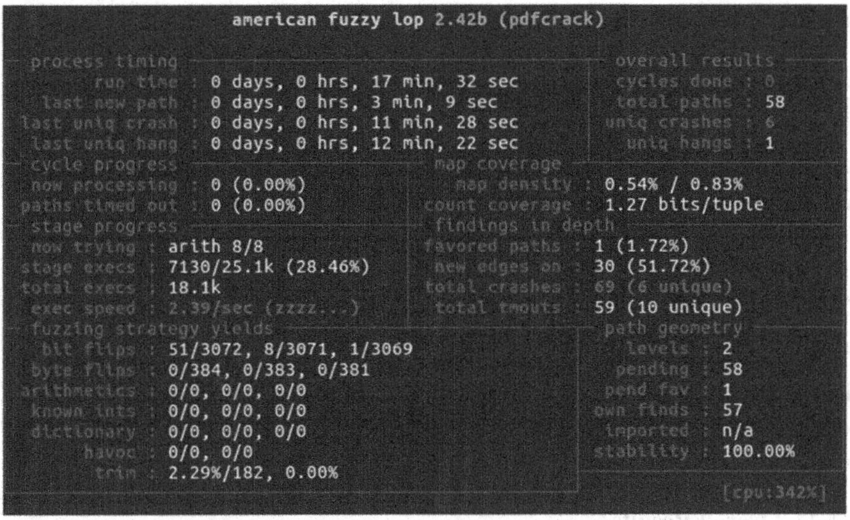

```
                    american fuzzy lop 2.42b (pdfcrack)
 process timing                                   overall results
          run time : 0 days, 0 hrs, 17 min, 32 sec    cycles done : 0
     last new path : 0 days, 0 hrs, 3 min, 9 sec      total paths : 58
   last uniq crash : 0 days, 0 hrs, 11 min, 28 sec    uniq crashes : 6
    last uniq hang : 0 days, 0 hrs, 12 min, 22 sec    uniq hangs : 1
 cycle progress                      map coverage
    now processing : 0 (0.00%)          map density : 0.54% / 0.83%
   paths timed out : 0 (0.00%)       count coverage : 1.27 bits/tuple
 stage progress                   findings in depth
         now trying : arith 8/8         favored paths : 1 (1.72%)
        stage execs : 7130/25.1k (28.46%)  new edges on : 30 (51.72%)
        total execs : 18.1k             total crashes : 69 (6 unique)
         exec speed : 2.39/sec (zzzz...)  total tmouts : 59 (10 unique)
 fuzzing strategy yields                        path geometry
          bit flips : 51/3072, 8/3071, 1/3069        levels : 2
         byte flips : 0/384, 0/383, 0/381           pending : 58
        arithmetics : 0/0, 0/0, 0/0                 pend fav : 1
         known ints : 0/0, 0/0, 0/0                 own finds : 57
         dictionary : 0/0, 0/0, 0/0                 imported : n/a
             havoc : 0/0, 0/0                       stability : 100.00%
              trim : 2.29%/182, 0.00%
                                                          [cpu:342%]
```

Fig. 2 Execution of AFL Fuzzer

If there is state transition involved in the executed path and not previously recorded then crashed and hangs are considered to be "unique". The names of the files in crashes and hangs are related with parent, and thus help in debugging.

3.6 Fuzzer Dictionary

By default, mutation engine of AFL fuzzer performs optimization for compact data formats such as images, multimedia, compressed data, and shell scripts. It is less suited for languages with most detailed and redundant verbiage, for example HTML, SQL, or JavaScript. To avoid the chaos of building syntax tools, afl-fuzz provides a way to seed the fuzzing process with an optional dictionary of language keywords, magic headers, or other special tokens associated with the targeted data type with AFL. New crashes, paths, and hangs are recorded in output directory (Fig. 2).

4   Implementation and Results

4.1   Execution of SAFAL Model

AFL fuzzer fires up and starts executing the program, AFL displays user interface
with number of metrics.

4.1.1   Processing Time

The processing timing describes the run time which specifies for how long the fuzzer
runs. It also mentions the time at which new paths, unique crashes and hangs are
explored. Every new crashes, paths, and hangs are recorded in output directory.

4.1.2   Overall Results

Overall results give the description of number of times the test cases were discovered
and then loop back to beginning. In Fig. 3, at this particular instance, total paths
(cases) that were found were 58 among which 6 were the unique crashes and unique
hangs were 1. Test cases, hangs, and crashes are recorded in output directory.

4.1.3   Map Coverage

Map coverage provides in and out of the target binary. Map density tells about the
number of branch tuples got hit and to proportion the bitmap can hold. Count coverage
it throws light on number of bits being covered for entire input corpus.

4.1.4   Findings in Depth

Findings in depth provide the details about the metrics that are of concern. It provides
with information of the number of test cases with better edge coverage, total number
of crashes and among them which crashes are at priority and the total timeouts.

4.1.5   Path Geometry

In path geometry:

I. The first part is levels which keep track of depth of path. The initial test cases
    are considered at "level 1". The derived test cases from fuzzing are considered
    at "level 2" and so forth.

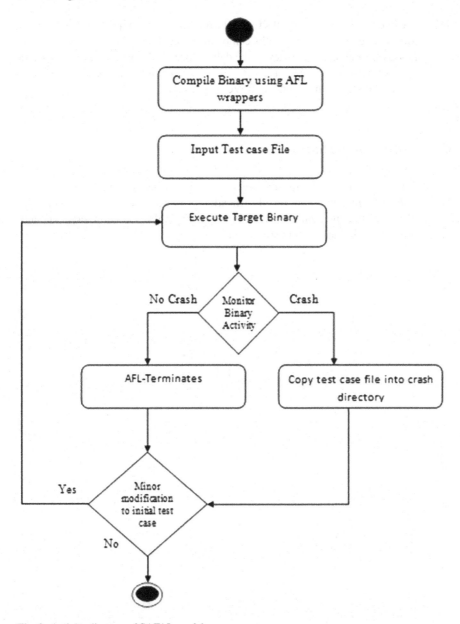

Fig. 3 Activity diagram of SAFAL model

II. Pending fav fields tells about the number of inputs of which fuzzing is yet to be done.

III. Favored field provides with statistics of number of entries want to enter the queue cycle and non-favored entries has to wait for couple of cycles.

IV. Own finds tells about new paths found during fuzzing and if parallelized fuzzing
     process is being going than paths can be imported from other fuzzers.
V.   Stability measures consistency of the data is fuzzed. If the behavior of program
     is the same as input data then stability is 100%. When value goes down, purple
     color is shown which affects the fuzzer negatively.

4.2   Crashes and Hangs

After the execution of the fuzzer, target binary is crashed and the more times the
crashes occur there is potential for identification of flaws. The crashes are stored in
output directory in a file named crashes. For each crash, the operation, position, and
source file at which crash occurs are stored.

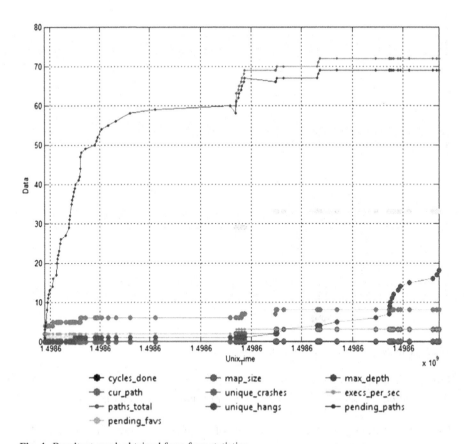

Fig. 4 Resultant graph obtained from fuzz statistics

```
Program received signal SIGABRT, Aborted.
0x00007ffff7a43418 in raise () from /lib/x86_64-linux-gnu/libc.so.6
1: x/i $rip
=> 0x7ffff7a43418 <raise+56>:    cmp    $0xfffffffffffff000,%rax
rax            0x0       0
rbx            0x7ffff7ff5000   140737354092544
rcx            0x7ffff7a43418   140737348121624
rdx            0x6       6
rsi            0xb7b0    47024
rdi            0xb7b0    47024
rbp            0x423e56 0x423e56
rsp            0x7fffffffd938   0x7fffffffd938
r8             0x627340 6452032
r9             0xffffffff00000000      -4294967296
r10            0x8       8
r11            0x246     582
r12            0xe8      232
r13            0x423e80 4341376
r14            0x54      84
r15            0x3a7     935
rip            0x7ffff7a43418   0x7ffff7a43418 <raise+56>
eflags         0x246     [ PF ZF IF ]
cs             0x33      51
ss             0x2b      43
ds             0x0       0
es             0x0       0
fs             0x0       0
gs             0x0       0
```

Fig. 5 Memory corruption

After running the triage script, each crash file in the output directory which passes through the script and hence crash data is printed. In the figure, target binary experienced a crash, i.e., segmentation fault for each crash. This means memory corruption has occurred (Figs. 4 and 5).

The exploitability of crashes and hangs may be ambiguous. Afl-fuzz tried to address by providing a crash exploration mode where a known-faulting test case is fuzzed in a manner similar to the normal operation of the fuzzer, but with a constraint that causes any non-crashing mutations to be thrown away. The table below describes about the hangs that occurred.

5   Conclusion and Future Work

The SAFAL model proposed uses AFL which is a powerful fuzzer, leverages on source code and binaries which founded potential vulnerabilities. This is the first step in exploit development. The model explored bugs in software and provided all the required metrics. It explored each executed path and on every path the possible vulnerability is founded. Stack-based buffer overflow is the software vulnerability and the model proposed to exploit this vulnerability in each executed path. Processing speed of fuzzer is fast and in short time, it explored 72 paths in which there were 8 crashes and 18 hangs. For each crash, bugs are found, and memory address and the type of vulnerability are determined by the model.

For future work, the model can be applied to real-world softwares. Models in machine learning can be applied to this fuzzer to determine the efficiency of occurence of bugs and vulnerability, and classify the vulnerabilities.

References

1. Chen, L.-H.: A robust kernel-based solution to control-hijacking buffer overflow attacks. J. Inf. Sci. Eng **27**(3), 869–890 (2011)
2. DeMott, J.: The evolving art of fuzzing. In: DEF CON 14 (2016)
3. Oehlert, P.: Violating assumptions with fuzzing. IEEE Secur. Priv., 58–62 (2005)
4. Cui, B.: Whirling Fuzzwork: a taint-analysis-based API in-memory fuzzing framework. Soft. Comput. **12**(21), 3401–3414 (2016)
5. Mouzarani, M., Sadeghiyan, B., Zolfaghari, M.: A smart fuzzing method for detecting heap-based buffer overflow in executable codes. In: 21st Pacific Rim International Symposium on. IEEE (2015)
6. Chen, Y.: Taming compiler fuzzers. ACM SIGPLAN Notices 48, no. 6 (2013)
7. Banks, G.: SNOOZE: toward a Stateful NetwOrk prOtocol fuzZEr. In: ISC 4176 (2006)
8. Rawat, S.: Vuzzer: Application-aware evolutionary fuzzing. In: Proceedings of the Network and Distributed System Security Symposium (NDSS) (2017)
9. Chen, K., Zhang, Y., Liu, P.: Dynamically discovering likely memory layout to perform accurate fuzzing. IEEE Trans. Reliab. **65**(3), 1180–1194 (2016)
10. Kim, J.-H., Ma, M.-C., Park, Jae-Pyo: An analysis on secure coding using symbolic execution engine. J. Comput. Virol. Hacking Tech **3**(12), 177–184 (2016)
11. Mouzarani, M., Sadeghiyan, B., Zolfaghari, M.: Smart fuzzing method for detecting stack-based buffer overflow in binary codes. IET Software **10**(4), 96–107 (2016)

Quantum Behaved Particle Swarm Optimization Technique Applied to FIR-Based Linear and Nonlinear Channel Equalizer

Rashmi Sinha, Arvind Choubey, Santosh Kumar Mahto and Prakash Ranjan

Abstract A novel application of quantum behaved particle swarm optimization technique (QPSO) in developing an adaptive channel equalizer based on finite impulse response (FIR) filter is presented. Equalizers form an inherent part of receiver in communication system that help in eliminating distortions in the received data to counterbalance the interference and nonlinearity that appears in practical channels using a simple and proficient optimization algorithm QPSO. Two examples of linear and nonlinear channels are undertaken. The results obtained are compared with those achieved by genetic algorithm (GA), standard PSO, and the conventional least mean square (LMS) methods. Extensive simulation carried out at two values of additive white Gaussian noise (AWGN), 20 and 30 dB reveals that the proposed approach to design the adaptive channel equalizer has an improved performance in terms of both mean square error (MSE) and bit error rate (BER).

Keywords Bit error rate (BER) · Genetic algorithm (GA) · Mean square error (MSE) · Particle swarm optimization (PSO) · Signal-to-noise ratio (SNR)

R. Sinha (✉) · A. Choubey · S. K. Mahto · P. Ranjan
Department of Electronics and Communication Engineering,
National Institute of Technology Jamshedpur, Jamshedpur, India
e-mail: rsinha.ece@nitjsr.ac.in

A. Choubey
e-mail: achoubey.ece@nitjsr.ac.in

S. K. Mahto
e-mail: ec51236@nitjsr.ac.in

P. Ranjan
e-mail: 2014rsec001@nitjsr.ac.in

© Springer Nature Singapore Pte Ltd. 2019   37
S. K. Bhatia et al. (eds.), *Advances in Computer Communication and Computational Sciences*, Advances in Intelligent Systems and Computing 759,
https://doi.org/10.1007/978-981-13-0341-8_4

1   Introduction

Tremendous rise in the demand of Internet Technology (IT) and multimedia applications has led to an unfulfilling urge for developing efficient and reliable data transfer techniques over communication channels. An adaptive equalizer installed at the front end of receiver is the most popular solution to mitigate the effects these aberrations and obtain reliable data transmission [1]. The existing adaptive equalizers that employ linear combiner structure using LMS algorithm and its different variants perform unsatisfactory in the presence of nonlinear distortions, as it cannot converge to the global minimum solution for multimodal and nonuniform objective function [2]. A deluge of adaptive equalizers using soft computing techniques is available in literature which works very well even under nonlinear and noisy channel conditions. Few popular techniques among them are GA [3], BFO [4], AIS [5], DE [6], PSO [7], and its variants [8].

Simplicity, easy coding and more competent performance exhibited by PSO has made it widely popular. Many variants of PSO are introduced in literature [9]. The PSO algorithm is initialized with a population of random solutions each generation is updated to reach an optimum. However, it has no evolution operators such as crossover and mutation as in GA. The potential solution called particles fly through the problem space by following individual own experience and current best particles. Therefore, it can fetch good performance as compared to other evolutionary paradigms. Confinement of particle to a finite sampling space deteriorates its global capability and convergence is not guaranteed in many cases. Inertia weight was introduced into the original PSO to enhance its performance in [10]. An alternative version of PSO was proposed in [11] where the parameter, constriction factor was added to regulate the restriction on velocities. Several modifications of PSO have been suggested [12–16].

Quantum behaved PSO introduced in [17, 18] was inspired from quantum mechanics. Quantum state is transformed or collapsed into classical state to measure particle position. The search of each particle begins in a direction starting from the individual current position and ends at a point that lies in between particle's individual best position pbest and particle's global best position gbest. Therefore, QPSO is preferred to standard PSO algorithm as far as search capability is concerned. Wide range of continuous optimization problems using QPSO has been successfully solved by researchers belonging to different technological communities, especially in field of electromagnetic [19], online system identification [20], image processing, electromagnetics, and microelectronics. It will undoubtedly continue to dominate the field of optimizing engineering problems.

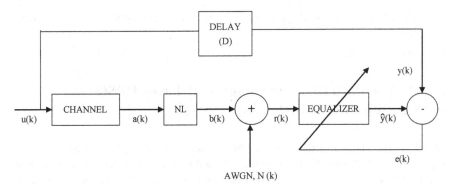

Fig. 1 Equalization in a digital communication scenario

Here, QPSO technique is successfully implemented in designing linear and nonlinear FIR-based adaptive channel equalizer under noisy conditions, which is yet an unexplored application. Two examples of linear and nonlinear channels are considered. The results obtained are compared to those achieved by GA, standard PSO, and the conventional LMS. Extensive simulation carried out at two values of AWGN (20 and 30 dB) reveals that the proposed approach to design the adaptive channel equalizer has an improved performance.

The remaining paper is arranged as follows. The basic concept of channel equalizer is discussed in Sect. 2 briefly. Section 3 contains PSO and QPSO theory. The proposed equalizer using QPSO is explained. Extensive simulation work and discussion are recorded in Section. The paper is finally concluded in Sect. 6.

2 Adaptive Channel Equalizer

It is pivotal to mitigate the interference effect of communication to achieve reliable digital transmission in today's crowded communication channels. This is where adaptive equalizers find its utility [1]. A typical digital communication scenario is illustrated in Fig. 1. The nonlinearity introduced by the channel is represented by NL.

The NL distorted output b(k) of the channel is $b(k) = f(a(k))$, where $f(\cdot)$ is the nonlinear function. The channel is assumed to be affected by AWGN, $N(k)$ with variance (σ^2). Input to equalizer is $r(k) = b(k) + N(k)$. Output equation of the equalizer, $\hat{y}(k)$ is compared with the desired signal $y(k)$ and the difference of these two signals forms the error signal $e(k)$ at the kth sample which is given as

$$e(k) = y(k) - \hat{y}(k); \qquad (1)$$

where $0 < k < M$, D and M represent the time delay and number of input samples, respectively.

3   The Standard PSO and Quantum Behaved QPSO Algorithms: Revisited

Eberhart and Kennedy [9] proposed PSO, which is an algorithm based on socio-psychological behavior of swarms which is open to adaptation. A population adapts and returns to the previously obtained best position. The flight of each particle is function of previous best, pbest and global best, gbest.

3.1   Standard PSO Algorithm

Each individual in an R-dimensional population space of swarm consisting of M individuals is assumed to be volumeless. For ith particle, position vector is represented as $X_i(t) = (X_{i1}(t), X_{i2}(t) \ldots X_{iR}(t))$ and velocity vector as $V_i(t) = (V_{i1}(t), V_{i2}(t), \ldots V_{iR}(t))$.

Movement of the particle is represented [12] as

$$V_{ij}(t+1) = \omega * V_{ij}(t) + C1 * \text{rand}() * \left(P_{ij}(t) - X_{ij}(t)\right)$$
$$+ C2 * \text{rand}() * \left(P_{gi}(t) - X_{ij}(t)\right) \tag{2}$$

for $i = 1, 2, \ldots M$; $j = 1, 2, \ldots R$. inertia weight, ω is tuned very carefully for good performance, and C1 and C2 are the acceleration constants. rand() is a uniformly distributed random number in the range [0, 1]. The personal best position of the ith particle is given by $Pi = (Pi1, Pi2 \ldots PiR)$, Vector $Pg = (Pg1, Pg2 \ldots PgR)$ is the global best position. The position of ith particle is then updated according to [9],

$$X_{ij}(t+1) = X_{ij}(t) + V_{ij}(t+1). \tag{3}$$

Vi vector is clamped to the range $[-V_{max}, V_{max}]$ so that particle's chance of abandoning the search space reduces.

3.2 QPSO Algorithm

The trajectory of a particle in the swarm cannot be determined by position vector X_i and velocity vector V_i following the Newtonian mechanics as is the case in PSO. This is due to fact that principle of uncertainty state that position and velocity of the particle cannot be determined simultaneously [17]. Therefore, the quantum state of a particle is represented by wave function $\Psi(X, t)$ [18] given by

$$|\Psi(X, t)|^2 = \frac{1}{L} \exp(-2\|P - X\|/L) \tag{4}$$

L is control parameter, the learning inclination point (LIP) or creativity or imagination and P is the center of delta potential well. Pbest of each particle is recorded and compared to other particles in the group to react gbest at each iteration. Random numbers $\varphi1$ and $\varphi2$ help in setting center of gravity of delta potential well, which is the only local attractor of each particle represented by

$$P = (\varphi_1 P_i + \varphi_2 P_g)/(\varphi_1 + \varphi_2) \tag{5}$$

so that local optimal point (Pi) or pbest converges to global optimal point (Pg) or gbest at $t \to \infty$.

$$L = 2 * \alpha * |P - X(t)| \tag{6}$$

where α is control parameter for delta potential field. Search space is transformed to solution space to fetch the new position

$$X_i(t + 1) = P + L*|P_i - X(t)| * \ln\left(\frac{1}{u}\right) \qquad \text{if } u > 0.5$$

$$X_i(t + 1) = P - L*|P_i - X(t)| * \ln\left(\frac{1}{u}\right) \qquad \text{if } u < 0.5 \tag{7}$$

where u is a random number distributed in range [0, 1]. The condition that must be satisfied to convergence is

$$\lim_{t \to \infty} L(t) = 0 \tag{8}$$

That is, when $L \to 0$, then $X \to P$ at $t \to \infty$. *pbest* is replaced by new position if it has better knowledge.

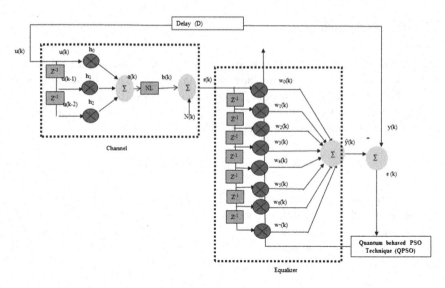

Fig. 2 Configuration of QPSO-based channel equalizer []

4 QPSO-Based Channel Equalizer

An adaptive channel equalizer consisting of finite impulse response filter (FIR) with QPSO algorithm employed to adjust the parameters (coefficients) of equalizer is illustrated in Fig. 2. A three-tap FIR filter h(z) taken as the channel output of channel at the kth instant is

$$a(k) = \sum_{L=0}^{2} h(L) * u(k - L) \tag{9}$$

where $h = [h_0, h_1, h_2]$ that represents the coefficient vector of linear channel filter h(z) and $u(k) = [u(k), u(k - 1), u(k - 2)...u(k - n)]$ is the binary input vector applied to the channel.

The output b(k) of nonlinear channel is

$$b(k) = f(a(k)) \tag{10}$$

Received output to the equalizer is

$$r(k) = b(k) + N(k) \tag{11}$$

The estimated output of the equalizer is given by

Table 1 Steps to implement channel equalizer using QPSO technique

Step#	Description
1.	Define the solution space
2.	Initialize position of particles (tap weights), X_i and velocities V_i inside R(=8) dimensional solution space
3.	Calculate cost function using (13)
4.	Perform the comparison between current and previous best cost function, Ji If $Ji(X_i) < Ji(pbest_i)$; then $pbest_i = X_i$ If $Ji(X_i) < Ji(gbest_i)$; then $gbest_i = X_i$
5.	Appraise P according to (5) and X_i using (7a) and (7b)
6.	Check stopping criteria otherwise repeat from step # 4

$$\hat{y}(k) = \sum_{Q=0}^{7} r(k - Q) * w(k) \tag{12}$$

where $w = [w_0, w_1 \ldots w_7]$ is the weight vector and length of equalizer is Q. Here, the cost/objective function J for ith particle is given by

$$J_i(k) = \frac{1}{N} \sum_{k=1}^{N} e_{ji}^2(k); \quad 1 < i < M \tag{13}$$

where e_{ji} is the jth error for the ith particle. This work is to minimize (13) by adjusting the weight vector iteratively using QPSO algorithm. The methodology to implement QPSO in the equalizer is described stepwise in Table 1.

5 Simulation Results and Discussions

Performance of the channel equalizer is investigated through extensive simulation carried out on two sets of channel under linear and nonlinear conditions at 20 and 30 dB SNR.

Example 1: CH1 NL0: Linear channel: $0.2090 + 0.9950 \, z^{-1} + 0.2090 \, z^{-2}$

Example 2: CH2 NL0: Linear channel: $0.2600 + 0.9300 \, z^{-1} + 0.2600 \, z^{-2}$

Nonlinear function $\tanh(\bullet)$ is incorporated in the above channel to make it into nonlinear channel, CH1 NL1 and CH2 NL1 (Table 2).

Table 2 QPSO, PSO, GA, and LMS parameters

Parameters	QPSO	Standard PSO	GA	LMS
Population size	80	80	80	–
Iteration cycles	200	200	200	200
Crossover rate	–	–	0.9	–
Mutation rate	–	–	0.5	–
Selection probability	–	–	1/3	–
Acceleration constant, C_1	2	2	–	–
Acceleration constant, C_2	2	2	–	–
Control parameter, α	0.6	–	–	–
Inertia weight, ω	–	0.9–0.4	–	–
Step size	–	–	–	0.08

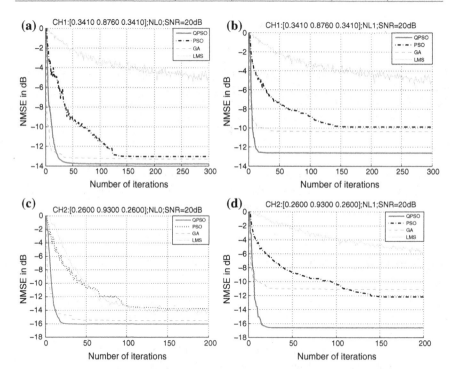

Fig. 3 **a–d** Convergence plot of CH1NL0, CH1NL1 and CH2NL0, CH2NL1 at 20 dB SNR

5.1 Convergence Characteristics

Comparative convergence characteristics at 20 dB and 30 dB SNR are depicted in Fig. 3 and Fig. 4, respectively, and the numerical results are summarized in Table 3 and Table 4. QPSO works better than standard PSO, GA, and LMS for both linear and nonlinear channel.

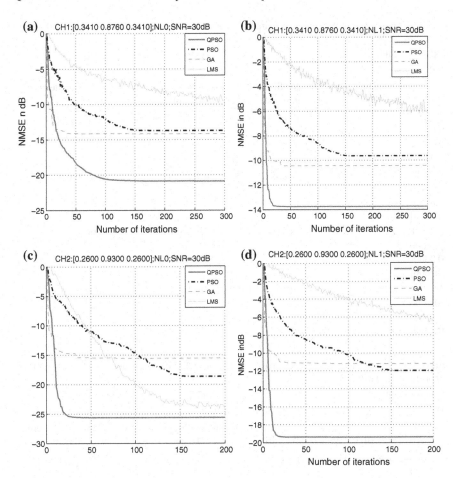

Fig. 4 **a–d** Convergence plot of CH1NL0, CH1NL1, and CH2NL0, CH2NL1 at 30 dB SNR

Normalized mean square error (NMSE) obtained by QPSO for SNR equal to 20–16.57 dB at 25 iterations where it is −12.13 dB at 152 iterations by standard PSO and −11.07 dB at 89 iterations by GA for CH2 NL1. LMS algorithm converges extremely slowly (beyond 200 iterations) or does not converge at all, for nonlinear channels. When SNR is increased to 30 dB, NMSE −25.55 dB at 35 iterations as compared to −18.62 dB at 157 iterations by standard PSO, −15.52 dB at 79 iterations by GA and −23.64 dB at 179 iterations by LMS for CH2 NL0. However, GA converges faster in case of the linear channel, CH1 NL0 (SNR = 30 dB) at 43 iterations against 157 iterations by QPSO and standard PSO and 270 iterations by LMS.

Table 3 Comparison of convergence characteristics of QPSO-based channel equalizer with other algorithms at 20 dB

Channel type	QPSO		PSO		GA		LMS	
	NMSEin (−dB)	Iterations	NMSE in (−dB)	Iterations	NMSE in (−dB)	Iterations	NMSE in (−dB)	Iterations
CH1NL0: [0.3410 0.8760 0.3410]	13.75	62	13.02	156	13.26	150	5.006	269
CH1NL1: [0.3410 0.8760 0.3410]	12.61	29	9.855	163	10.29	36	Do not converge	
CH2NL0: [0.2600 0.9300 0.2600]	16.03	35	13.73	150	15.50	80	13.99	126
CH2NL1: [0.2600 0.9300 0.2600]	16.57	25	12.13	152	11.07	89	Do not converge	

Table 4 Comparison of convergence characteristics of QPSO-based channel equalizer with other algorithms at 30 dB

Channel type	QPSO		PSO		GA		LMS	
	NMSE in (−dB)	Iterations	NMSE in (−dB)	Iterations	NMSE in (−dB)	Iterations	NMSE in (−dB)	Iterations
CH1NL0: [0.3410 0.8760 0.3410]	**20.83**	157	13.68	157	14.1	**43**	5.968	270
CH1NL1: [0.3410 0.8760 0.3410]	**13.71**	**26**	9.62	152	10.43	35	Do not converge	
CH2NL0: [0.2600 0.9300 0.2600]	**25.55**	**35**	18.62	157	15.52	79	23.64	179
CH2NL1: [0.2600 0.9300 0.2600]	**19.36**	23	11.94	150	11.131	31	Do not converge	

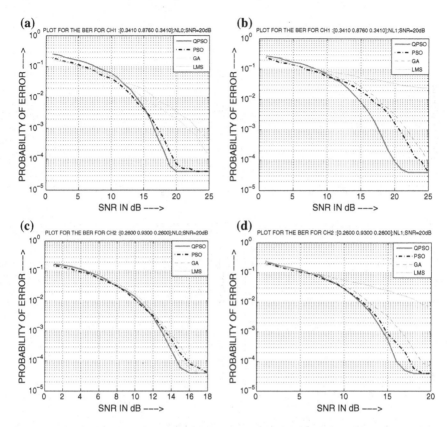

Fig. 5 a–d BER plot of CH1 NL0, CH1 NL1, and CH2 NL0, CH2 NL1 at 20 dB SNR

QPSO is capable of converging rapidly to below −16 dB in most cases which is faster than other algorithms considered in this paper.

5.2 BER Performance

BER is the true performance indicator for equalization task. Comparative BER performance at 20 dB and 30 dB SNR is shown in Fig. 5 and Fig. 6, respectively. QPSO shows an improved and consistent performance up to 3–4 dB better than standard PSO and GA, for nonlinear channel CH1 NL1 and 5 dB better than GA for linear channel CH1 NL0. Performance comparison is summarized in Tables 5 and 6. It confirms that an adaptive nonlinear equalizer trained by QPSO produces better overall BER performance.

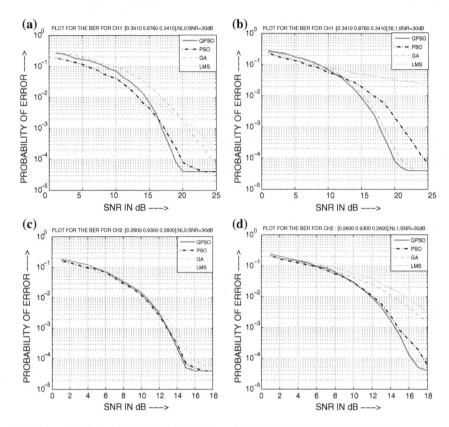

Fig. 6 a–d BER plot of CH1 NL0, CH1 NL1, and CH2 NL0, CH2 NL1 at 30 dB SNR

Table 5 BER improvement of QPSO over other algorithms at 10^{-3} and 20 dB SNR

Channel	Gain in SNR (dB) over standard PSO	Gain in SNR (dB) over standard GA	Gain in SNR (dB) over standard LMS
CH1NL0: [0.3410 0.8760 0.3410]	0	5	6
CH1NL1: [0.3410 0.8760 0.3410]	3	4	Out of range
CH2NL0: [0.2600 0.9300 0.2600]	0	1	0
CH2NL1: [0.2600 0.9300 0.2600]	1	3	Out of range

Table 6 BER improvement of QPSO over other algorithms at 10^{-3} and 30 dB SNR

Channel	Gain in SNR (dB) over standard PSO	Gain in SNR (dB) over standard GA	Gain in SNR (dB) over standard LMS
CH1NL0: [0.3410 0.8760 0.3410]	0	4	5
CH1NL1: [0.3410 0.8760 0.3410]	3	0	Out of range
CH2NL0: [0.2600 0.9300 0.2600]	0	1	0
CH2NL1: [0.2600 0.9300 0.2600]	1	4	Out of range

6 Conclusion

A novel based channel equalizer based on QPSO is proposed. The supremacy of this methodology is established by comparing with other popular and successful methods such as LMS, standard PSO, and GA. The simulation results demonstrate the improved performance of QPSO is obtained without increase in the computation complexity compare to other reported evolutionary algorithm.

References

1. Haykin, S.: Communication Systems. Wiley (2008)
2. Widrow, B., Stearns, S.D.: Adaptive Signal Processing, vol. 491, p. 1. Englewood Cliffs, NJ, Prentice-Hall, Inc. (1985)
3. Merabti, H., Massicotte, D.: Nonlinear adaptive channel equalization using genetic algorithms. In: 2014 IEEE 12th International New Circuits and Systems Conference (NEWCAS). IEEE (2014)
4. Majhi, B., Panda, G., Choubey, A.: On the development of a new adaptive channel equalizer using bacterial foraging optimization technique.In: 2006 Annual IEEE India Conference. IEEE (2006)
5. Nanda, S.J., Panda, G., Majhi, B.: Improved identification of Hammerstein plants using new CPSO and IPSO algorithms. Expert Syst. Appl. **37**(10), 6818–6831 (2010)
6. Karaboga, N., Cetinkaya, B.: Design of digital FIR filters using differential evolution algorithm. Circuits Syst. Signal Process. **25**(5), 649–660 (2006)
7. Yogi, S., et al.: Equalization of digital communication channels based on PSO algorithm. In: 2010 IEEE International Conference on Communication Control and Computing Technologies (ICCCCT). IEEE (2010)
8. Al-Awami, A.T., et al.: A new modified particle swarm optimization algorithm for adaptive equalization. Digital Signal Process. **21**(2), 195–207 (2011)
9. Eberhart, R.C., Kennedy, J.: A new optimizer using particle swarm theory. In: Proceedings of the Sixth International Symposium on Micro Machine and Human Science, vol. 1 (1995)
10. Shi, Y., Eberhart, R.: A modified particle swarm optimizer. In: The 1998 IEEE International Conference on Evolutionary Computation Proceedings, 1998. IEEE World Congress on Computational Intelligence. IEEE (1998)

11. Clerc, M.: The swarm and the queen: towards a deterministic and adaptive particle swarm optimization. In: Proceedings of the 1999 Congress on Evolutionary Computation, 1999. CEC 99, vol. 3. IEEE (1999)

12. Eberhart, R.C., Shi, Y.: Comparing inertia weights and constriction factors in particle swarm optimization. In: Proceedings of the 2000 Congress on Evolutionary Computation, 2000, vol. 1. IEEE (2000)

13. Hu, X., Eberhart, R.C.: Adaptive particle swarm optimization: detection and response to dynamic systems. In: Proceedings of the IEEE Congress on Evolutionary Computation, CEC2002. IEEE (2002)

14. Bratton, D., Kennedy, J.: Defining a standard for particle swarm optimization. In: 2007 IEEE Swarm Intelligence Symposium. IEEE (2007)

15. Janson, S., Middendorf, M.: A hierarchical particle swarm optimizer and its adaptive variant. IEEE Trans. Syst. Man Cybern. Part B (Cybern.) $35(6)$, 1272–1282 (2005)

16. Krohling, R.A., dos Santos Coelho, L.: PSO-E: particle swarm with exponential distribution. In: 2006 IEEE International Conference on Evolutionary Computation. IEEE (2006)

17. Sun, J., Feng, B., Xu, W.: Particle swarm optimization with particles having quantum behavior. Cong. Evol. Comput. (2004)

18. Yang, S., Wang, M., Jiao, L.: A quantum particle swarm optimization. IEEE Cong. Evol. Comput. 1, 320–324 (2004)

19. Dos Santos Coelho, L., Alotto, P.: Global optimization of electromagnetic devices using an exponential quantum-behaved particle swarm optimizer. IEEE Trans. Magn. $44(6)$, 1074–1077 (2008)

20. Gao, F., Tong, H.Q.: Parameter estimation for chaotic system based on particle swarm optimization. Acta Phys. Sinica $55(2)$, 577–582 (2006)

Deep ConvNet with Different Stochastic Optimizations for Handwritten Devanagari Character

Mahesh Jangid and Sumit Srivastava

Abstract In this paper, we present a deep learning model to recognize the handwritten Devanagari characters, which is the most popular language in India. This model aims to use the deep convolutional neural networks (DCNN) to eliminate the feature extraction process and the extraction process with the automated feature learning by the deep convolutional neural networks. It also aims to use the different optimizers with deep learning where the deep convolution neural network was trained with different optimizers to observe their role in the enhancement of recognition rate. It is discerned that the proposed model gives a 96.00% recognition accuracy with fifty epochs. The proposed model was trained on the standard handwritten Devanagari characters dataset.

Keywords Handwritten character recognition · Deep learning · Convolutional neural network · Stochastic optimization · Gradient-based learning

1 Introduction

Handwritten character recognition (HCR) can be explained as the system of automatically recognizing the characters from a digitalized or scanned handwritten document. This system consists of primarily preprocessing, segmentation, feature extraction, classification, and postprocessing phases. Although there are a lot of challenges, can be seen at each phase of HCR system and many researches have been contributed a lot towards the solution of these challenges. The style of writing of same document by different writer is different that makes HCR system more challenging and difficult for research community. There are different types of OCR model such as template-based [1], feature-based [2], and hybrid-based systems [1]. In the last decade, feature based systems have been paid more attention than others. Moment-based features [3], gradient direction features [4], shadow-based features [5], graph-based features

M. Jangid (✉) · S. Srivastava
School of Computing & Information Technology, Manipal University Jaipur, Jaipur, India
e-mail: mahesh_seelak@yahoo.co.in

© Springer Nature Singapore Pte Ltd. 2019
S. K. Bhatia et al. (eds.), *Advances in Computer Communication and Computational Sciences*, Advances in Intelligent Systems and Computing 759,
https://doi.org/10.1007/978-981-13-0341-8_5

[6], etc., are few examples. Entire features are handcrafted features and explicitly designed. Some researchers have also proposed different models which do not require the crafted and designed features. The model itself has a capability to generate the feature from raw image for further processing. Artificial neural network [7], hidden markov model [8], and deep learning [9] are the examples of these models.

Despite the development in feature-based systems in the last decade, today deep learning based approaches have drawn the more attention from researcher's community to develop the efficient and accurate HCR system. In spite of handwritten character recognition, the deep learning approaches are being used in many areas like speech, natural language processing, computer vision applications, etc. Deep learning concept occurred from the study on artificial neural networks [10]. A comprehensive survey of deep neural architectures and their applications can be found in [11].

2 Previous Work

Ample works on handwritten Devanagari character recognition have been reported in the past years. There are numerous research [1, 3, 12, 13] works which have been discussed the history of handwritten Devanagari character and numeral recognition. The first research work has been published by Sethi and Chatterjee [14] in 1976. They recognized the handwritten Devanagari numerals by a structure approach which found the existence and their positions of horizontal and vertical line segment, D-curve, C-curve, left slant, and right slant. A directional chain code based feature extraction technique has been used by Sharma [2]. A bounding box of a character sample is divided into blocks and computed 64-D direction chain code features from each divided block and then applied a quadratic classifier for the recognition of 11270 samples. They reported 80.36% accuracy on handwritten Devanagari characters. Deshpande [15] used the same chain code features with regular expression to generate an encoded string from character and improved the recognition accuracy by 1.74%. A two-stage classification approach for handwritten characters has been reported by Arora [16], she used structural properties of characters like shirorekha, spine in the first stage and, in another stage, used intersection features. These features further fed to a neural network for the classification. She also defined the method for finding the shirorekha properly. This approach has been tested on 50,000 samples and obtained 89.12% accuracy. In the paper [17], Arora has combined the different features like chain code, four side views and shadow based. These features have been fed in multilayer perceptron neural network to recognize the 1500 handwritten Devanagari characters and obtained 89.58% accuracy.

A fuzzy model based recognition approach has been reported by Hanmandlu [5]. The features have been extracted by the box approach which divided the character into 24 boxes (6 × 4 grid) and a normalized vector distance for each box has been computed except the empty boxes. A reuse policy is also used to enhance the speed of the learning of 4750 samples and obtained 90.65% accuracy. This paper [18] computed shadow features, chain code features and classified the 7154 samples using two multilayer perceptrons and a minimum edit distance method for handwritten Devanagari characters. They reported 90.74% accuracy. Kumar [19] has tested five different features named Kirsch directional edges, chain code, directional distance distribution, gradient and distance transform on the 25,000 handwritten Devanagari characters and reported 94.1% accuracy. During the experiment, he found that the gradient feature outperforms the remaining four features with the SVM classifier and Kirsch directional edges feature was the least performer. A new kind of features was also created that computed total distance in four directions after computing the gradient map and neighborhood pixels weight from the binary image of the sample. In the paper [20], Pal has applied the mean filter 4 times before extracting the direction gradient features that have been reduced using Gaussian filter. They used modified quadratic classifier on 36,172 samples and reported 94.24% accuracy using cross-validation policy. Pal [4] has further extended his work with SVM and MIL classifier on the same database and obtained 95.13% and 95.19% recognition accuracy respectively.

Despite the high recognition rate achieved by existing methods, there is still space for the improvement of handwritten Devanagari character recognition.

3 Proposed Recognition Strategy

The recognition of handwritten Devanagari character has been done by deep convolution neural network (DCNN) which is biologically inspired machine learning strategy. The process of object recognition like face recognition, etc., have mainly two steps named as feature extraction and the classification steps likewise the DCNN has these two steps in its architecture as shown in Fig. 1. The convolution (Conv) step works like a feature extraction which is handcrafted and requires a lot of skills to observe them but the DCNN automatically finds the best features for the model. The fully connected (FC) step works like a classification step.

Fig. 1 The conceptual deep convolution neural network architecture

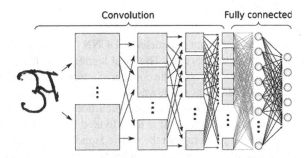

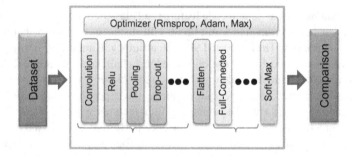

Fig. 2 The schematic diagram of deep convolution neural network architecture

3.1 Deep Convolution Neural Network

A DCNN contains the sequence of convolution, relu, pooling, drop-out, flatten, fully connected and soft-max layers, and each layer has a connection with the previous layer as shown in Fig. 2. DCNN is performed convolution in place of the general matrix multiplication as perform in simple neural network, thereby the complexity of the network can be reduced. The raw images are directly fed to the deep network. There are three important factors in the learning process of a DCNN: sparse interaction, parameter sharing, and equivariant representation [21]. Spare interaction plays a key role to reduce the computation burden during the learning phase of DCNN. The kernels or filter are being used and the value of kernels is kept smaller than the intensity value of raw input images. Further, the parameters are also being shared so that in spite of learning, a separate set of parameters at each location, learning can be done with only one set of them. That helps in getting better performance of the DCNN. The next important factor of DCNN is equivariance representation which means that whenever there are changes in the input, the changes are also realized in the output as well.

3.2 RMSProp

One of the important parameter of DCNN is learning rate which is further divided in local and global learning. The local learning rate helps in faster training and better convergence of DCNN. The concept of adaptive local learning is used by Hinton [22] called as RMSProp which was designed for stochastic gradient descent. RMSProp is an updated version of rprop which does not work with mini-batches. Rprop is the same as gradient but it is also divided by the size of the gradient but RMSProp keeps a moving average of the squared gradient for each weight and further it divides the gradient by square root of mean square value. The first moving average of the squared gradient is given by

$$Av_t := \gamma Av_{t-1} + (1 - \gamma)(\nabla Qw)^2 \tag{1}$$

where γ is the forgetting factor and ∇Qw is the derivative of the error, and Av_{t-1} is previously adjustment value. The weights are updated as per the following equation:

$$w_{t+1} := w_t - \frac{\alpha}{\sqrt{Av_t}} \nabla Qw \tag{2}$$

where w is the previous weight and w_{t+1} is the updated weight and α is the global learning rate.

3.3 Adam and AdaMax

Adam (adaptive moment estimation) [23] is another optimizer for DCNN that needs first-order gradient with small memory and computes adaptive learning rates for different parameters. This method has proven to be better than the RMSprop and rprop optimizers. The rescaling of the gradient is dependent on the magnitudes of parameter updates. The adam does not need a stationary objective and works with sparse gradients. It also contains a decay average of past gradients M_t.

$$M_t = B_1 M_{t-1} + (1 - B_1)G_t \tag{3}$$
$$V_t = B_2 V_{t-1} + (1 - B_2)G_t^2 \tag{4}$$

where M_t and V_t are calculated first and second moment of the gradients and these values are biased towards zero when the decay rates are small thereby bias correction does first and second moments estimates:

$$\check{M}_t = \frac{M_t}{1 - B_1^t} \tag{5}$$

$$\check{V}_t = \frac{V_t}{1 - B_2^t} \tag{6}$$

As per authors of adam, the default values of B_1 and B_2 were fixed at 0.9 and 0.999. Empirically, they have shown its work in practice and a good choice as an adaptive learning method.

Adamax is an extension of adam, where in place of L^2 norm, L^P norm based update rule has been followed.

3.4   Detailed Description of Deep Convolution Neural Network

The schematic diagram of deep convolution neural network has shown in Fig. 2. A deep network may contain the multiple sections of similar layers. In our case, there were multiple sections of convolution, Relu, and pooling layers. The first layer called input layer which takes the raw images that were first normalized in 64×64 dimension. The convolution layer performed convolution operation over the image that requires the kernel size $K \times K$ and the number of different kernel Kr_n. The outcome of convolution layer called feature map will be $64 \times 64 \times Kr_n$ size if no overlap and padding have set already otherwise, the size of feature map may reduce. The pooling layer has followed after convolution layer that acts as downsamplers and enables us to reduce dimensionality by performing the max, min or mean operators on feature map. In our experiment, the block size for pooling layer has kept 4×4. The deep neural network has a lot of neuron in each layer that introduces the overfitting problem. There are many techniques that have been observed in literature. The dropout is one of the best ways to keep our model away from the overfitting problem.

4   Experimental Results

Many set of experiments were performed to compare the recognition rate of handwritten Devanagari characters with different setups of DCNN and also with variation in the selection of optimizer. The experiments were executed on the ParamShavak supercomputing system having 2 multicore CPUs with each CPU consists of 12 cores along with two number of accelerator cards. This system has 64 GB RAM with CentOs 6.5 operating system. The handwritten Devanagari dataset [2] has 36,172 samples and 75% (27,129) of samples were used for training and remaining 25% of samples were used for the testing (validation) purpose. The deep neural network model were coded in Python using Keras a high level neural network API that uses Theano Python library. The basic preprocessing tasks like background elimination, gray normalization and image resizing were done in Matlab.

The representation of the architecture is followed as described in [9]. The architecture of the first model was 64C4-4P2-64C4-4P2-64C4-4P2-1000FC-47SM that applied on handwritten Devanagari characters without padding and with padding layer before convolution layer to observe the impact of it on this model. From Fig. 3 and Table 1, it can clearly be observed that the DCNN model performed better with padding used for training data as well as validation data. In our further experiments, we used padding layer as well and the DCNN will be tested up to 50 epochs.

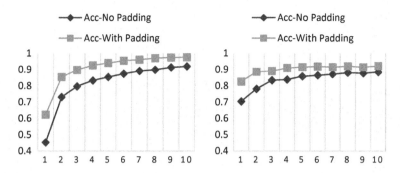

Fig. 3 The left graph has shown the rate of convergence (Accuracy) of DCNN model with training data and right graph has shown the rate of convergence (Accuracy) of DCNN model on validation data

Table 1 Accuracy reported during DCNN model were executed with and without padding layer

Epoch	Training		Validation	
	Accuracy no padding	Accuracy with padding	Accuracy no padding	Accuracy with padding
1	0.4528	0.6233	0.7046	0.8264
2	0.7316	0.8548	0.7812	0.8865
3	0.7966	0.8980	0.8342	0.8910
4	0.8327	0.9263	0.8382	0.9090
5	0.8541	0.9405	0.8582	0.9151
6	0.8747	0.955	0.8655	0.9194
7	0.8916	0.9613	0.8721	0.9154
8	0.8990	0.9707	0.8810	0.9206
9	0.9135	0.9747	0.8782	0.9151
10	**0.9191**	**0.9760**	**0.8851**	**0.9223**

The aforesaid DCNN model was trained and tested with different optimizers to find the best optimizer out of adam, adamax, and RMSProp. Figure 4 shows the accuracy of DCNN model at different optimizer and it can be clearly seen that RMSProp out formed the adam and adamax optimizers. Adam optimizer has also given the better accuracy than the adamax. The only maximum and minimum value of accuracy at each optimizer is reported in Table 2 and the maximum accuracy obtained is 96% on the handwritten Devanagari character recognition. This work is also compared with the previously reported works on the handwritten Devanagari characters as shown in Table 3 and the highest recognition accuracy is found by this work (Fig. 5).

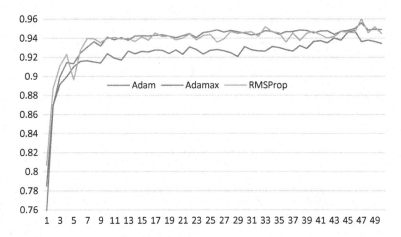

Fig. 4 Accuracy (rate of convergence) of DCNN model at different optimizers

Table 2 Max and Min value of accuracy at different optimizer

Accuracy	Adam	Adamax	RMSProp
Maximum value	0.9558	0.9479	**0.9600**
Minimum value	0.7585	0.7851	**0.8067**

Table 3 Comparison of recognition accuracy by other researchers

S.No	Accuracy obtained	Feature; classifier	Method proposed by	Data size
1	80.36	Chain code; quadratic	Sharma [24]	11,270
2	82	Chain code; RE and MED	Deshpande [16]	5000
3	89.12	Structural; FFNN	Arora [5]	50,000
4	90.65	Vector Dist.; fuzzy sets	Hanmandlu [18]	4750
5	94.10	Gradient; SVM	Kumar [20]	25,000
6	95.19	Gradient; MIL	Pal [25]	36,172
7	95.24	GLAC; SVM	Jangid [2]	36,172
8	**96.00**	Masking, SVM	Proposed work	36,172

5   Conclusion

This paper applied the deep convolution neural network on the handwritten Devana-gari characters. The ISIDCHAR dataset was tested only with shallow technique where feature extraction and classification steps were used. Deep learning approach has been applied for the first time on this dataset and 96% accuracy which is more

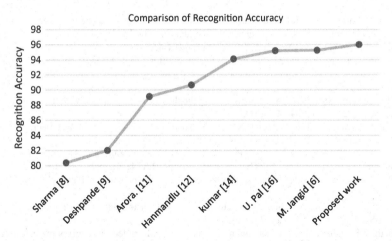

Fig. 5 Improvement in accuracy of handwritten Devanagari characters

than the previously reported accuracy has been obtained. During the experiment, it has been found that the padding layer and pooling layer play the important role in the improvement of convergence rate. The experiments were done with different optimizers of deep convolution neural network and RMSProp was found as the best optimizer for aforementioned model.

Acknowledgements The authors are thankful to the ISI, Kolkata to provide dataset and the Manipal University Jaipur to provide the supercomputing facility without this facility deep learning concept was not possible on the handwritten Devanagari characters.

References

1. Trier, Ø.D., Jain, A.K., Taxt, T.: Feature extraction methods for character recognition-a survey. Pattern Recogn. **29**(4), 641–662 (1996)
2. Jangid, M., Srivastava, S.: Gradient local auto-correlation for handwritten Devanagari character recognition. In: 2014 International Conference on High Performance Computing and Applications (ICHPCA), pp. 1–5. IEEE (2014)
3. Kan, C., Srinath, M.D.: Invariant character recognition with Zernike and orthogonal Fourier-Mellin moments. Pattern Recogn. **35**(1), 143–154 (2002)
4. Pal, U., Sharma, N., Wakabayashi, T., Kimura, F.: Off-line handwritten character recognition of devnagari script. In: Ninth International Conference on Document Analysis and Recognition, 2007. ICDAR 2007, vol. 1, pp. 496–500. IEEE (2007)
5. Arora, S., Bhattacharjee, D., Nasipuri, M., Basu, D.K., Kundu, M., Malik, L.: Study of different features on handwritten Devnagari character. In: 2009 2nd International Conference on Emerging Trends in Engineering and Technology (ICETET), pp. 929–933. IEEE (2009)
6. Bunke, H., Riesen, K.: Recent advances in graph-based pattern recognition with applications in document analysis. Pattern Recogn. **44**(5), 1057–1067 (2011)
7. Rajavelu, A., Musavi, M.T., Shirvaikar, M.V.: A neural network approach to character recognition. Neural Netw. **2**(5), 387–393 (1989)

8. Ayyalasomayajula, K.R., Nettelblad, C., Brun, A.: Feature evaluation for handwritten character recognition with regressive and generative Hidden Markov Models. In: International Symposium on Visual Computing, pp. 278–287. Springer International Publishing (2016)
9. Roy, S., Das, N., Kundu, M., Nasipuri, M.: Handwritten isolated Bangla compound character recognition: a new benchmark using a novel deep learning approach. Pattern Recogn. Lett. **90**, 15–21 (2017)
10. Hinton, G.E., Salakhutdinov, R.R.: Reducing the dimensionality of data with neural networks. Science **313**(5786), 504–507 (2006)
11. Liu, W., Wang, Z., Liu, X., Zeng, N., Liu, Y., Alsaadi, F.E.: A survey of deep neural network architectures and their applications. Neurocomputing **234**, 11–26 (2017)
12. Jayadevan, R., Kolhe, S.R., Patil, P.M., Pal, U.: Offline recognition of Devanagari script: A survey. IEEE Trans. Syst. Man Cybern. Part C: Appl. Rev. **41**(6), 782–796 (2011)
13. Pal, U., Chaudhuri, B.B.: Indian script character recognition: a survey. Pattern Recogn. **37**(9), 1887–1899 (2004)
14. Jangid, M.: Devanagari isolated character recognition by using statistical features. Int. J. Comput. Sci. Eng. **3**(2), 2400–2407 (2011)
15. Sethi, I.K., Chatterjee, B.: Machine recognition of hand-printed Devnagri numerals. IETE J. Res. **22**(8), 532–535 (1976)
16. Deshpande, P.S., Malik, L., Arora, S.: Fine classification & recognition of hand written Devnagari characters with regular expressions & minimum edit distance method. JCP **3**(5), 11–17 (2008)
17. Arora, S., Bhatcharjee, D., Nasipuri, M., Malik, L.: A two stage classification approach for handwritten Devnagari characters. In: International Conference on Conference on Computational Intelligence and Multimedia Applications, 2007, vol. 2, pp. 399–403. IEEE (2007)
18. Hanmandlu, M., Ramana Murthy, O.V., Madasu, V.K.: Fuzzy model based recognition of handwritten Hindi characters. In: 9th Biennial Conference of the Australian Pattern Recognition Society on Digital Image Computing Techniques and Applications, pp. 454–461. IEEE (2007)
19. Arora, S., Bhattacharjee, D., Nasipuri, M., Basu, D.K., Kundu, M.: Recognition of non-compound handwritten Devnagari characters using a combination of MLP and minimum edit distance (2010). arXiv:1006.5908
20. Kumar, S.: Performance comparison of features on Devanagari handprinted dataset. Int. J. Recent Trends **1**(2), 33–37 (2009)
21. Bengio, I.G.Y., Courville, A.: Deep Learning. Book in preparation for MIT Press (2016)
22. Hinton, G.: Slide 6, Lecture Slide 6 of Geoffrey Hinton's Course. http://www.cs.toronto.edu/~tijmen/csc321/slides/lecture_slides_lec6.pdf. Accessed 19 July 2017
23. Kingma, D., Ba, J.: Adam: a method for stochastic optimization (2014). arXiv:1412.6980
24. Sharma, N., Pal, U., Kimura, F., Pal, S.: Recognition of off-line handwritten Devnagari characters using quadratic classifier. In: Computer Vision, Graphics and Image Processing, pp. 805–816. Springer, Berlin, Heidelberg (2006)
25. Pal, U., Wakabayashi, T., Kimura, F.: Comparative study of Devnagari handwritten character recognition using different feature and classifiers. In: 10th International Conference on Document Analysis and Recognition, 2009. ICDAR'09, pp. 1111–1115. IEEE (2009)

Speech Recognition of Punjabi Numerals Using Convolutional Neural Networks

Thakur Aditi and Verma Karun

Abstract Achieving accuracy for speech recognition has been a huge obstacle in the domain of Natural Language Processing and the model used predominantly for this is GMM-HMM. But, now with the boom of deep learning, it took primacy over the earlier model. With the advancement in the parallel processing and usage of the GPU power, deep learning has set forth results that have outperformed the GMM-HMM. This paper evaluates the performance of deep learning algorithm—Convolutional Neural network (CNN) on dataset comprising of audio (.wav) files capturing the recital of numerals from 0 to 100 in Punjabi language. The accuracy of the network is evaluated for two datasets that are with and without noise reduction. The model gives better results than the baseline GMM-HMM showing a reduction of error rate by 3.23% for data with noise reduction and by 3.76% for data without noise reduction.

Keywords Speech recognition · Deep learning · CNN
Convolutional · Dropout · Max-pooling

1 Introduction

Automatic Speech Recognition allows to transcribe the human speech into words. There are numerous challenges in this field because in the real-world implementation, the ASR systems should be noise robust and should be able to handle the variability of speech signals due to different styles of speaking.

The Gaussian Mixture Model (GMM) has been regarded as one of the powerful models for speech recognition. The Hidden Markov Model (HMM) has been successful in modeling the temporal behavior of speech signals using a sequence of states, each of which is associated with a particular probability distribution of observations

T. Aditi (✉) · V. Karun
Computer Science and Engineering Department, Thapar University, Patiala, Punjab, India
e-mail: adity2792@gmail.com

V. Karun
e-mail: karun.verma@thapar.edu

© Springer Nature Singapore Pte Ltd. 2019
S. K. Bhatia et al. (eds.), *Advances in Computer Communication and Computational Sciences*, Advances in Intelligent Systems and Computing 759,
https://doi.org/10.1007/978-981-13-0341-8_6

[1]. The GMM model recognizes the speech signal associated with each of the HMM states. In 2012, a paper was presented by G. Hinton et al. regarding the deep neural networks in speech recognition. The main findings were that the feedforward neural network can be used to evaluate how perfectly each state fits into a frame/frames. This takes many frames of coefficients as input and as an output, it gives out the posterior probabilities over HMM states [2].

GMM-HMM has remained the state of art model for a long time. But, these systems have been susceptible to background noise and channel distortions. However, GMM-HMM based models are faster and easier to train but DNN is more accurate classifiers. DNN can operate extremely well on complex data that has spatial arrangements whereas GMM-HMM based model tends to operate better on flat data and cannot use high dimensional features.

Deep learning is a re-branded concept that has been in existence since 1960. In the past, it did not overpower the existing models of that time because of the lack of the tremendous volume of data that is required to train it and the amount of computing power that is required. Now with the technological advancement, deep learning is evolving as the state of the art. The deep nets are the modified versions of the neural nets with much more layers and added functionality. There are many types of deep net models available, each one having a signature functionality. The key is that deep nets are able to break the complex pattern into series of simpler patterns. These are inspired by the structure of our human brains. These decipher patterns just like deep nets do in layers. Deep neural networks could model complex higher order statistical structure effectively but training deep neural nets is difficult [3].

The concept of deep nets evolved slowly and steadily. In 1988, Rumelhart et al. stated a new mechanism called as backpropagation [4, 5]. This mechanism adjusts the weights of the edges by comparing the actual output and the desired output. And thus in these ways, the hidden units learn about the data and represent important features. In 1989, Yann Le Cun et al. at the AT&T Labs applied backpropagation in order to recognize handwritten zip code [6]. In the 1990s, the deep nets began to lose its hold and then in 2006, the concept was reintroduced as deep learning.

The major breakthrough paper was presented by G. Hinton et al. stating that the weights for the deep net model if initialized using the wake sleep greedy algorithm resulted in better performance [7]. The selection of a nonlinear activation function impacts the performance on a huge scale and the weights should not just be random but the scale should vary according to the layers [8].

Convolutional neural networks have also been used in classifying environmental and urban sound sources [9]. In 2017, a paper was presented by Zhang et al. for end-to-end speech recognition systems based on CNN. This paper presents a method to train CNN in a manner which adds more expressive power and better generalization. In order to build deep convolutional and recurrent networks batch normalization, network-in-network principles, convolutional LSTMs, and residual connections have been applied [10]. Context-dependent systems have also shown significant improvement on a larger dataset of 300 h of data (without speaker adaptation) using deep net up to nine layers. Multiple layers structure allow the higher layer feature detectors to reuse the lower layer feature detectors [11].

2 Convolutional Neural Network

CNN is an extension of the basic neural network making it more powerful to handle the data that can have spatial patterns. CNN works efficiently for image recognition but has also shown good results for speech recognition. The input that is fed to the CNN which consists of [width * height * channel]. As the dataset is audio so instead of width and height, there are bands and frames [bands * frames * channel].

Convolutional networks capture local "spatial" patterns in data. These are great at finding patterns but if the data remains useful after swapping any of the columns with each other than the convolutional net cannot be used as it does not form any pattern. So, convolutional nets can work well with image and anything that has a pattern or can be represented in that format such as for speech signal processing and sentence processing.

2.1 Layer Architecture

Convolutional Layer. The sliding is defined by the displacement and the projection is the net product between two functions. For the case of vectors, the filter slides on and a filter dot product is computed along with the actual data values and as a result, we obtain the convolved feature. The method that has been used in our model is conv2D as the sound data is depicted as two-dimensional data.

Pooling Layer. The pooling that has been used here is max-pooling. In max-pooling, the largest number is taken in a vector neighborhood. The largest numbers in a convolved vector are those where the convolutional filter forms the best matching inputs. From this point of view, it can be said that convolutions contribute to feature matching and pooling gives the best match detection. The combination of both convolutional layer and pooling layer is what gives CNN a great shift invariant representations. In the study shown by Yanmin Qian et al. the most optimised strategies are investigated for pooling, padding and input feature map selection [12].

Fully Connected Layer. This layer is the basic neural network layer that is fully connected. High-level features of data are represented by the convolutional layer as output. This output is further flattened and connected to the output layer by using the fully connected layer.

2.2 Activation Function and Optimizer

Rectified Linear Unit (ReLU). This computational unit performs normalization by removing all negative values and changing it to 0. The ReLU function is $f(x) = \max(x, 0)$, these help in speeding up the training process. ReLU is computed after the convolution layer. It is a nonlinear activation function and it also does not suffer from the vanishing gradient problem. But if the learning rate is set too high then the ReLU units might die if a large gradient is flowing through the ReLU unit hence, the learning rate needs to be tuned.

Softmax. It is used as a classifier at the end of the neural network. The output of each unit is squashed in between a range of 0–1 and along with that it sets all outputs in a manner so that the sum total of the outputs is equal to 1. The output of this Softmax function gives us the probabilities of each class being true for an input.

Adam Optimizer. It is a replacement optimization algorithm for stochastic gradient descent for training deep learning models.

2.3 Mechanisms in CNN

Forward Propagation. It is the way the data propagates in the forward direction. The input data is provided to the first convolutional layer. Before processing the data, the weights and biases are set for each node and edge. Further, the data is processed and can be pushed into multiple convolutional and pooling layers. The activation function work in order to normalize the features in between different layers. At last, the intermediate features are passed on to the fully connected layer to obtain the final results.

Backpropagation. The principle behind backpropagation is that the error in the final answer is used to determine how much the network needs to adjust. This error drives a process called gradient descent. It is calculated from the output layers towards the initial layers, moving in the opposite direction of the forward propagation. The training process utilizes gradient. Gradient is the rate at which cost changes with respect to change in weight or bias. Gradient at a layer is the product of all the Gradient at prior layer. The fundamental problem that the DNN had faced at its earlier stages was vanishing gradient. When the gradient is large the net will train quickly but when the gradient is small the net will train slowly. The gradients are much smaller in the earlier layers, as a result, the earlier layers are slowest to train. Deeper neural networks, especially deeper autoencoders, are known to be difficult to train with backpropagation alone [13].

Dropout. It is a method of regularization in which randomly selected neurons are ignored during training. This implies their contribution to the activation of downstream neurons is temporally neglected on the forward pass and any weight updates are not applied to the neuron on the backward pass. The size of the values being propagated forward needs to be increased. This has to be done in proportion to the number of values being turned off. As a result, the network becomes less sensitive to the specific weights of neurons and leads to a network that most probably will not overfit the training data and gives a better generalization.

3 Methodology

3.1 Dataset

The data set comprises of an audio recording of the spoken words, i.e., numerals from 0 to 100 in Punjabi. This recording has 101 samples for each person. The total samples obtained are 101 * 30 = 3030. Two data sets are prepared, the first data set contains audio files with noise reduction and the second data set contains audio files without noise reduction. The data is divided into different ratios to compare the performance. It is divided into training, testing, and validations.

Data Preprocessing.

Noise Reduction.

The removal of noise does not affect the quality of the audio but it just reduces the noise component. The spectral subtraction method has been used to reduce the noise component.

Silence Removal.

Silence removal is a technique to remove silence before and after the isolated word. This helps in extracting out just the speech part. End point detection is a technique of identifying speech parts in audio signal. Inputs which contain speech surrounded by silence are easier to work with in terms of energy comparison.

3.2 Framing Window

Librosa package in python has been used for the audio analysis. This package provides some core functionality like loading audio files and provides various methods for analyzing audio. The audio has been loaded and decoded as time series, which is represented as a one-dimensional Numpy floating point array. Time series are decomposed in terms of sinusoid presence. The sampling rate by default is approximately 22 kHz but it can be set by setting the parameter sr = 16,000 (16 kHz). The sampling rate is defined as the number of samples of audio carried per second. For each audio clip, the sound wave needs to be windowed to a particular size. Because of windowing, we need to take overlapping segments to make up for the attenuated parts of the input.

3.3 Feature Extraction

For the selected window size melspectogram is computed and then mapped directly onto the mel scale. Melspectogram is a representation of power spectrum of sound. Then further logamplitude of the melspectogram is computed. This is done to scale the melspectogram in a stable format.

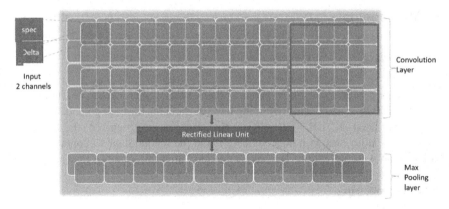

Fig. 1 The stack of convolutional layer, ReLU, and max-pooling layer

3.4 Experimental Setup

In order to train a CNN, an architecture has to be decided, which involves deciding the number and the size of layers and the filter size. Also, it involves parameter tuning (learning rate, dropout, window size, stride size, and number of neurons for the fully connected layers).

- In the implementation, audio data is used and two channels are used. One being the log-scaled melspectogram values of the recordings (16,000 Hz) and the other being its delta values with the window size being 1024 and bands = 40 and frames = 21.
- A layer stack is created using Convolution, ReLU layer, and pooling layer. In this model, deep stack architecture is followed. A total of 4 stacks are used.
- All the Convolutional ReLU layers and max-pooling layers (Fig. 1) consist of filter size (20×20) and stride size (2×2) for the first two layers and then for the last two layers, the filter size of (2×2) and stride size of (1×1).
- The first layer Convolutional ReLU layer consisted of 30 filters.
- The second layer Convolutional ReLU layer consisted of 50 filters.
- The third layer Convolutional ReLU layer consisted of 70 filters.
- The fourth layer Convolutional ReLU layer consisted of 90 filters.
- The final layer is the fully connected layer. The values present here are the probabilities or values corresponding to each class that helps in identifying to which class a particular input belongs to. It can be said that this is the final voting layer.
- Fully Connected Layer Stack contains two fully connected layers. The first layer has a dropout of 25% and the second layer has no dropouts. The number of neurons used in the fully connected layer is 512.
- Learning rate = 0.001 and epochs = 9000
- Output Layer: The output layer maps the output to the 101 classes (0–100) and feeds the final output to the Softmax Unit which converts the final output values to

a range of 0–1 and sum total of 1 thus representing the outputs as the probabilities for the predicted class.

4 Results

We collected 3030 samples of audio files by recording recital of Punjabi counting from 1 to 100. Then this data was fed to CNN in order to recognize the speech. The following results were drawn from the study.

- Table 1 represents the division of data into training, testing and validation. It also shows the different testing and validation accuracy for different divisions of data with noise reduction.
- Table 2 represents the division of data into training, testing, and validation. It also shows the different testing and validation accuracy for different divisions of data without noise reduction.
- Table 3 represents the comparison of the results obtained using CNN (my work) and GMM-HMM (existing work) [14] for the data with noise reduction and without noise reduction.
- Table 4 represents the error improvement rates for the data with and without noise reduction.

Table 1 The results obtained for dataset with noise reduction

	Training samples	Validation samples	Testing samples	Testing accuracy (%)	Validation accuracy (%)
60/20/20	1818	606	606	92.43	94.12
50/20/30	1515	606	909	88.66	92.07
40/20/40	1212	606	1212	84.38	88.49

Table 2 The results obtained for dataset without noise reduction

	Training samples	Validation samples	Testing samples	Testing accuracy (%)	Validation accuracy (%)
60/20/20	1818	606	606	88.56	85.39
50/20/30	1515	606	909	83.01	81.67
40/20/40	1212	606	1212	78.93	80.23

Table 3 The results obtained for dataset with noise reduction

Model	Accuracy (%)	Error (%)
GMM-HMM (With noise reduction)	89.20	10.80
CNN (With noise reduction)	92.43	7.57
GMM-HMM (Without noise reduction)	84.80	15.20
CNN (Without noise reduction)	88.56	11.44

Table 4 Reduction in error rate

	Reduction in error rate (%)
Recognition with noise reduction	3.23
Recognition without noise reduction	3.76

5 Conclusion and Future Work

It can be concluded from the results that the CNN model performs better than the GMM-HMM model for the present dataset. The reduction in the error rate in case of the dataset without noise reduction is greater than the dataset with noise reduction. So we can conclude that it can perform well with data that is near to real world (without noise reduction). Also when the data is raw it has complex patterns and for this kind of data, the deep net starts to outperform all other algorithms. The extra baggage is that deep nets take much longer to train, but the GPU can train these neural nets faster following the parallel processing. Convolutional networks capture local "spatial" patterns in data. If the data cannot be made to look like an image Convolutional Nets are less useful.

The scope of future work, first include bridging the gap between the current accuracy and the maximum accuracy by expanding the structure of the architecture and deploying it on the system which has more computation power. This can also be done by trying out variation of this model with other deep net models in a hybrid format. Second, this model can be implemented on a larger dataset and thereby tuning the parameters to achieve better results. A comparison of CNN model with GMM-HMM for a specific dataset is used in this experiment. In future, more annotated data can be appended to get a robust model with low error rate.

Also, the computational cost of CNN is high and to deal with this separated filter architecture can be used. It basically means that convolutional layer is represented as a linear combination of separable filters with the help of a concept known as filter group's concept, which allows one to get separable convolution filters by some standard methods [15]. The robustness can be improved by filtering the speech signal using the Wiener filter technique and then feeding it into the CNN [16]. Not only this many other ways be used to improve the computation cost. However, for the fast implementation of the model, GPU power can be used in an optimized way.

References

1. Hamid, A.O., Mohamed, A., Jiang, H., Deng, L., Penn, G., Yu, D.: Convolutional neural networks for speech recognition. IEEE/ACM Trans. Audio Speech Lang. Process. **22**(10), 1533–1545 (2014)
2. Hinton, G., Deng, L., Yu, D., Dahl, G., Mohamed, A., Jaitly, N., Senior, A., Vanhoucke, V., Nguyen, P., Sainath, T., et al.: Deep neural networks for acoustic modeling in speech

recognition: the shared views of four research groups. IEEE Sig. Process. Mag. **29**(6), 82–97 (2012)

3. Jaitly, N., Nguyen, P., WSenior, A., Vanhoucke, V.: Application of pretrained deep neural networks to large vocabulary speech recognition. In: Interspeech, pp. 2578–2581 (2012)

4. Rumelhart, D., Hinton, G., Williams, R. et al.: Learning representations by back-propagating errors. Cognitive Modeling (1988)

5. Rumelhart, D., Hinton, G., Williams, R.: Learning internal representations by error propagation. Technical Report, California Univ San Diego La Jolla Inst for Cognitive Science (1985)

6. LeCun, Y., Boser, B., Denker, J., Henderson, D., Howard, R., Hubbard, W., Jackel, L.: Back-propagation applied to handwritten zip code recognition. Neural Comput. **1**(4), 541–551 (1989)

7. Hinton, G., Osindero, S., Teh, Y.: A fast learning algorithm for deep belief nets. Neural Comput. **18**(7), 1527–1554 (2006)

8. Glorot, X., Bengio, Y.: Understanding the difficulty of training deep feedforward neural networks. In: Proceedings of the Thirteenth International Conference on Artificial Intelligence and Statistics, pp. 249–256 (2010)

9. Piczak, K.: Environmental sound classification with convolutional neural networks. In: 2015 IEEE 25th International Workshop on Machine Learning for Signal Processing (MLSP), pp. 1–6 (2015)

10. Zhang, Y., Chan, W., Jaitly, N.: Very deep convolutional networks for end-to-end speech recognition. In: 2017 IEEE International Conference on Acoustics, Speech and Signal Processing (ICASSP), pp. 4845–4849 (2017)

11. Yu, D., Seide, F., Li, G.: Conversational speech transcription using context-dependent deep neural networks. In: Interspeech, pp. 437–440 (2011)

12. Qian, Y., Bi, M., Tan, T., Yu, K.: Very deep convolutional neural networks for noise robust speech recognition. IEEE/ACM Trans. Audio Speech Lang Process **24**(12), 2263–2276 (2016)

13. Dahl, G.E., Yu, D., Deng, L., Acero, A.: Context-dependent pre-trained deep neural networks for large-vocabulary speech recognition. IEEE Trans. Audio Speech Lang. Process. **20**(1), 30–42 (2012)

14. Mittal, S., Verma, K.: Speaker independent isolated word speech to text conversion using HTK, Ph.D. thesis (2014)

15. Limonova, E., Sheshkus, A., Nikolaev, D.: Computational optimization of convolutional neural networks using separated filters architecture. Int. J. Appl. Eng. Res. **11**(11), 7491–7494 (2016)

16. Palaz, D., Collobert, R., et al.: Analysis of CNN-based speech recognition system using raw speech as input. Technical Report, Idiap (2015)

Fault Diagnosis of Bearings with Variable Rotational Speeds Using Convolutional Neural Networks

Viet Tra, Jaeyoung Kim and Jong-Myon Kim

Abstract Feature extraction is of great significance in the traditional bearing fault diagnosis methods which detect incipient defects by identifying peaks at characteristic defect frequencies in the envelope power spectrum or by developing discriminative models for features extracted from the signals through time and frequency domain analysis. This issue not only requires expert knowledge to design discriminative features but also requires appropriate feature selection algorithms to eliminate irrelevant and redundant features that degrade the performance of the methods. In this paper, we present a convolutional neural networks (CNNs) based method to automatically derive optimal features which not require any feature extraction or feature selection. The proposed method identifies bearing defects efficiently by processing feature extraction and classification into a single learning machine and automatically extracts optimal features from these energy distribution maps to diagnose various single and compound bearing defects and yields better diagnostic performance compared to state-of-the-art AE-based methods.

Keywords Convolutional neural networks · Stochastic diagonal
Levenberg–Marquardt algorithm · Bearing fault diagnosis

1 Introduction

Rolling element bearings are crucial components of rotating machines, and they are the leading cause of failure in induction motors, where they account for 51% of total failures [1]. These failures can cause unscheduled and costly shutdowns. The early

V. Tra · J. Kim · J.-M. Kim (✉)
School of IT Convergence, University of Ulsan, Ulsan 680-749, South Korea
e-mail: jongmyon.kim@gmail.com

V. Tra
e-mail: traviet.vt@gmail.com

J. Kim
e-mail: kjy7097@gmail.com

© Springer Nature Singapore Pte Ltd. 2019   71
S. K. Bhatia et al. (eds.), *Advances in Computer Communication and Computational Sciences*, Advances in Intelligent Systems and Computing 759,
https://doi.org/10.1007/978-981-13-0341-8_7

detection of these defects is helpful in preventing such abrupt shutdowns and keeping costs low.

Various techniques are based on the analysis of vibration acceleration signals [2, 3] or current signals [4, 5] for the purpose of bearing fault detection. In the low rotational speeds, acoustic emission (AE)-based techniques are more effective due to the ability to capture low energy acoustic emissions [6–8]. In this paper, we use AE signals to diagnose bearing defects.

AE-based methods have predominantly utilized feature extraction-based techniques for bearing fault diagnosis that usually consists of three steps: signal acquisition, extraction of features from the acquired signal, and classification of various defect types based on the extracted features. Feature extraction is of great significance in these methods, as discriminative features lead to better discriminative models and hence good diagnostic performance. Thus, diagnostic performance of these discriminative models depends upon the quality of the features. In this paper, we propose a method using convolutional neural networks (CNNs) to extract distinguishing features from the energy distribution maps of the spectrum of an AE signal, and then we use those automatically learned features to diagnose various bearing defects. Although CNNs have been traditionally used for automatic representation learning and classification of images, the strong local structure of the AE signal spectrum from the high correlation between the amplitudes of nearby frequencies justifies their use in bearing fault diagnosis.

To improve the diagnosis performance of CNNs-based fault diagnosis system, we use the stochastic diagonal Levenberg–Marquardt (S-DLM) algorithm, which was first proposed by LeCun. It is a fast and efficient method to train convolutional neural networks (CNNs). The empirical results have been verified that the S-DLM algorithm helps CNNs to converge quickly when compared with stochastic gradient descent (S-GD) algorithm, and thereby enhances the performance of fault diagnosis system. The contributions of this paper are as follows:

(1) We propose a new method for diagnosing bearing using CNNs that does not require any feature extraction or feature selection.
(2) We propose the use of energy distribution maps of the AE signal spectrum to diagnose bearing defects.

The rest of this paper is organized as follows: Sect. 2 presents the experimental testbed and the data acquisition system used to collect the acoustic emission data. Section 3 introduces CNNs using LeNet-5 architecture and analyzes the stochastic diagonal Levenberg–Marquardt algorithm. Section 4 presents the proposed method and experimental results. Finally, Sect. 5 concludes this paper.

2 The Experimental Testbed and Seeded Defect Acoustic Emission Data

The experimental test bed [6] which is used to generate acoustic emission (AE) data for seeded bearing defects at variable operating speeds is shown in Fig. 1a. There are two shafts in this setup, a drive end shaft (DES) and a non-drive end shaft (NDES). The operating speed is measured using a displacement transducer installed on the NDES. The AE data is collected from bearings fastened to the NDES that are seeded with single and multiple combined defects. The acoustic emissions are captured using a wideband AE sensor that is coupled to the bearing housing at the NDES at a distance of 21.48 mm. The AE signals are recorded at a sampling rate of 250 kHz using a PCI-2 data acquisition system, as shown in Fig. 1b.

AE signals are recorded for a normal, defect-free bearing and bearings seeded with seven types of localized defects. The seven seeded defect types contain both single and multiple combined defects [7]: outer raceway crack (BCO), inner raceway crack (BCI), roller crack (BCR), inner and outer raceway cracks (BCIO), outer and roller cracks (BCOR), inner and roller cracks (BCIR), and inner, outer, and roller cracks (BCIOR), as shown in Fig. 2.

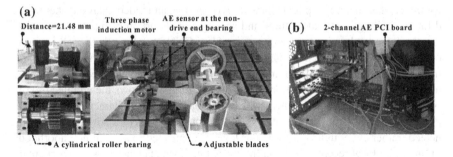

Fig. 1 **a** Machinery fault simulator [6]. **b** Data acquisition system [6]

Fig. 2 Examples of single and compound seeded bearing defects considered in this study, **a** BCI, **b** BCO, **c** BCR, **d** BCIO, **e** BCIR, **f** BCOR, and **g** BCIOR [7]

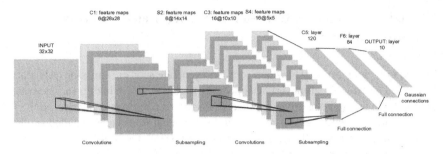

Fig. 3 Architecture of LeNet-5, convolutional neural networks [9]

3 The Architecture of CNNs and Backpropagation Through CNNs Using a Stochastic Diagonal Levenberg–Marquardt Algorithm

3.1 Overview of CNNs and LeNet-5

CNNs are feedforward artificial neural networks (ANNs) that have been widely used for image classification, optical character recognition, and video analysis where it is difficult to design optimal features and scaling and distortion of inputs can significantly degrade classification performance [13]. CNNs combine three architectural concepts, using local receptive fields instead of fully connected layers, weight sharing, and spatial or temporal subsampling to achieve a reasonable degree of invariance to shifting, scaling, and distortion in images [9]. Figure 3 shows the architecture of typical CNNs, the LeNet-5, which was initially proposed for character recognition [9]. It is different from typical ANNs as the consecutive layers are not fully connected, rather every unit in a particular layer is connected to a small neighborhood of units or a local receptive field in the preceding layer. The neurons in a particular layer extract simple features from the receptive fields in the previous layers, and this process continues in subsequent layers, which combine those simple features to learn higher order representations of the input.

3.2 Training CNNs Using a Stochastic Diagonal Levenberg–Marquardt Algorithm

Many authors have claimed that second-order methods should be used instead of gradient descent for neural network training. Second-order learning algorithms generally consider both gradient and curvature information in optimization and compute the unique update step sizes (learning rates) for each weight based on weight-specific information. Thus, they are also considered as local adaptive algorithms [11]. In the classical second-order methods such as the Newton method, the weights of units are computed by [14]:

$$\Delta w = \eta_g \left(\frac{\partial^2 E}{\partial w^2} \right)^{-1} \frac{\partial E}{\partial w} = \eta_g H(w)^{-1} \frac{\partial E}{\partial w} \qquad (1)$$

where $\eta_g \in (0, 1)$ is the global learning rate and $(\partial^2 E / \partial w^2)^{-1}$ is the inverse of Hessian matrix that used to compute individual learning rate (local learning rate) for weights. To obtain the local learning rate for individual weights, we have to take the inverse of the Hessian, as shown in (1), which is clearly a computationally expensive task and makes them impractical for large networks. It requires $O(N^3)$ operations per update, where N is the number of parameters [11]. To address the drawbacks of the classical second-order method, we need to find an appropriate approximation method to reduce the complexity of the Hessian computation.

In this paper, the stochastic diagonal Levenberg–Marquardt method (S-DLM) is proposed for updating learnable weights that approximates the Hessian matrix by dropping the off-diagonal terms, thereby the individual learning rate for each connection weight w_{ij} is determined by diagonal elements of Hessian matrix [15]:

$$\eta_{ij} = \frac{\eta_g}{\left(\frac{\partial^2 E}{\partial w_{ij}^2} \right) + \mu} \qquad (2)$$

Thereby, the adjustment $\Delta w_{ij}(n)$ for each corresponding connection weight becomes

$$\Delta w_{ij} = \frac{\eta_g}{\left(\frac{\partial^2 E}{\partial w_{ij}^2} \right) + \mu} \left(\frac{\partial E^p}{\partial w_{ij}} \right), \qquad (3)$$

where $h_{kk} = (\partial^2 E / \partial w_{ij}^2)$ is a running estimate of the diagonal second derivative with respect to w_{ij} (the k-th diagonal element of Hessian matrix). Unlike most other second-order acceleration methods for backpropagation, the proposed method focuses on s stochastic mode. It uses a diagonal approximation of the Hessian. Similar to the classical Levenberg–Marquardt algorithm, it uses a "safety" factor μ to prevent the step sizes from getting too large when the second derivative estimates are small.

In Eq. (3), the instantaneous gradient of synaptic weight $(\partial E^p / \partial w_{ij})$ is efficiently calculated by backpropagation using a stochastic gradient descent method [10]. The running estimates of the diagonal second derivative with respect to w_{ij}; $\partial^2 E / \partial w_{ij}^2$ is computed over the training samples as follows:

$$\left(\frac{\partial^2 E}{\partial w_{ij}^2} \right)_{new} = (1 - \gamma) \left(\frac{\partial^2 E}{\partial w_{ij}^2} \right)_{old} + \gamma \left(\frac{\partial^2 E^p}{\partial w_{ij}^2} \right) \qquad (4)$$

where γ is a small constant that determines the amount of memory which is being used, $(\partial^2 E^p / \partial w_{ij}^2)$ is the instantaneous second derivative with respect to w_{ij}, which can be effectively obtained via backpropagation which is well known [15, 16].

4 Experimental Results and Discussion

4.1 Fault Signature Pool Configuration and Energy Distribution Maps

In this study, we implement two experiments to evaluate the proposed method. The first experiment evaluates the effectiveness of the S-DLM algorithm by comparing with the stochastic gradient descent training (S-GD) algorithm, while the second experiment compares the performance of the proposed method with traditional AE-based methods under variable operating speeds.

In the first experiment, we use three datasets. For each dataset, AE signals of normal, defect-free bearing, and bearings seeded with seven types of localized defects of 3 mm crack size are recorded at one of operating speeds of 300, 400, and 500 revolutions per minute (RPM) [6, 7, 17], as given in Table 1. For every operating speed, 90 AE signals of each bearing condition are recorded for a duration of 5 s each. Hence, each dataset contains a total of $\sum_{Fault\_classes=1}^{8} 90 \langle AE\ Signals \rangle$, i.e., 720 AE signals. For the purpose of training and testing the CNNs, each dataset is divided into training and testing subsets. For each dataset, we use a half of AE signals to train model and a remaining half to test model.

For the second experiment, we use AE signals that bearings seeded with seven types of localized defects are recorded of two sizes (3 and 12 mm). The recorded

Table 1 Description of the datasets used to evaluate the proposed method in experiment 1

Single and compound seeded bearing failures		Rotational speed (RPM)	Crack size		
			Length (mm)	Width (mm)	Depth (mm)
Dataset 1	Training dataset	300	3	0.35	0.30
	Testing dataset	300		0.35	0.30
Dataset 2	Training dataset	400	3	0.35	0.30
	Testing dataset	400		0.35	0.30
Dataset 3	Training dataset	500	3	0.35	0.30
	Testing dataset	500		0.35	0.30

Table 2 Description of the datasets used to evaluate the proposed method in experiment 2

Single and compound seeded bearing failures		Rotational speed (RPM)	Crack size		
			Length (mm)	Width (mm)	Depth (mm)
Dataset 4	Training dataset	300, 400, 500	3	0.35	0.30
	Testing dataset	250, 350, 450		0.35	0.30
Dataset 5	Training dataset	300, 400, 500	12	0.49	0.50
	Testing dataset	250, 350, 450		0.49	0.50

AE signals are divided into two datasets, one for each crack size. For each dataset, we divide into training and testing subsets. The training subset includes AE signals acquired at shaft speeds of 300, 400, and 500 RPM, while the testing subset includes the AE signals recorded at shaft speeds of 250, 350, and 450 RPM [6, 7, 17], as given in Table 2. Hence, each dataset contains $N_{RPM} \times N_{Classes} \times N_{Signals}$, or 4320 AE signals, where $N_{RPM} = 6$, $N_{Classes} = 8$, and $N_{Signals}$ is the total number of AE signals recorded for each bearing condition at each shaft speed, $N_{Signals} = 90$.

As discussed in Sect. 2, the experimental testbed is used to acquire acoustic emission data, which is essentially a 1-D signal captured at ultrahigh sampling rates. Mining the raw AE signal for distinctive features would require very large CNNs. Hence, we propose to convert the raw AE signals into an equivalent 2-D representation that shows the distribution of energies in various frequency bands of the AE signal, which is then used as input to the CNNs. The energy distribution map is generated by first multiplying the AE signal by a Hanning window function and then computing its FFT. The spectrum of the AE signal is then split into an appropriate number of frequency bands, and the root mean square (RMS) value of each of these bands is calculated. The RMS value gives an approximation of the energy carried by each band [12]. These RMS values are arranged in the form of a 2-D array that shows the distribution of energies across the entire spectrum of the AE signal. Examples of these maps are shown in Fig. 4.

4.2 Verification of the Stochastic Diagonal Levenberg–Marquardt Algorithm

In this experiment, we show the effectiveness of the stochastic diagonal Levenberg—Marquardt (S-DLM) algorithm by comparing with the stochastic gradient descent (S-GD) algorithm in terms of convergence time and convergence ability.

Figure 5 shows mean square errors (MSE) of training samples and misclassification error rate (MCR) of testing samples of datasets in Table 1. The learning curves

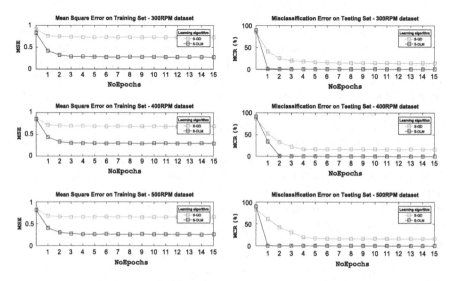

Fig. 4 The 2-D energy distribution maps that are used as input in CNNs

Fig. 5 Training mean square error and testing misclassification error of datasets in Table 1

(MSE graphs) in Fig. 5 show that the S-DLM model is better than S-GD since it utilizes the second-order (curvature) information.

4.3 Verification of the CNNs-Based Methodology for a Bearing Fault Diagnosis Application

The key difference between the CNN-based method and feature extraction-based methods is the manner of extracting features. While the performance for detecting various bearing failures using conventional methods depends highly on selecting reliable feature set, the CNN-based method can automatically extract the optimal

Table 3 Average classification accuracies and sensitivities for single and multiple combined bearing of variant bearing speed datasets

Datasets	Methodologies	Average sensitivity of each fault types								ACA (%)
		BCI	BCO	BCR	BCIO	BCIR	BCOR	BCIOR	BNC	
Dataset 4	[6]	19.6	47.4	75.1	17.0	59.6	30.7	10	3.33	32.8
	Developed method	66.6	100	100	100	89.2	99.2	99.2	99.6	94.2
Dataset 5	[6]	7.03	70	66.6	79.6	5.92	44.8	74.0	62.9	51.3
	Developed method	100	100	91.8	98.1	99.2	99.2	100	99.2	98.4

feature from original signals. Under variant shaft speed condition, since the characteristics of the signal are unstable, features extracted by feature extraction-based methods are weak in describing the signal, leading to false diagnosis.

To validate the effectiveness of features extracted by the CNNs, a comparison is provided between the CNN-based method and one of AE-based methods [6]. In this method, various statistical measures of the time and frequency domain AE signals are used as features, along with statistical quantities calculated through complex envelope analysis. The extracted features are then evaluated and selected using outlier-insensitive hybrid feature selection (OIHFS). Finally, the selected features are used to detect bearing defects by the K-NN classifier.

Table 3 presents the diagnostic performance of the proposed method and the method in [6] for each dataset in Table 2 in terms of the average classification accuracy (ACA). In addition, sensitivity, which is a useful metric for evaluating the diagnostic performance of each bearing condition, is also provided. It is clear from the results given in Table 3 that the proposed method delivers better diagnostic performance than the method in [6]. The average classification accuracy of the proposed method is 94.2% and 98.4% using dataset 4 and dataset 5, respectively, in Table 2, which shows significant improvement when compared to the method in [6] by 61.4% and 47.1%.

5 Conclusions

This paper proposed a CNN-based method that merges feature extraction and classification tasks into a single learner and has great ability to automatically extract optimal features from the original AE signal. In this study, we applied the stochastic diagonal Levenberg–Marquardt (S-DLM) to train CNNs. The empirical results showed that the proposed CNN-based method provides better performance in diagnosing both single and combined bearing defects when compared with the state-of-the-art conventional method in various bearing speed conditions.

Acknowledgements This work was supported by the Korea Institute of Energy Technology Evaluation and Planning (KETEP) and the Ministry of Trade, Industry & Energy (MOTIE) of the Republic of Korea (Nos. 20162220100050, 20161120100350, 20172510102130). It was also funded in part by The Leading Human Resource Training Program of Regional Neo industry through the National Research Foundation of Korea (NRF) funded by the Ministry of Science, ICT and Future Planning (NRF-2016H1D5A1910564), and in part by the Basic Science Research Program through the National Research Foundation of Korea (NRF) funded by the Ministry of Education (2016R1D1A3B03931927).

Conflict of Interest The authors declare that there is no conflict of interest regarding the publication of this manuscript.

References

1. Thorsen, O.V., Dalva, M.: Failure identification and analysis for high-voltage induction motors in the petrochemical industry. IEEE Trans. Ind. Appl. **35**(4), 810–818 (1999)
2. Jin, X., Zhao, M., Chow, T.W.S., Pecht, M.: Motor bearing fault diagnosis using trace ratio linear discriminant analysis. IEEE Trans. Ind. Electron. **61**(5), 2441–2451 (2014)
3. Seshadrinath, J., Singh, B., Panigrahi, B.K.: Investigation of vibration signatures for multiple fault diagnosis in variable frequency drives using complex wavelets. IEEE Trans. Power Electron. **29**(2), 936–945 (2014)
4. Lau, E.C.C., Ngan, H.W.: Detection of motor bearing outer raceway defect by wavelet packet transformed motor current signature analysis. IEEE Trans. Instrum. Meas. **59**(10), 2683–2690 (2010)
5. Frosini, L., Bassi, E.: Stator current and motor efficiency as indicators for different types of bearing faults in induction motors. IEEE Trans. Ind. Electron. **57**(1), 244–251 (2010)
6. Kang, M., Islam, R., Kim, J., Kim, J.-M., Pecht, M.: A hybrid feature selection scheme for reducing diagnostic performance deterioration caused by outliers in data-driven diagnostics. IEEE Trans. Ind. Electron. **63**(5), 3299–3310 (2016)
7. Kang, M., Kim, J., Wills, L., Kim, J.-M.: Time-varying and multi-resolution envelope analysis and discriminative feature analysis for bearing fault diagnosis. IEEE Trans. Ind. Electron. **62**(12), 7749–7761 (2015)
8. Nguyen, P., Kang, M., Kim, J.M., Ahn, B.H., Ha, J.M., Choi, B.K.: Robust condition monitoring of rolling element bearings using de-noising and envelope analysis with signal decomposition techniques. Expert Syst. Appl. **42**(22), 9024–9032 (2015)
9. Lecun, Y., Bottou, L., Bengio, Y., Haffner, P.: Gradient-based learning applied to document recognition. In: Proceedings of the IEEE, vol. 86, no. 11, pp. 2278–2324 (1998)
10. Bouvrie, J.: Notes on convolutional neural networks (2006)
11. LeCun, Y., Bottou, L., Orr, G., Muller, K.: Efficient backprop. Neural Networks: Tricks of the Trade, vol. 1524, pp. 9–50 (1998)
12. Abu-Mahfouz, I.A.: A comparative study of three artificial neural networks for the detection and classification of gear faults. Int. J. Gen. Syst. **34**(3), 261–277 (2005)
13. LeCun, Y., Cortes, C., Burges, C.J.: MNIST handwritten digit database. AT&T Labs (2010)
14. Haykin, S.: Neural Networks and Learning Machines. 3rd edn, pp. 122–229. Prentice Hall, New York (2009)
15. Becker, S., Le Cun, Y.: Improving the convergence of back-propagation learning with second order methods. In: Proceedings of the Connectionist Models Summer School, pp. 29–37 (1988)

16. LeCun, Y.: Generalization and network design strategies. In: Pfeifer, R., Schreter, Z., Fogelman, F., Steels, L. (eds.) Connectionism in Perspective Zurich. Elsevier, Switzerland (1989)

17. Tra, V., Kim, J., Khan, S.A., Kim, J.-M.: Incipient fault diagnosis in bearings under variable speed conditions using multiresolution analysis and a weighted committee machine. J. Acoust. Soc. Am. **142**(1), EL35–EL41 (2017)

Fault Diagnosis of Multiple Combined Defects in Bearings Using a Stacked Denoising Autoencoder

Bach Phi Duong and Jong-Myon Kim

Abstract Bearing fault diagnosis is an inevitable process in the maintenance of rotary machines. Multiple combined defects in bearings are more difficult to detect because of the complexity in components of acquired acoustic emission signals. To address this issue, this paper proposes a deep learning method that can effectively detect the combined defects in bearings. The proposed deep neural network (DNN) is based on the stacked denoising autoencoder (SDAE). In this study, the proposed method trains single faulty data while it efficiently classifies multiple combined faults. Experimental results indicate that the proposed method achieves an average accuracy of 91% although it only has single mode fault information.

Keywords Bearing fault diagnosis · Combined fault classification · Deep neural networks · Stacked denoising autoencoder

1 Introduction

The condition-based maintenance (CB) is an inevitable process of industrial machines, as it not only helps in avoiding the economic losses and unscheduled downtime but also increases the safety and efficiency of the production environment. Defects in bearings frequently occur in rotating machines, as they are vulnerable to degradation processes. Thus, bearing fault diagnosis is an essential task in rotary machinery. Bearing faults are usually divided into two categories: the first one is single mode faults and the other is combined faults. A single mode fault in bearings includes outer raceway fault (BCO), inner raceway fault (BCI), and roller element fault (BCR), whereas combined faults are those in which bearing have faults in more than one component simultaneously. The combined faults include outer and inner

B. P. Duong · J.-M. Kim (✉)
School of Electrical Engineering, University of Ulsan, Ulsan 680-749, South Korea
e-mail: jongmyon.kim@gmail.com

B. P. Duong
e-mail: duongbachphi@gmail.com

© Springer Nature Singapore Pte Ltd. 2019
S. K. Bhatia et al. (eds.), *Advances in Computer Communication and Computational Sciences*, Advances in Intelligent Systems and Computing 759,
https://doi.org/10.1007/978-981-13-0341-8_8

raceways defect (BCOI), outer raceway and roller defect (BCOR), inner raceway and roller defect (BCIR), outer raceway—inner raceway—roller defect (BCOIR). Due to the complexity inherent in the combined fault analysis, combined fault identification and classification is very difficult.

In recent years, several machine learning techniques have been developed for bearing fault diagnosis of rotary machines, such as the artificial neural network (ANN) [1, 2], support vector machine (SVM) [3], and K-nearest neighbor (K-NN) [4]. Kang et al. [4] and Nguyen et al. [5] presented effective fault diagnosis schemes, which involve feature extraction from both single and combined modes. The extracted features were provided to a machine learning algorithm as input for classification. In these algorithms, both single and combined fault data were trained to discriminate each fault. In contrast to these conventional methods, this study employs one of the deep neural networks (DNNs) [6], called stacked denoising autoencoders [7], and it only requires single fault information to detect multiple combined faults. The DNN-based method can automatically extract useful features from the raw bearing signal and identifies each fault of single and combined faults efficiently.

The main contribution of this study is to develop a method that classifies the combined bearing failures with only single fault information using deep neural networks, where the DNN is pretrained by a stacked denoising autoencoder.

The remaining paper is structured as follows. Section 2 presents the background of different types of bearing faults, and Sect. 3 describes the architecture of the DNN. Section 4 provides the results of the proposed method, and Sect. 5 concludes this paper.

2 Different Types of Bearing Faults

A bearing failure usually occurs by mechanical cracks, poor lubrication, and corrosion due to the increased load stress. The defect caused by poor lubrication increases friction, which results in increased metal to metal contact and enlarges the plastic deformation due to local contact phenomenon. It also increases the temperature, helping in speeding up the deterioration process. The high humidity in the environment can cause the bearing corrosion [8]. When the defect occurs in the bearing, it induces anomalous patterns in signal amplitude. Particularly, the defect in outer or inner raceway produces impulses when the rolling elements pass over the defect area. The impulse stimulates the bearing's vibration at its resonance frequency and is modulated by the characteristic frequencies [9]. These impulses repeat periodically depending on the structure of the bearing and rotating speed. The frequencies of these impulses are known as defect frequencies which associate with different types of failures. Figure 1 depicts the signal for eight different types of bearing faults. As compared to the signal of defect-free bearing (DFB), seven different fault signals show the appearance of abnormal impulses. The seven seeded defect types contain both single and multiple combined defects: outer raceway crack (BCO), inner raceway crack (BCI), roller crack (BCR), inner and outer raceway cracks (BCIO), outer

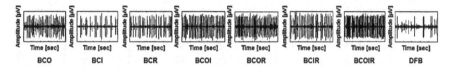

Fig. 1 Acoustic emission signal for each bearing condition

and roller cracks (BCOR), inner and roller cracks (BCIR), and inner, outer, and roller cracks (BCIOR).

The defect frequencies correlate with ball-pass frequency for the outer raceway (BPFO), ball-pass frequency for the inner raceway (BPFI) and twice ball spin frequency ($2 \times$ BSF). They are calculated as (1)–(3)

$$BPFO = \frac{nf_r}{2}\left(1 - \frac{d}{D}\cos\phi\right) \tag{1}$$

$$BPFI = \frac{nf_r}{2}\left(1 + \frac{d}{D}\cos\phi\right) \tag{2}$$

$$BSF = \frac{Df_r}{2d}\left[1 - \left(\frac{d}{D}\cos\phi\right)^2\right] \tag{3}$$

where f_r is the shaft speed, n is the number of rolling element, ϕ is the angle of the load from the radial plane, d is roller diameter, and D is pitch diameter.

These low-frequency periodic impulses are amplitude modulated to higher frequency because of the modulation with higher frequency carrier signal. Therefore, envelope analysis [10] is applied as a demodulation technique to detect these impulses by removing the carrier signal. First, the Hilbert transform is applied to the time domain signal [11] as

$$\tilde{x}(t) = H[x(t)] = \frac{1}{\pi}\int\limits_{-\infty}^{\infty}\frac{x(\tau)}{t-\tau}d\tau \tag{4}$$

where $\tilde{x}(t)$ is the Hilbert transform of $x(t)$. Hence, the analytic signal is calculated by $z(t) = x(t) + i\tilde{x}(t)$, where $i = \sqrt{-1}$. The envelope signal is given by $e(t) = |z(t)| = \sqrt{x^2(t) + \tilde{x}^2(t)}$. And the envelope power spectrum is computed by squaring the absolute value of the Fast Fourier Transform of the envelope signal. Figure 2 shows envelope power spectra of BCO, BCI, and BCOI.

Figure 2a, b shows the envelope power spectra of the faults in outer and inner raceway cases, respectively. Figure 2c shows the envelope spectrum of a combined defect on both outer and inner raceways. In the envelope spectrum of combined mode bearing failure, the appearance of both BPFO and BPFI illustrates that the acoustic emission (AE) signal of combined mode BCOI contains the harmonics of both BCO and BCI. The same phenomenon was observed when analyzing the envelope spectra

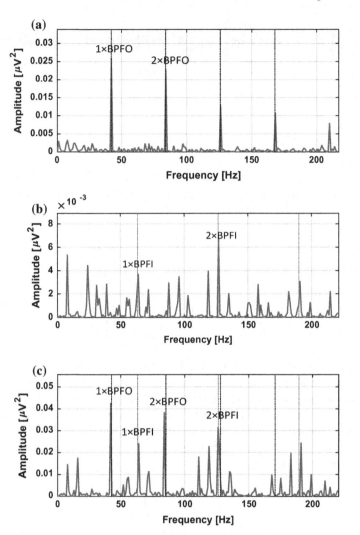

Fig. 2 Envelope spectrum of signals: **a** Defect on outer raceway **b** Defect on inner raceway **c** Defect on outer and inner raceways

of other combined failures such as BCOR, BCIR, and BCOIR. In this study, the proposed DNN is developed to detect the combined faults with single fault training information.

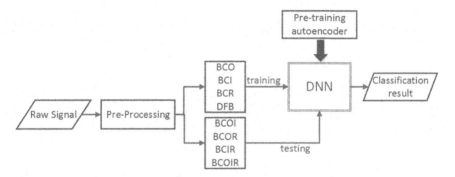

Fig. 3 Outline process of classification

Fig. 4 Autoencoder architecture

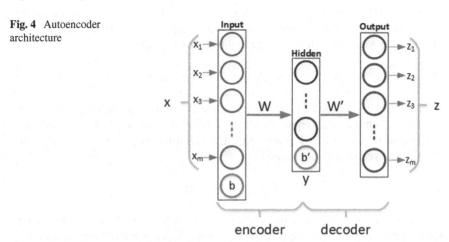

3   The Proposed Methodology

Figure 3 shows the proposed method, which depicts overall process during the training and testing of the DNN. The proposed method includes two phases: the pretraining and the fine-tuning, which are described more details below.

3.1   Pretraining Phase with Stacked Denoising Autoencoder

The traditional autoencoder is a type of neural networks that can learn a mapping from input X in dimension d to find a new representation Y in dimension d'. The learning procedure of autoencoder consists of two steps: encoder and decoder. Figure 4 illustrates the architecture of the original autoencoder.

The encoder finds deterministic mapping f_θ that transforms the input vector x to the representation vector y, such as

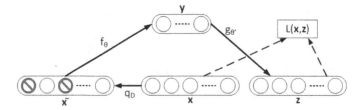

Fig. 5 Denoising autoencoder architecture

$$y = f_\theta(x) = s(Wx + b),\tag{5}$$

where $\theta = \{W, b\}$, W is a $d' \times d$ weight matrix and b is bias vector of dimension d', and $s(\cdot)$ is nonlinear mapping. The decoder maps back the representation y to reconstruct vector z which has the same dimension with the vector x, such as

$$z = g_{\theta'}(y) = \varphi(W'y + b'),\tag{6}$$

where $\theta' = \{W', b'\}$, and $\varphi(\cdot)$ is affine mapping or nonlinear mapping. The reconstruction z is not exactly equal to the input x, but it exists a distribution such that z is similar to x with a high probability. The reconstruction error is yielded as the distance between x and z, such as

$$L(x, z) = \|x - z\|^2\tag{7}$$

The training autoencoder minimizes the reconstruction error by finding the higher level representation of the input x. However, the dimension d' of representation y can be equal (or larger) than the dimension d of x. In this case, the representation y can achieve perfect reconstruction simply by "copying" the input vector without discovering useful information. To avoid this problem, a denoising autoencoder is implemented to establish the higher level of representation. The underlying idea of the denoising autoencoder is that the higher level of representation is more stable and robust if it is yielded from the corrupted input. In the denoising autoencoder, the input x is corrupted into $\tilde{x}$ by a stochastic mapping, such as

$$\tilde{x} \sim q_D(\tilde{x}|x)\tag{8}$$

Then, the noising input $\tilde{x}$ is mapped by the autoencoder $y = f_\theta(\tilde{x}) = s(W\tilde{x} + b)$ and reconstructed by $z = g_\theta(y)$. Figure 5 shows the procedure of the denoising autoencoder. The weights (W, W') and the biases (b, b') are trained to obtain the reconstruction vector z as much similar as possible to the original input x. This minimizes the reconstruction error $L(x, z) = \|x - z\|^2$. Parameters of the autoencoder are randomly initialized and optimized by a stochastic gradient descent technique.

The denoising autoencoders are stacked to form a stacked denoising autoencoder, which generates the structure of DNN. Thereby, each hidden layer of DNN is

Table 1 Encoding the output label of DNN

Types of bearing faults	Labels		
	O	I	R
Normal	0	0	0
Outer	1	0	0
Inner	0	1	0
Roller	0	0	1
Outer + inner	1	1	0
Outer + roller	1	0	1
Inner + roller	0	1	1
Outer + inner + roller	1	1	1

pretrained as one autoencoder. The output y of the first autoencoder feeds as the input for pretraining the next autoencoder. The output from the highest layer can be used as input to a supervised learning algorithm in the fine-tuning phase. After the creation of DNN, its fine-tuning is done using a gradient descent algorithm in a supervised manner.

3.2 Fine-Tuning for Classification

In this study, our dataset consists of eight different classes including BCO, BCI, BCR, BCOI, BCOR, BCRI, BCOIR, and DFB. Each class is independent and not mutually exclusive. For example, one AE signal can contain both outer fault and inner fault at the same time. It is classified as the defect in both the inner raceway and the outer raceway. For this purpose, the design labels for each class must show the relationship between classes. In the normal case of classification, the one-hot encoder is used to encode the labels of the training set. The one-hot is a group of bits in which the number of bits is equal to the number of classes. When a sample data belongs to one class, the value of the bit related to that classes is made 1 in one-hot encoding, while the value of other bits is kept as 0. Since this type of label encoding cannot provide the relationship between the classes, for example of a single inner raceway fault and a combined inner and outer raceway fault, this study proposes a new encoding for a normal, defect-free bearing and bearings seeded with seven types of localized defects. In this encoding, a group of 3 bits is used to encode eight classes. Each bit presents one of three values inner (I), outer (O), roller (R). The bits flip to 1 if the signal contains the failure. Accordingly, the defect-free bearing (normal) has 3 bits with values of 0, the bearing with the defect on outer-inner-roller (BCOIR) has the 3 bits having values of 1. Thereby, the output of DNN has three nodes. The detailed label of the classes is illustrated in Table 1.

Similar to other supervised learning techniques, in the fine-tuning phase, the loss function is used to calculate the difference between the prediction value and the target value. In order to properly design encoding labels of the proposed method, the loss function is also designed to show the relationship between classes. In our bearing fault classification, classes are independent and not mutually exclusive. In other words, one can perform multi-label classification, where a bearing signal can contain both an outer fault and an inner fault at the same time. Therefore, the more suitable loss function for the classification is the cross-entropy loss, which compares the log-odds of the data belonging to either of the classes. In this study, we only consider the probability of a data point belongs to a class. The real values are first fed into the sigmoid function $\sigma(x)$ that squashes the real values output to $[0, 1]$ and the model prediction becomes $\hat{y}(x) = \sigma(W^T x + b)$. Then, the cross-entropy loss is defined by the sigmoid function as

$$L_\theta(y, \hat{y}) = \frac{1}{M} \sum_{m=1}^{M} \left[-y^{(m)} \log \hat{y}(y^{(m)}) - (1 - y^{(m)}) \log(1 - \hat{y}(y^{(m)})) \right] \quad (9)$$

Finally, the DNN is trained to optimize entire weight and bias using the back-propagation algorithm.

4 Experimental Setup and Results

4.1 Data Acquisition

To evaluate the proposed method, data is collected from the test rig of the laboratory, as shown in Fig. 6. A three-phase motor is coupled with the drive-end shaft of the gearbox. The gearbox transmits torque to the non-drive-end shaft. The cylindrical roller element bearing is installed at both ends of each shaft. The non-drive-end shaft is connected with a belt and pulley to rotate the blade as the loaded condition of a bearing. The displacement transducer is placed on the non-drive-end shaft to measure the rotating speed of the bearing. An acoustic emission (AE) sensor is installed on the bearing housing to record AE signals stating bearing health condition.

Signals for seven classes of faults consist of both single and combined modes (BCO, BCI, BCR, BCOI, BCOR, BCIR, and BCOIR). A healthy or defect-free bearing is used as the reference for the baseline measurement. Eight different types of signals regarding bearing health condition were recorded under different rotational speeds (300, 350, 400, 450, and 500 rpms). Each signal contains 1-s signal and collected at sampling rate 250 kHz for each bearing defect in this study.

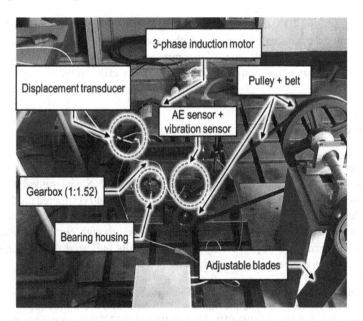

Fig. 6 Laboratory experimental set up to collect AE data [12]

Table 2 Results of the proposed method

# Training sample	500 input		1000 input	
	Single (%)	Combined (%)	Single (%)	Combined (%)
600	75–76	73–74	80–81	76–77
2000	90–91	88–89	94–95	90–91

4.2 Experimental Results

Each 1-s signal is a vector with 250,000 values corresponding to the sampling rate 250 kHz. It is segmented into 500 segments and 1000 segments. In each segment, the root means square value of the amplitude is calculated as the input of DNN. Therefore, DNN is provided with two different sizes of inputs, one is 500, and the other is 1000. In this experiment, for the case of 500 inputs, the DNN is set up with three hidden layers, and the number of nodes in each layer is (600, 300, 50). The SDAE activation function is a sigmoid function, and the affine function is used in the deactivation phase. The stacked denoising autoencoder with a stochastic gradient descent algorithm is used to minimize the reconstruction error. For the case of 1000 inputs, the network is set to 5 hidden layers with the number of nodes (1400, 600, 400, 100, 50). The rest of the structure parameters of DNN is set as same as with the case of 500 inputs. The results of DNN are depicted in Table 2.

The results show that the DNN classifies single mode bearing faults with the accuracy of 94–95% and combined faults with the accuracy of 90–91% with 2000 training sample and 1000 input size. There is significant variation in the classification accuracy when the number of training samples and input data change. This is because during training for each fault type, a large number of samples are available, and the proposed model can extract informative features more effectively.

5 Conclusion

This paper presented a stacked denoising autoencoder for reliable fault diagnosis, which is one of the deep neural networks (DNNs). The DNN was trained with only single fault samples and tested with data with both single and combined mode faults. Due to the layered architecture, DNN can automatically learn features single mode fault data and classify multiple combined faults even though combined fault data are not used during training. The experimental results indicated that the proposed method achieves a classification accuracy of 91%.

Acknowledgements This work was supported by the Korea Institute of Energy Technology Evaluation and Planning (KETEP) and the Ministry of Trade, Industry & Energy (MOTIE) of the Republic of Korea (Nos. 20162220100050, 20161120100350, 20172510102130). It was also funded in part by The Leading Human Resource Training Program of Regional Neo industry through the National Research Foundation of Korea (NRF) funded by the Ministry of Science, ICT and Future Planning (NRF-2016H1D5A1910564), and in part by the Basic Science Research Program through the National Research Foundation of Korea (NRF) funded by the Ministry of Education (2016R1D1A3B03931927).

References

1. Ben Ali, J., Fnaiech, N., Saidi, L., Chebel-Morello, B., Fnaiech, F.: Application of empirical mode decomposition and artificial neural network for automatic bearing fault diagnosis based on vibration signals. Appl. Acoust. **89**, 16–27 (2015)
2. Prieto, M.D., Cirrincione, G., Espinosa, A.G., Ortega, J.A., Henao, H.: Bearing fault diagnosis by a novel condition-monitoring scheme based on statistical-time features and neural networks. IEEE Trans. Ind. Electron. **30**(8), 3398–3407 (2013)
3. Konar, P., Chattopadhyay, P.: Bearing fault detection of induction motor using wavelet and support vector machines (SVMs). Appl. Soft Comput. **11**(6), 4203–4211 (2011)
4. Kang, M., Kim, J., Wills, L., Kim, J.-M.: Time-varying and multi-resolution envelope analysis and discriminative feature analysis for bearing fault diagnosis. IEEE Trans. Ind. Electron. **62**(12), 7749–7761 (2015)
5. Nguyen, P.H., Kim, J.-M.: Multifault diagnosis of rolling element bearings using a wavelet kurtogram and vector median-based feature analysis. Shock Vibr. (2015)
6. Bengio, Y., Delalleau, O.: On the expressive power of deep architectures. In: ALT'11 Proceedings of the 22nd International Conference on Algorithmic Learning Theory, pp. 18–36. Berlin (2011)
7. Vincent, P., Larochelle, H., Lajoie, I., Bengio, Y., Manzagol, P.: Stacked denoising autoencoders: learning useful representations in a deep network with a local denoising criterion. J. Mach. Learn. Res. **11**, 3371–3408 (2010)

8. McInerny, S.A., Dai, Y.: Basic vibration signal processing for bearing fault detection. IEEE Trans. Educ. **46**(1), 149–156 (2003)
9. Stack, J.R., Harley, R.G., Habetler, T.G.: An amplitude modulation detector for fault diagnosis in rolling element bearings. IEEE Trans. Ind. Electron. **51**(5), 1097–1102 (2004)
10. Yang, Y.: A signal theoretic approach for envelope analysis of real-valued signals. IEEE Access **5**, 5623–5630 (2017)
11. Kang, M., Kim, J., Kim, J.-M.: High-performance and energy-efficient fault diagnosis using effective envelope analysis and denoising on a general-purpose graphics processing unit. IEEE Trans. Power Electron. **30**(5), 2763–2776 (2015)
12. Kang, M., Ramaswami, G.K., Hodkiewicz, M., Cripps, E., Kim, J.-M., Pecht, M.: A sequential K-nearest neighbor classification approach for data-driven fault diagnosis using distance- and density-based affinity measures. LNCS **9714**, 253–261 (2016)

Cancer Prediction Through a Bacterial Evolutionary Algorithm Based Adaptive Weighted Fuzzy C-Means Approach

M. Sangeetha, N. K. Karthikeyan, P. Tamijeselvy and M. Nachammai

Abstract Clustering of tumor plays a significant part in classifying malignancies from carcinoma genetic data and hence is introduced to deal among the classification problem. It is used in critical applications like cancer treatment for diagnosis and prognosis, analysis of gene expression and related areas. In earlier research works, various tumor clustering schemes were presented based on the single clustering systems and implemented successfully to a variety of biomolecular data for cancer class detection. But, it suffered from some drawbacks similar to starvation of stability, accuracy, and robustness. The ensemble grouping schemes are introduced to overcome these limitations. In this research work, efficient dimensionality reduction is done by using Enhanced Independent Component Analysis (EICA) and effective Feature Selection (FS) using Geometric Particle Swarm Optimization (GPSO) are used. Finally, the competent Adaptive Weighted Fuzzy C-Means Clustering (AWFCM) with metaheuristic optimization scheme of Bacterial Evolutionary Algorithm (BEA) is also proposed. It improves the performance of tumor clustering of biomolecular data. The performance of various ensemble clustering schemes is evaluated based on the efficient feature selection methods.

Keywords Bi-clustering Bayesian principal component analysis (Bi-BPCA)
Weighted fuzzy C-Means · Bacterial evolutionary algorithm

M. Sangeetha (✉)
IT Department, SKCT, Coimbatore, India
e-mail: m.sangeetha@skct.edu.in; godsan2003@gmail.com

N. K. Karthikeyan
CSE & IT Department, CIT, Coimbatore, India
e-mail: karthiaish1966@gmail.com

P. Tamijeselvy
CSE Department, SKCT, Coimbatore, India
e-mail: p.tamijeselvy@skct.edu.in

M. Nachammai
CS Department, SKACAS, Coimbatore, India
e-mail: nachammaimuthu1611@gmail.com

© Springer Nature Singapore Pte Ltd. 2019   95
S. K. Bhatia et al. (eds.), *Advances in Computer Communication and Computational Sciences*, Advances in Intelligent Systems and Computing 759,
https://doi.org/10.1007/978-981-13-0341-8_9

1 Introduction

The prime challenge in disease diagnosis is accuracy in prediction. Most helpful tool for assessing the presence and state of Leukemia is the microarray technology. Before performing any analysis, however, the microarray data has to go through several steps of preprocessing. Microarray experiments are employed to calculate the interpretation measures of genes and then use the data for analysis. The standard dataset may have missing values or noisy data. These data are to be analyzed using data extracting techniques. The data has to be preprocessed to enhance the accuracy of the result. Data preprocessing is an essential step in the data mining process. The goal is to examine the preprocessing and gene selection on Leukemia gene clustering. The primary goal is to explore the potential of AWFCM with BEA using microarray gene expression Leukemia dataset in cancer diagnosis. The aim of clustering a microarray data is to simplify analysis, thereby enabling the user to make further decisions.

2 Related Works

Data Mining has the great potential to handle the healthcare systems. Thus data mining plays a significant part in reducing the cost. For large data sets Bayesian approach functions fastly and provides the exact result [1]. Data mining methods are utilized to increase the essence of prognostication of disease [2]. Different methods of data mining have been utilized to identify the similar patterns. Healthcare and therapeutic fields are possessing the huge amount of data but they are not accurately utilized for making decisions. Leukemia is a dangerous disease that may cause death earlier. Microarray profiling of cancer cells is suggested as a more informative approach to diagnosis [3]. Leukemia cells are typically self-sufficient to grow and have an increased activity enabling a larger number of cell divisions [4]. As Leukemia grows, it becomes capable of rapidly affecting the blood cells at proximity, and spreading to all the blood cells within weeks or months. Acute Lymphoblastic Leukemia and Acute Myelogenous Leukemia diagnostic task type from Gene Expression Model Selector database [5] which is publicly available as open access. This dataset provides information about microarray genes of both normal and cancer cells. J. C. Dunn [6] developed the Fuzzy C-Means clustering procedure, and it is an advancement of K-measure algorithm. In fuzzy clustering, it is a vital one and permitting the entire data as an element of any cluster. Dudoit et al. [7] utilized the various approaches like nearest neighbor classifiers, categorization trees and Linear Discriminant Analysis for clustering the cancer cells and noncancer cells. Shi and Malik [8] presented the Normalized Cuts clustering algorithm and it examines the information as a mixture of nodes and graph. These nodes and graph indicate the edges and data points. LCE is extended from the Hybrid Bipartite Graph Formulation which was an important strategy [9]. This has been implied as a process that depends on graph to enhance the combination of cluster matrix, rather than the current $P \times N$

Binary Matrix. In fuzzy clustering, it is a vital one and permits entire data to be an element of any group. Various clustering schemes like Adaptive Weighted Fuzzy C-Means Clustering-Bacterial Evolutionary procedure (AWFCM-BEA), Hybrid Fuzzy Cluster Ensemble Frameworks (HFCEF) [10], Link-based Cluster Ensemble (LCE) scheme, Random Subspace technique and K-Means (RSKM) [11] are used. Out of all these clustering schemes, AWFCM-BEA provides better results.

3 Motivation

The main objective of the proposed work is as follows:

i.   To efficiently group the cancer cells and noncancer cells using clustering techniques.
ii. To evaluate and compare the results using various clustering methods.
iii. To find the missing values and also to eliminate the redundant values.

The remaining section of paper is categorized as below. Section 4 specifies the data set. Section 5 explains the scheduled method that specifies preprocessing methods and various clustering strategies. Section 6 includes the analysis of clustering experiments and resolution is provided in Sect. 7.

4 Data Set

In this research work, the Leukemia dataset (www.gems-system.org) reported by Golub et al. (2007) has been utilized. The gene dataset possesses information about more than 40 patients discomforted on behalf of Acute Lymphoblastic Leukemia and more than 22 patients discomforted from Acute Myeloid Leukemia. The bone marrow specimens of more than 70 patients received while diagnosing Leukemia have been used.

5 Proposed Method

The primary goal of this proposed work for classifying leukemia gene expression is to develop a method for the intention of leukemia cancer cell investigation utilizing gene classification method inclusive of feature selection process. Multiple strategies are available to identify the cancer cell investigation in an automatic way. The proposed Leukemia gene classification process makes use of Enhanced preprocessing Independent Component Analysis (EICA) method and Bi-PCA with Geometric Particle Swarm Optimization (GPSO) based feature selection and classification. After the completion of preprocessing and feature selection, the data set is given as input to various clustering algorithms such as Hybrid Fuzzy Cluster Ensemble Framework, Link-Based Cluster Ensemble Framework, Random Subspace Technique, and K-Means, Bacterial Evolutionary Algorithm with Adaptive Weighted Fuzzy Clustering Means Strategy.

5.1 Preprocessing

The standard dataset may have in it or some omitted data. The entire data has to be investigated by utilizing mining techniques. The dataset has to be preprocessed to improve the accuracy of the result. Data preprocessing is an important step in the data extracting process. Initially, the data set may have omitted data and redundant data. Bi-clustering Bayesian Principal Component Analysis has been utilized to investigate the omitted data. The redundant data has been removed with the help of Enhanced Independent Component Analysis. After the completion of preprocessing feature selection has been used. This aims to remove the unwanted features that are not needed by the users. For selecting the best features from the data set, Particle Swarm Optimization and Geometric Particle Swarm Optimization have been used.

5.2 Hybrid Fuzzy Cluster Ensemble Framework Ensemble with FCM and Ncut Algorithms

The Hybrid Fuzzy Cluster Ensemble Framework involves Affinity Propagation (AP) algorithm to carry out clustering on the sample elements. Subsequently, it chooses a specimen from every group at random that is provided as the foundation sample, along with a group of foundation samples. The mentioned work is repeated 'B' times, which produces 'B' foundation specimen groups. Second, the fuzzy membership objective is implemented to detain the association among the specimens in the real dataset 'P'. Thus, the foundation specimens and a fuzzy matrix are obtained. Ultimately, the concurrence task is intended to review the downy matrices and get the concluding results.

5.3 Link-Based Cluster Ensemble Approach

Link-Based Cluster Ensemble is extended from the Hybrid Bipartite Graph Formulation strategy and a graph-based consensus process is applied to improve cluster association matrix, rather than the current 'P × N' Binary Matrix. It is used to improve the cluster association matrix by decreasing the number of missing entries. Initially, 'M' foundation groups are created to generate a cluster ensemble. Second, Refined

---

Input: Group of N vectors, K Number of Groups
Output: K Groups
STEP 1: Choose 'K' points as the centroids of cluster.
STEP 2: Allocate every point in the vector close to cluster depending on the Euclidean measure between cluster centroid and data point.
STEP 3: Recompute every centroid of the cluster as the moderate of the point in a group and is computed as below Equation:

$$v_i = (1/c_i) \sum_{j=1}^{c_i} x_i$$

Where c_i indicates the count of data particles in i^{th} cluster.

STEP 4: Recompute the measure among each data point and attain the newly cluster centres.
STEP 5: Reiterate steps 2 and 3 till there is no movement of centroids.

Fig. 1 K-means procedure

cluster-linked Matrix (RM) is set up depending on an established similarity algorithm known as Weighted Connected-Triples. Finally, the ultimate data separation is generated using the Spectral Graph Partitioning scheme as a consent task.

5.4 Random Subspace Technique and K-Means (RSKM)

Initially, the dataset is preprocessed by means of enhanced independent component analysis method which helps to identify the missed data and also to remove the redundant values. Thus, EICA method is used for reducing dimensions. The best easiest unsupervised training strategy is K-means (KM) clustering, and it is used to find solutions for the popular clustering dilemma. K-means procedure is given in Fig. 1. 'K' indicates the predefined clusters numbers, and it defines the closeness of data points. In this process, initially, the cluster centroids are chosen randomly. It does not accurately give the unique cluster result. Therefore, user selected centroids are specified for good results. The closest centroids are grouped as clusters by finding the distance among data points and the cluster centroid.

Random Subspace Technique and K-Means approach is proposed to determine the fundamental types of cancers in some datasets.

5.5 Enhanced Ensemble Clustering of BEA with AWFCM for Cancer Diagnosis

Cluster analysis is a useful tool for determining co-expressed genes of gene expression data and it partitions data into clusters depending on specific features. Genetic expression analysis faces some issues. To overcome the issues, in this research work,

EICA and AWFCM with BEA algorithm are proposed for dimensionality reduction and clustering. The objective function of AWFCM is optimized using BEA, and it yields better clustering PU. The input is preprocessed Fuzzy C-Means clustering algorithm is an enhancement of K-measure algorithm. FCM is given in Algorithm 1. In fuzzy clustering, data must be a member of any cluster. The clustering block, similar features are grouped. This clustering stimulates the classifier to quickly and easily classify with more accuracy.

5.5.1 Bacterial Evolutionary Algorithm (BEA)

There are various optimization algorithms presented for optimization problems, motivated by the procedures of metaheuristic evolution. An individual is signified via a flow of numbers that may be binary digits or chromosome. This scheme is the extension of GA, which depends on bacterial growth other than eukaryotic. Germs distribute portions of their gene alternatives to implement clear crossover in genetic code that denotes the microbes can develop or narrow down. It is utilized in the microbial evolution and the genetic substitution operations. Then, crossover operation of GA is applied. Consequently, information is moved among various individuals. In the suggested scheme, multiple functions can be carried out in parallel and implemented in a parallel measuring environment.

Step-by-step evolutionary process is given below. It includes three steps.

Step 1: Create initial population.

An initial bacterium population of 'N_{ind}' bacteria $\{\xi_i, i \in I\}$ is created randomly $(I = \{1, \dots, N_{ind}\})$.

Step 2: Apply microbial crossover.

In order to find a universal maximum in AWFCM, it is essential to find the new portions of the search area that are not covered by the present population. It has been achieved by inserting newly created data to the microbes using crossover method.

Step 3: Apply Genetic shift.

Microbial transfiguration operation optimizes the germs in the population. Genetic shift has been applied to confirm that the data from microbes spreads among the entire population. The entire population is split into two portions: one with origin chromosome and the other with resulting chromosome.

5.5.2 Adaptive Weighted Fuzzy C-Means (AWFCM) Clustering

Basic FCM objective function is defined as given in Eq. 1.

$$\text{Obj}(U, V) = \sum_{i=1}^{n} \sum_{j=1}^{c} (\mu_{ij})^m \left|\left|x_i - v_j\right|\right|^2 \qquad (1)$$

Table 1 Rate of success of clustering schemes

Clustering schemes	AWFCM-BEA	HFCEF	LCE	RSKM
Rate of success or purity	65.65	55.7	49.54	47.48

here,

m	Real number>1
μ_{ij}	Degree of association of x_i in the group
x_i	ith of d-spatial data
v_j	Group centroid of d-dimension data
$\|x_i - v_j\|$	Measure of similarity among the estimated data and the group data.

The optimized value of the objective function is derived using BEA.

In the proposed scheme, once the tumor cancer region within a given gene expression is identified, an unsupervised clustering is performed by k-class AWFCM clustering of the tumor features and weighted mean concept. The number of clusters (k) is adaptively determined on a per-data basis, so as to partition the tumor tissue into subregions of relatively homogenous gray-level intensity clusters. In the proposed algorithm, the input is the smallest and largest number of clusters and the maximum number of iterations, and its output is the adaptive clustering results. The output of AWFCM clustering is known to be dependent on the overall distribution of the feature values in a gene expression. Biomolecular data may be thought of as an admixture of several different Gaussian distributions. The proposed scheme adaptively computes 'k' for each data based on the properties of features values within the tumor region. To accomplish this, the gene expression data of the tumor tissue is initially distributed to possess a zero-mean and unit SD.

6 Experimental Results and Conclusion

The above-said algorithms are implemented, and the results are validated using MAT-LAB 8.1 tool. The above clustering algorithms are performed separately without any added block. Among these four algorithms, the proposed AWFCM-BEA clustering provides better performance results. The grouping of features mostly reduce the relevant data from the dataset. Hence, this clustering reduces the count of characteristics in the data. Due to this dimensionality reduction process, the classifier is fast and more accurate.

Figure 2 indicates the rate of success of proposed AWFCM-BEA with the previous methods. The Leukemia 1 dataset is directly applied to the clustering block and the predicted output of the related success values are shown in Table 1.

Figure 3 indicates the NMI of suggested AWFCM-BEA clustering and existing methods. It is clear that the proposed AWFCM-BEA grouping works out good when related with other schemes without any preprocessing and FS. If preprocessing and

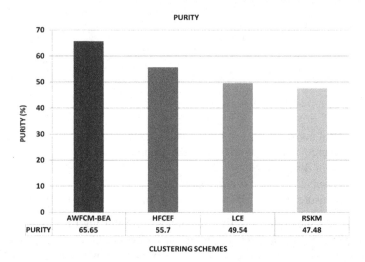

Fig. 2 Comparison of rate of success of clustering methods

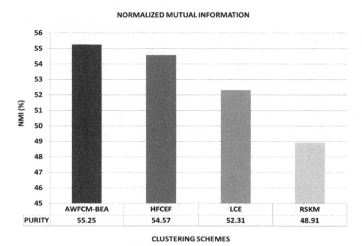

Fig. 3 Comparison of NMI of clustering methods

Table 2 NMI of clustering schemes

Clustering schemes	AWFCM-BEA	HFCEF	LCE	RSKM
Rate of success or purity	55.25	54.57	52.31	48.91

FS blocks are included, the result values will be improved compared to the direct application of clustering. The corresponding NMI values are shown in Table 2.

The results of the proposed AWFCM-BEA clustering are compared schemes with and without GPSO, and existing HFCEF, LCE, RSKM clustering algorithms for Leukemia cancer samples.

7 Conclusion

In this work, the proposed and existing clustering schemes for Leukemia gene clustering in cancer diagnosis system have been discussed. The performance of these schemes with respect to rate of success and NMI has been evaluated. It is seen that the proposed Adaptive Weighted Fuzzy C-Means with Bacterial Evolutionary Algorithm (AWFCM-BEA) with GPSO for cancer gene clustering yields better results than the existing methods on Leukemia dataset. In the future work, classification methods can be proposed to deal with gene classification in the cancer diagnosis system. In future, it will be focused on an additional feature extraction process for most accurate classification results in the proficient early cancer diagnosis.

References

1. Ahmad, P., Qamar, S., Rizvi, S.Q.A.: Techniques of data mining in health care—a review. Int. J. Comput. Appl. **120**(1), 0975–8887 (2015)
2. Khaleel, M.A., Pradham, S.K., Dash, G.N.: A survey of data mining techniques on medical data for finding locally frequent diseases. Int. J. Adv. Res. Comput. Sci. Softw. Eng. **3**(8) (2013). ISSN 2277 128X
3. Golub, T.R., et al.: Molecular classification of cancer: class discovery and class prediction by gene expression monitoring. Science **286**(5439), 531–537 (1999)
4. Hanahan, D., Weinberg, R.A.: The hallmarks of cancer. Cell **100**(1), 57–70 (2000)
5. Statnikov, A., et al.: GEMS: a system for automated cancer diagnosis and biomarker discovery from microarray gene expression data. Int. J. Med. Inf. **74**(7), 491–503 (2005)
6. Dunn, J.C.: A fuzzy relative of the ISODATA process and its use in detecting compact well-separated clusters. 32–57 (1973)
7. Dudoit, S., Fridlyand, J., Speed, T.P.: Comparison of discrimination methods for the classification of tumors using gene expression data. J. Am. Stat. Assoc. **97**(457), 77–87 (2002)
8. Shi, J., Malik, J.: Normalized cuts and image segmentation. IEEE Trans. Pattern Anal. Mach. Intell. **22**(8), 888–905 (2000)
9. Fern, X.Z., Brodley, C.E.: Solving cluster ensemble problems by bipartite graph partitioning. In: Proceedings of the Twenty-First International Conference on Machine Learning. ACM (2004)
10. Yu, Z., Chen, H., You, J., Han, G., Li, L.: Hybrid fuzzy cluster ensemble framework for tumor clustering from biomolecular data. IEEE/ACM Trans. Comput. Biol. Bioinf. **10**(3), 657–670 (2013)
11. Yu, Z., Wong, H.S., Wang, H.: Graph-based consensus clustering for class discovery from gene expression data. Bioinformatics **23**(21), 2888–2896 (2007)
12. Tsukasaki, K., et al.: Definition, prognostic factors, treatment, and response criteria of adult T-cell leukemia-lymphoma: a proposal from an international consensus meeting. J. Clin. Oncol. **27**(3), 453–459 (2009)

Cold Start Problem in Social Recommender Systems: State-of-the-Art Review

V. R. Revathy and S. Pillai Anitha

Abstract Recommender systems have become very important for the effective customer or user support in the field of online shopping, social networking, e-commerce, and so on. Collaborative filtering methods have been used by social networking sites for providing the most interesting service for a user by finding similar tastes from many other users. To generate top-N recommendations, we have to study and solve one of the major challenges faced with this method which is the cold start problem. This paper discusses different types of solutions to the cold start problem. Each researcher has provided different solutions to the cold start problem through different methodologies.

Keywords Recommender systems · Cold start · Cross-domain · Deep learning
Active node technique

1 Introduction

Recommender systems are used in different fields to provide good recommendations to users while they are searching online for some information. Social recommender systems are an interesting area of research because of the importance, expansion, consideration, and value of the social networks. The role that social recommender systems play in recommending the information according to a specific user's taste is great. Thus, social recommender systems are inevitable for the success of any business/organization.

Recommendations may be specific to the item in which users are interested in (for example, a movie or a product). In the case of a social network, there are different types of data and metadata which are shared or tagged. One of the major challenges

V. R. Revathy (✉) · S. P. Anitha
School of Computing Sciences, Hindustan Institute of Technology and Science, Chennai, India
e-mail: revathyvrajendran@gmail.com

S. P. Anitha
e-mail: anithasp@hindustanuniv.ac.in

© Springer Nature Singapore Pte Ltd. 2019
S. K. Bhatia et al. (eds.), *Advances in Computer Communication and Computational Sciences*, Advances in Intelligent Systems and Computing 759,
https://doi.org/10.1007/978-981-13-0341-8_10

that affects the performance of recommender system is cold start problem. It arises due to insufficient information about a user or an item. This could be because sufficient review or feedback about a product is not available. Cold start problems are mainly found in collaborative filtering. This paper reviews different solutions to the cold start problem. A recommender system architecture incorporating sentiment analysis is proposed as future work.

2   Literature Review: Solutions for Cold Start Problems

In order to find possible solutions to the cold start problem, we identified around 29 papers where the authors had discussed their work and how they encountered the cold start problem. Most of the selected papers discussed and experimentally showed a solution to the cold start problem. The cold start problem particularly refers to a situation when the information about a user or an item is very few or not available. Solutions to the cold start problem that each researcher arrived were classified into 10 categories based on the approach and the methodology they used. These categories include solutions based on cross-domain system and social network data, elicitation of user profile via applying association rules, demographic similarities, solutions via studying users' behavior on the website, user elicitation, matrix factorization, and building decision trees. Deep learning techniques were followed in some studies.

2.1   Cross-Domain Recommender Systems and Social-Network Data

Figure 1 shows a cross-domain recommender system. When the information about the user is insufficient, the system can use external sources (e.g., social networking sites) to use it as additional information. These systems aim to establish knowledge-based links between domains or to transfer knowledge from the source domain to the target domain [1]. That is, user information from some external sources is used to bootstrap the system. One popular example of cross-domain recommender systems is travel recommender systems which give personalized recommendations to tourists to unknown places with the help of social media geo-tagged photos [2–4]. Memon I. et al. designed an intelligent application to predict the user's travel interests based on geo-tagged images from social media with historical weather data. Personalized location recommendations for a new trip are generated using the historical last tourist visit to the place where the trip is planned. The places are geo-tagged with images to generate location-based recommendations. The cold start problem arises here when a new user is going on a trip. The new user plan trip is predicted using similar tourist trip planned to that particular location last time. Unrelated or irrelevant user preference recommendations are also generated to a new user while matching a similar user

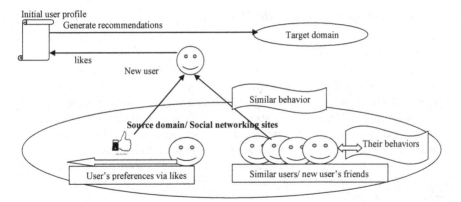

Fig. 1 Model of a cross-domain recommender system [1, 5]

profile from social networking sites. These irrelevant recommendations may affect the quality of results [4].

Shapira et al. put forward a remedy to the cold start problem that the system may collect information about a new user in a domain from his/her Facebook page and generate an initial user profile by extracting data relevant to the specific domain (if such information exists), or data related to other domains, as well as utilizing information about the behavior and preferences of the new user's friends. In this case, the authors used a dataset composed of unary Facebook likes (i.e., If liked unary 1, else 0) as a user preference. Popularity information from social networking sites gave more satisfying results to generate recommendations [5]. Shapira et al. introduced another cross-domain learning method called k-NN (k-Nearest Neighbor) source (k-NN-s) aggregation method for a new user problem in target domain. k-NN-s method learns user's friends' information and applies the information what it learnt to the target domain. The authors achieved significantly more accurate results by using aggregation based methods when the available user preferences are sparse [5].

The quality of recommendations in cross-domain platform depends on dataset density. The dataset density is very important for cold start user problems. The performance of cross-domain recommender system is sensitive to various features such as relativity of domains, the number of users, number of items, and overlapping users.

2.2 Enrichment of User Profile Through Association

Hridya and Mariappan [6] proposed a solution to the cold start problem by applying association rules and clustering technique. The new user profiles are enriched by applying association rules and frequent pattern taxonomy driven profiles are constructed. For newly added item, clustering technique was applied. Thus, Top-N recommendations were generated.

2.3 Demographic Similarities/User Interests

Naive Filterbots model. This model is based on Naive Filterbots algorithm an extended form of Filterbots algorithm [7]. The Naïve Filterbots algorithm injects pseudo users or bots into the system [8]. Naive Filterbots gives rating for items based on the attributes of items or users, e.g., average rate of some demographic similarities between users. The system treats the injected agents in the user-item matrix as any other existing users or items ratings (actual user-item ratings), and then standard collaborative filtering algorithms are applied to generate recommendations. Naive Filterbots used average ratings format [8, 9]. Predictions are done using user-based and item-based algorithms. Pearson correlation coefficient is used to find the similarity between users as follows:

$$S(a, b) = \frac{\sum_{i \in I_a \cap I_b} (r_{a,i} - \bar{r}_a) \cdot (r_{b,i} - \bar{r}_b)}{\sqrt{\sum_i (r_{a,i} - \bar{r}_a)^2} \cdot \sqrt{\sum_i (r_{b,i} - \bar{r}_b)^2}} \tag{3}$$

where $S(a, b)$ is the similarity between users a, and b, while $r_{a,i}$ and $r_{b,i}$ are the ratings of an item i which had been done by both users a and b. In addition, $\bar{r}_a$ represents the user a average rating for all items, and $\bar{r}_b$ represents the user b average rating for all items, and $I_a \cap I_b$ is the set of items that rated by both users a and b.

Triadic Aspect Model. This model depends on users' demographic information such as age, gender, and job [10]. The authors used a triadic aspect model suggested by Hofmann [11], which includes three-way co-occurrence demographic information like age, gender, and job. Lam X. N. et al. put forward an aspect model which uses the information of the current similar users and generates prediction ratings. Predictions are done by the triadic aspect model using the three features of the user (age, gender, and job). Similar matching features of existing users are found. The experimental results show that the approach produces satisfactory predictions with 284 different types of users. The performance with large number of users was not effective.

Model-based on classification. Lika et al. [12] proposed a three-phase model to tackle the new user cold start problem. In the first phase *classification*, using the demographic dataset A = {a1, a2, ..., an}, a list of users U = {u1, u2, ..., um}, and list of items I = {i1, i2, ..., ip}, a model is built. User preferences are also considered in this phase.

The built model is mapped to different categories based on the instances of some variables of set A and prediction variable. For each user *u* there is a category C with set of N neighbors. So category relates homogenous neighbors. The second phase is *similarity estimation* that measures the similarity between the users. Semantic similarity measure is applied and similar weights are found out. The third phase is *collaborative filtering* that combines the similarity measure and historical ratings of similar users to generate predictions. The performance of this model-based approach is good when the users are available in large number.

2.4 Browsing Targets

Embarak developed a privacy-protected model named as Active Node Technique (ANT) that generates recommendations purely based on the power of thinking of a user. The click streams of a new user visiting the website are collected and then the user sessions are created. The active web pages which users have clicked are active nodes. Active nodes and a newly added item are linked by estimating a logical/virtual weight. Estimation of this logical weight is based on the user's browsing target through a semantic link between the items, and the threshold of the items in the sessions [9]. The node recommendations are based on the virtual relationship between active node and the candidate items. Node recommendations applied to the dataset showed highest accuracy in coverage and precision rate. Batch recommendations are based on all super sessions of visited sessions. Both recommendations showed accuracy in novelty, when compared to the abovementioned models on demographic similarities [9]. This technique needs to prove scalability with increasing number of users, adaptivity to different system requirements and system performance [9].

2.5 Elicitation/Asking New Users to Provide Information

Eliciting information from users is a form of active learning. Rubens et al. [13] discussed the Active Learning (AL) where RSs will actively ask users to rate the items. Rating of items provides information about the personal preferences. Active learning also provides the personalization of recommendations. For this, the system is allowed to select the items for display (during the initial sign-up or in log-in) to the website and allowing the users to explore their desires through asking them to do rating. Fernández-Tobías et al. in their experiments using personalized AL methods, they got a considerably larger number of likes from the user rating elicitation [14]. These ratings were very useful for improving the recommendations. The accuracy and ranking quality for these personalized AL methods seem to be little effective when at least one rating or like was given (that is in the case of known users). Novelty and coverage yielded by these methods showed the best trade-offs.

2.6 Using Content Information to Combine with Collaborative Filtering

Combining the content information with collaborative filtering is a hybrid approach towards user cold start problem.

Two-split decision tree building. Sun D. et al. proposed an algorithm that consists of four-step procedures namely user *Clustering*, decision *Tree* building, new user classifying, *And* item *Predicting*, thus called in short as CTAP algorithm. Their approach to the cold start user problem through decision tree is a machine learning approach. The algorithm first creates clusters based on the user-item rating matrix; the number of clicks is used to evaluate similar preferences. Thus, the clustering step produces 'n' number of clusters in which a cluster will be with users of similar preference. Best attribute is chosen from the demographical information. Using the attributes and clusters, the decision tree is built [15]. The third step consists of new user classification. Novel user's demographic details are collected instantly [15]. These details are inputted as answers for the questions (in each node) of the inbuilt trained decision tree accurately and at last will reach a particular cluster where user's information matches. The last step does item predictions based on the user preferences. The algorithm showed robustness when compared to standard collaborative algorithms. This algorithm gave better performance for cold start users [15]. The performance of CTAP algorithm reduces with the increase in the ratings.

Multiple-split (questions) decision tree building. Sun M. et al. proposed an approach that collects user's taste information by asking multiple questions like an interview with minimum number of interaction with users [16]. This kind of approach for creating a user profile can help in generating accurate recommendations. Each node is connected to multiple splits which ask different questions. To minimize the cost of searching all the splits, a framework having less prediction loss is used to find a split. The decision tree learns a regressor inside each split to retrieve all the answers from the root node to the current visiting node. The algorithm shows outstanding performance in the prediction accuracy and user interaction efforts. The accuracy rate is comparatively very high with single-question decision trees [16].

2.7   User Similarity for Cold Start Items

The cold start item problem arises when the ratings for a newly added item is very less. Sarumathi M. et al. proposed a recommender system that generates recommendations based on user similarity for item correlation [17]. The user's profile will be listed with initial item preferences. Item-item similarity index is calculated and an expanded list is also prepared. The top 'n' items from this list are selected to generate recommendations.

2.8   Imputing Missing Values into the Input Matrix

The nonnegative matrix factorization with stochastic Gradient (NG) is good to deal with sparse matrices but computational time is big when handling large matrices. The study conducted by Ocepek et al. found that the pure matrix factorization algorithms perform better due to the imputation strategy. The Matrix Factorization (MF) by data fusion and nonnegative matrix factorization with alternating least squares was not giving satisfactory results. Matrix factorization can be applied to real-time domains, new item cold start problem, and RS in generating good recommendations. For the experiment, MF used different combinations of parameters. The combination of 25 and 10% of nearest neighbors, and choosing most frequent value attributes and imputation of a missing value for the user (i.e., for cold start user) with attribute's mean value delivered better performance than just applying the raw matrix factorization approaches [18]. The imputation step can also reduce prediction error.

2.9 Deep Learning

Collaborative deep learning can be used for alleviating cold start problem [19, 20]. The introduction of Deep Belief Network (DBN) was the first work that proved the deep learning network can be trained to get better AI models [21]. Now there are different types of deep models. Deep Neural Network (DNN) is formed by Multilayer Perceptron (MLP) with many intermediate layers (hidden layers) [21]. A DNN with a large number of hidden layers performs better. These networks follow effective parameter initialization methods. A DNN called Stacked Denoising Autoencoder (SDAE) is used to retrieve the content information of items. Time-Based Single-Valued Decomposition (Time-SVD++) is a state-of-the-art collaborative filtering model that can be applied to convert the content information into prediction ratings for the new items or cold start item [22].

For the complete and the incomplete cold start problems, Wei J. et al. proposed Integrated Recommendation models with Collaborative filtering and Deep Learning (IRCD)-CCS and IRCD-ICS, respectively. The IRCD-CCS model is trained with the rating matrix by giving attribute values of non-complete cold start items. Wei J. et al. developed multiview deep neural network model by integrating the data sources from multiple domains. The semantic feature mapping can be learnt from this network model and the user cold start problem can be alleviated. The multiview model is the improved version of Deep Structured Semantic Model (DSSM) where the feature dimension vectors from multiple domains are passed as input through two different neural networks for two different inputs [22]. The structure of the model is such that recommendations can be done easily through a pivot vector X_i to the users' feature dimension and then produces views for each heterogeneous item. The idea behind this is to find a unique mapping for users' features, which specifically converts the whole user features into a semantic space that emulates the heterogeneous items the users are interested into multiple views/domains [22]. This transformation through mapping a user feature into item view will solve the problem of those domains which has insufficient or no information.

2.10   Algorithmic Approaches

Son L. H. reviewed the related studies on user cold start problem and classified them
into three groups and gave an experimental evaluation of algorithms that comes
under each category, such as Modified Intuitionistic Possibilistic Fuzzy Geographi-
cally Weighted Clustering-for Cold Start problem (MIPFGWC-CS), New Heuristic
Similarity Model (NHSM), Fuzzy Association Rules and Multiple-level Similarity
(FARAMS) and Hybrid User-based Fuzzy Collaborative Filtering (HU–FCF) [23].
The *first* classification was making use of demographic information. MIPFGWC-CS
algorithm comes under this category [23]. The main limitation of this algorithm is the
unavailability of source information during selection that happens due to the absence
of recording corresponding social network account.

The *second* classification makes use of the idea of improving the approaches
to match similar users. The studies [15, 16] come under this category. Liu H. et al.
introduced a refined version of the similarity measure that showed better performance
[24]. The accuracy of NHSM is better than the MIPFGWC-CS algorithm. The limitation of NHSM algorithm is that it will give inaccurate results on similarity. The *third*
classification uses a hybrid approach. Fuzzy Association Rules and Multiple-level
Similarity (FARAMS) is a hybrid approach of fuzzy set theory and association rule
mining technique towards collaborative filtering [23, 25]. Cross-Level Association
RulEs (CLARE) was included in their approach to combine different user preference
ratings and content information of items. Experimental result shows that the fuzzification in FARAMS leads to inaccurate predictions when membership values are not
correct [23].

Another algorithm that comes under the *third* classification is HU-FCF (Hybrid
User-based Fuzzy Collaborative Filtering) [26]. Fuzzy similarities between the users
are found out from the demographic information and integrated with the similarities
on user-based preferences such as ratings and final rating similarities are calculated
[23]. The HU-FCF algorithm does predictions if the demographical values are available. This is because there is no historical rating for a new user, thus similarities
cannot be found out and final ratings cannot be predicted. This is one of the limitations of the HU-FCF algorithm [26]. NHSM algorithm employed a better similarity
metric when compared to the metrics other (FARAMS and HU-FCF) algorithms
because it is not using any fuzzy similarity measures. If the demographic information and historical ratings are available HU-FCF will give better performance. The
hybrid approach towards NHSM will give more accurate results.

Son L. H. proposed the improved version of HU-FCF, HU-FCF++ [27].
MIPFGWC-CS and NHSM give prediction only when either the information
about users or the historical ratings are available. To overcome this drawback,
Son L. H. combined HU-FCF and NHSM algorithms [23, 26]. HU-FCF++ considers
the demographical information such as attributes of the user and prior ratings of the
similar users for an item. The exact number of clusters cannot be estimated using
MIPFGWC-CS. To handle this problem a procedure, FACA-DTRS (Fast and Auto-

matic Clustering Algorithm based on a Decision-Theoretic Rough Set) model works quickly to find out extreme number of clusters [28].

To resolve the irrelevant user similarities from degrading the performance the combination of fuzzy metric in MIPFGWC-CS and Association Rule Mining (ARM) was used [23]. ARM will also tackle the drawback of fuzzification in FARAMS that leads to inaccurate results. Thus, the HU-FCF++, a novel generalized method by combining MIPFGWC-CS, NHSM, and HU-FCF a resolving method was proposed [27]. HU-FCF++ shows better accuracy rates when compared to the approaches that only use demographic information. The drawback of the HU-FCF++ is that it takes more computational time when compared to the other algorithms.

3   Conclusion and Future Work

The cold start problem can be solved using different kinds of approaches as discussed above. This review has gone through most of the experimentally proven solutions that are discussed by different researchers. Some approaches that gave better recommendations suffer from time complexity. Data from social media sites can be used to improve the quality of relevant recommendations. If we are applying the sentiment analysis to the social media data, more relevant information required for high-quality recommendations can be extracted. User choices and feedbacks can be extracted by reviewing the comments of a particular item.

In our future work, we like to propose an intelligent approach to the ones suggested by Arain Q. A. et al. that handles cold start user problem in improving relevancy of recommendations. With the help of sentiment analysis, this approach shall be able to study a new user's choice for an event and other parameters such as user preference, user behavior, sentiments, economic status, pattern of user's activity, the purpose of user's activity, user's plan for an event, and so on by extracting the hidden information from users social media profile by preserving privacy of the user. This approach can be suggested as a modification of the architecture developed by Arain Q A et al. shown in the Fig. 2. The process of reviewing the user text information from user's social media profile can be added to the user profile building phase of this architecture. The extraction of the abovementioned data of new user's preference is the next phase. This can be done by interacting with the user during sign-up. After this, the new user profile can be built. A collaborative filtering algorithm that combines the results of sentiment analysis phase and user similarities is needed. Historical ratings of similar users can also be considered for better results.

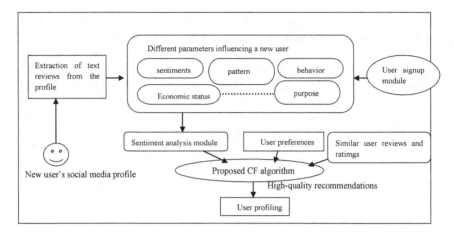

Fig. 2 Modified recommendation framework [29]

References

1. Cantador, I., Fernández-Tobías, I., Berkovsky, S., Cremonesi, P:. Cross-domain recommender systems. Recommender Systems Handbook, pp. 919–959. Springer, US (2014)
2. Memon, I., Chen, L., Majid, A., Lv, M., Hussain, I., Chen, G.: Travel recommendation using geo-tagged photos in social media for tourist. Wireless Pers. Commun. **80**(4), 1347–1362 (2015)
3. Arain, Q.A., Memon, H., Memon, I., Memon, M.H., Shaikh, R.A., Mangi, F.A.: Intelligent travel information platform based on location base services to predict user travel behavior from user-generated GPS traces. Int. J. Comput. Appl. 1–14 (2017)
4. Memon, M.H., Li, J.P., Memon, I., Arain, Q.A.: GEO matching regions: multiple regions of interests using content based image retrieval based on relative locations. Multimedia Tools Appl. **76**(14), 15377–15411 (2017)
5. Shapira, B., Rokach, L., Freilikhman, S.: Facebook single and cross domain data for recommendation systems. User Model. User Adap. Interaction **23**(2–3), 211–247 (2013)
6. Hridya, S., Mariappan, A.K.: Addressing cold start problem in recommender systems using association rules and clustering technique. In: 2013 International Conference on Computer Communication and Informatics (ICCCI), pp. 1–5 (2013)
7. Good, N., Schafer, B., Konstan, J., Borchers, A., Sarwar, B., Herlocker, J., Riedl, J.: Combining collaborative filtering with personal agents for better recommendations. In: AAAI-'99 Conference, pp. 439–446 (1999)
8. Park, S.T., Pennock, D., Madani, O., Good, N., DeCoste, D.: Naïve filterbots for robust cold-start recommendations. In: 12th ACM SIGKDD International Conference on Knowledge Discovery and Data Mining, pp. 699–705. ACM (2006)
9. Embarak, O.H.: A method for solving the cold start problem in recommendation systems. In: 2011 International Conference on Innovations in Information Technology (IIT), pp. 238–243. IEEE (2011)
10. Lam, X.N., Vu, T., Le, T.D., Duong, A.D.: Addressing cold-start problem in recommendation systems. In: 2nd International Conference on Ubiquitous Information Management and Communication, pp. 208–211. ACM (2008)
11. Hofmann, T.: Probabilistic latent semantic indexing. In: 22nd Annual International ACM SIGIR Conference on Research and Development in Information Retrieval, pp. 50–57, Berkeley, California, USA (1999)

12. Lika, B., Kolomvatsos, K., Hadjiefthymiades, S.: Facing the cold start problem in recommender systems. Expert Syst. Appl. **41**(4), 2065–2073 (2014)
13. Rubens, N., Elahi, M., Sugiyama, M., Kaplan, D.: Active learning in recommender systems. Recommender Systems Handbook, pp. 809–846. Springer, US (2015)
14. Fernández-Tobías, I., Braunhofer, M., Elahi, M., Ricci, F., Cantador, I.: Alleviating the new user problem in collaborative filtering by exploiting personality information. User Model. User Adap. Interaction **26**(2–3), 221–255 (2016)
15. Sun, D., Li, C., Luo, Z.: A content-enhanced approach for cold-start problem in collaborative filtering. In: 2011 2nd International Conference on Artificial Intelligence, Management Science and Electronic Commerce (AIMSEC), pp. 4501–4504. IEEE (2011)
16. Sun, M., Li, F., Lee, J., Zhou, K., Lebanon, G., Zha, H.: Learning multiple-question decision trees for cold-start recommendation. In: Sixth ACM International Conference on Web Search and Data Mining, pp. 445–454. ACM (2013)
17. Sarumathi, M., Singarani, S., Thameemaa, S., Umayal, V., Archana, S., Indira, K., Devi, M.K.: Systematic approach for cold start issues in recommendations system. In: 2016 International Conference on Recent Trends in Information Technology (ICRTIT), pp. 1–7. IEEE (2016)
18. Ocepek, U., Rugelj, J., Bosnić, Z.: Improving matrix factorization recommendations for examples in cold start. Expert Syst. Appl. **42**(19), 6784–6794 (2015)
19. Wang, H., Wang, N., Yeung, D.Y.: Collaborative deep learning for recommender systems. In: 21th ACM SIGKDD International Conference on Knowledge Discovery and Data Mining, pp. 1235–1244. ACM (2015)
20. Elkahky, A.M., Song, Y., He, X.: A multi-view deep learning approach for cross domain user modeling in recommendation systems. In: 24th International Conference on World Wide Web, pp. 278–288. ACM (2015)
21. Zheng, L.: A survey and critique of deep learning on recommender systems (2016). bdsc.lab.uic.edu/docs/survey-critique-deep.pdf
22. Wei, J., He, J., Chen, K., Zhou, Y., Tang, Z.: Collaborative filtering and deep learning based recommendation system for cold start items. Expert Syst. Appl. **69**, 29–39 (2017)
23. Son, L.H.: Dealing with the new user cold-start problem in recommender systems: a comparative review. Inf. Syst. **58**, 87–104 (2016)
24. Liu, H., Hu, Z., Mian, A., Tian, H., Zhu, X.: A new user similarity model to improve the accuracy of collaborative filtering. Knowl. Based Syst. **56**, 156–166 (2014)
25. Leung, C.W.K., Chan, S.C.F., Chung, F.L.: An empirical study of a cross-level association rule mining approach to cold-start recommendations. Knowl. Based Syst. **21**(7), 515–529 (2008)
26. Son, L.H.: HU–FCF: a hybrid user-based fuzzy collaborative filtering method in recommender systems. Expert Syst. Appl. **41**(15), 6861–6870 (2014)
27. Son, L.H.: HU-FCF++: a novel hybrid method for the new user cold-start problem in recommender systems. Eng. Appl. Artif. Intell. **41**, 207–222 (2015)
28. Yu, H., Liu, Z., Wang, G.: An automatic method to determine the number of clusters using decision-theoretic rough set. Int. J. Approximate Reasoning **55**(1), 101–115 (2014)
29. Zheng, X., Luo, Y., Xu, Z., Yu, Q., Lu, L.: Tourism destination recommender system for the cold start problem. KSII Trans. Internet Inf. Syst. **10**(7) (2016)

Traffic Surveillance Video Summarization for Detecting Traffic Rules Violators Using R-CNN

Veena Mayya and Aparna Nayak

Abstract Many a times violating traffic rules leads to accidents. Many countries have adopted systems involving surveillance cameras at accident zones. Monitoring each frame to detect the violators is unrealistic. Automation of this process is highly desirable for reliable and robust monitoring of traffic rules violations. With deep learning techniques on GPU, the violation detection can be automated and performed in real time on surveillance video. This paper proposes a novel technique to summarize the traffic surveillance videos that uses Faster Regions with Convolutions Neural Networks(R-CNN) to automatically detect violators. As the proof of concept, an attempt is made to implement the proposed method to detect the two-wheeler riders without helmet. Long duration videos can be summarized into very short video that includes details about only rules violators.

Keywords Convolution neural network · Deep learning · Helmet detection R-CNN · Surveillance video

1 Introduction

Most of road the accidents happen due to violating the traffic rules such as wrong side driving, driving over the speed limit, riding without wearing a helmet, etc. As per survey carried out by Sutikno et al. [17], most of the road accidents happen with two-wheeler riders who do not wear helmet. Wearing helmet substantially reduces the accident impact. Even though helmet is mandatory as per traffic rules, two-wheeler riders violate it. Monitoring manually traffic surveillance 24 × 7 videos, to detect traffic rule violators is very inefficient, time consuming, and prone to errors.

V. Mayya (✉) · A. Nayak
Department of Information & Communication Technology,
Manipal Institute of Technology, MAHE, Manipal, India
e-mail: veena.mayya@manipal.edu

A. Nayak
e-mail: aparna.nayak@manipal.edu

© Springer Nature Singapore Pte Ltd. 2019
S. K. Bhatia et al. (eds.), *Advances in Computer Communication and Computational Sciences*, Advances in Intelligent Systems and Computing 759,
https://doi.org/10.1007/978-981-13-0341-8_11

117

Recently state-of-the-art results are achieved using Convolution Neural Networks (CNN) based techniques. Using CNN, powerful visual features can be extracted without handcrafting the features. CNN has been adopted in many image/ video processing techniques such as classification [19], colorization [9], object detection [15], restoration [7], semantic segmentation [19], and video summarization [13]. Recently state-of-the-art object detection network is achieved using Faster Regional CNN [15]. R-CNN used Region Proposal Network (RPN) which is a fully convolutional network that simultaneously predicts object bounds and corresponding scores at each position. In comparison with traditional CNNs, Faster R-CNN can detect the violators in very short duration. Due to this, R-CNN is more suitable for processing videos in real time.

2 Literature Survey

Supervision of video is indispensable in order to monitor any kind of criminal or anti law activity. CCTV video surveillance systems are installed in major countries and public video surveillance is used as a primary tool to monitor traffic conditions. It is highly impossible to keep track of the surveillance videos 24×7. Automation of such system is required to fire an alarm or highlight, whenever there is a violation of traffic rule.

Baran et al. [1] have implemented the traffic surveillance to detect make and model of the vehicle as well as the color of the vehicle. Their model makes use of the smart camera which is not suitable for the existing systems. Existing cameras would be difficult to replace with new smart cameras because of cost.

Romuere Silva et al. [16] have proposed a method to detect whether the vehicle is motorcycle or not. The entire work they have divided into four parts such as region of interest, background detection, dynamic objects segmentation, and vehicle classification. For feature extraction of images, the algorithms SURF, HAAR, HOG and LBP were used as descriptors. Multilayer Perceptron, Support Vector Machines, and Radial-Based Function Networks were used for image classification. Comparison is made with different classification algorithms used with different feature descriptors and the results show that 97.63% accuracy is found with LBP descriptor and SVM classifier.

Maharsh Desai et al. [5] have proposed method to detect the helmet of two-wheeler riders. Background subtraction and Hough descriptors are used detect the curve, lines, etc. It is just proposed method where accuracy is not mentioned.

Sutikno et al. [17] have implemented the helmet detection of motorcycle riders. They have used back-propagation neural network with 40 neurons in the hidden layer. Almost 150 images are considered for experiment among which 75 images are with helmet and another 75 images are without helmet. Up to 86.3% accuracy is achieved with their technique wherein the same experiment with SVM has resulted 85% as proposed by Chiverton [4]. The number of images considered was very less which may not be applicable for the real-time scenarios as the captured videos consist of background such as other vehicles, road, etc.

Dahiya et al. [6] proposed the system to detect the bike riders without helmet in surveillance videos. Here, they have preprocessed the data using background subtraction and applied SIFT, HOG, LBP feature extraction techniques and classified using DCSVM. They could achieve 93% accuracy for detection of biker riders without helmet. This method cannot be adopted for generic surveillance videos captured at the public traffic, as handcrafted feature extraction techniques are used which fails to efficiently extract the visual features of small regions in the whole frame that are used to identify the violators.

Recently, Vishnu et al. [18] have proposed the helmet detection of motor cycle using convolution neural network where they have achieved 92.87%. As the preprocessing step, Gaussian background subtraction is applied to detects the motorcycle riders. Top one-fourth part of the motorcyclist images are cropped and then Alexnet CNN model [11] is used for violators detection. In the proposed method, preprocessing, cropping, separation, etc., are avoided by using Faster Regional CNN [15].

Compared to all the literature work done in the traffic surveillance, the proposed method gives better accuracy for the helmet detection use case. The training and detection time using R-CNN is substantially less as compared with other CNN models. Here, [11] is applied on each object's region in an image and "background" class is used distinguish areas with or without objects. The proposed model can be easily adopted to detect any kind of violators with minimal changes.

3   Proposed Method

Overall traffic surveillance summarization is as shown in Fig. 1, where the video is parsed to generate summarized video of only traffic violators. Figure 2 depicts the block diagram of the proposed surveillance video summarization system to automatically detect the traffic rules violators. As the preprocessing step, all the bounding boxes of violators are annotated using simple BBox-Label-Tool [14]. An xml file for each image is generated in the format similar to PASCAL VOC [8] format. Algorithm 1 depicts the steps carried out during preprocessing phase.

The CCTV surveillance videos of two hour duration are collected from one of the private circle near MIT, Manipal campus. One hour of the video is used for training. Along with this some of the Google images were crawled. In total 500 images were used to train the model for each class. Algorithm 1 details about the steps to generate the training data set and the network details is depicted in Algorithm 2. The R-CNN model used for training with three classes, viz., background, helmet, noHelmet, is shown in Fig. 3. Here only modified layers, as compared with Zeiler and Ferguss (ZF) net [20], are highlighted.

Fig. 1 High level system diagram

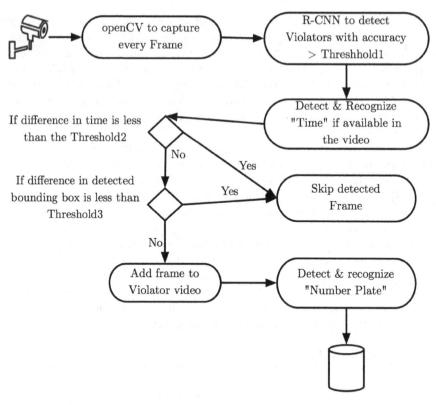

Fig. 2 Control flow diagram of surveillance summarization system

---

Algorithm 1 Steps To Generate Training Dataset
___

1: Collect around 500 images of with two-wheeler riders with and without helmet. Some images captured from the surveillance camera is included to train the model.
2: Generate the XML files for annotations in PASCAL VOC [8] format
___

Algorithm 2 RCNN training the network to detect rules violators
___

1: Update the R-CNN training layers(prototxt) to include only three classes, i.e., background, withHelmet, withoutHelmet
2: Train the model using GPU with pretrained weights from ZF faster RCNN [20]
___

The proposed method is evaluated to summarize the detection of two-wheeler riders who did not wear helmet. The steps carried out to summarize the surveillance video to detect two-wheeler riders without helmet are depicted in Algorithm 3. Here every 15th frame is read using openCV library [2] as the frame rate was 15 FPS. The frame is tested using the proposed method to detect the violators. Threshold1 is considered 0.98, i.e., if the region detected belongs to noHelmet class with confidence

```
name: "ZF"              layer {                 layer {                    layer {
layer {                   name: "cls_score"        name: 'roi-data'            name: "bbox_pred"
  name: 'input-data'      type: "InnerProduct"     type: 'Python'             type: "InnerProduct"
  type: 'Python'          bottom: "fc7"            bottom: 'rpn_rois'         bottom: "fc7"
  top: 'data'             top: "cls_score"         bottom: 'gt_boxes'         top: "bbox_pred"
  top: 'im_info'          param { lr_mult: 1.0 }   top: 'rois'                param { lr_mult: 1.0 }
  top: 'gt_boxes'         param { lr_mult: 2.0 }   top: 'labels'              param { lr_mult: 2.0 }
  python_param {          inner_product_param {    top: 'bbox_targets'        inner_product_param {
    module:                 num_output: 3          top: 'bbox_inside_weights'   num_output: 12
'roi_data_layer.layer'    weight_filler {          top: 'bbox_outside_weights'  weight_filler {
    layer:                    type: "gaussian"     python_param {               type: "gaussian"
'RoIDataLayer'              std: 0.01}               module:                    std: 0.001}
    param_str:            bias_filler {           'rpn.proposal_target_layer'  bias_filler {
"'num_classes': 3"}}        type: "constant"        layer: 'ProposalTargetLayer'  type: "constant"
                            value: 0}}}             param_str:                   value: 0}}}
                                                  "'num_classes': 3"}}
```

Fig. 3 R-CNN model used for detection o violators

of 98% then only such frames will be included in the output video. This drastically reduces the false alarms. Threashold3 is set as 5 pixels. This is used to reduce the number of detected bounding boxes around the violators, i.e., if the detected boxes are too close to each other then it would not be marked on the output frame. Similarly annotations can be performed to mark other violators and also to identify the time details (normally shown at the corner of CCTV footage). Only the *num_classes* and *num_output* parameters need to be updated depending on the kind of violators to be identified. *num_output* should be four times the *num_classes* parameter while using R-CNN model.

Algorithm 3 Algorithm for Summarizing surveillance Video

1: **procedure** MAINMODULE
2:    Read frames from video using openCV. Here every 15th frame is read.
3:    Create a output video with height and width as of input first Frame
4:    Use trained RCNN model to test the current image to detect the boxes and scores for the current image
5:    **if** current class is noHelmet **then**
6:        Get all the regions detected by RCNN
7:        **for** each of the regions **do**
8:            **if** The score for noHelmet class is greater than Threshold2 (0.98) **then**
9:                **if** Region does not fall in close proximity of previously identified region Threshold3 (5 pixels range) **then**
10:                    Draw the bounding box using opencv library API
11:                    Add the frame to the output Video
12:    Close the output video

---

4   Results

Param Shavak system is used to implement and test the proposed method. Figure 4 shows the frames of input 5 minutes video used to test the proposed system. Here every 50th frame is extracted to avoid duplicates. Only few frames are shown due

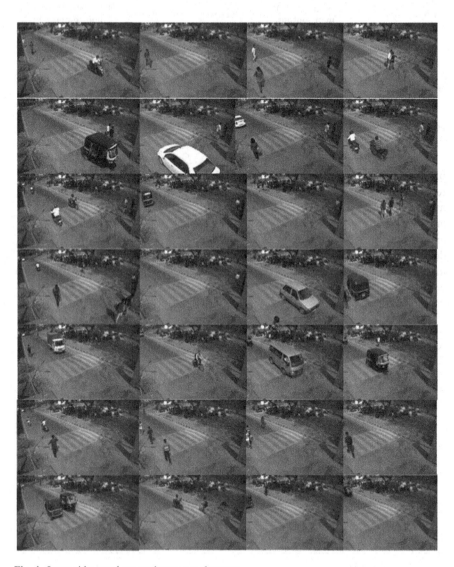

Fig. 4 Input video used to test the proposed system

to size constraint. The summarized video length is only 25 seconds, the frames of resultant video is shown in Fig. 5. Here the detected violators are clearly marked by the proposed system. Even the violators sitting behind the two-wheeler riders are also detected by the proposed method. The false alarm rate is less than 5%. Few bicycle riders are detected as violators, but this would not affect accuracy as only when the bicycle riders come nearer to the CCTV camera view then it would be identified as violators. It can be seen that frames that include two-wheeler riders with helmet are rightly identified by the proposed system and are skipped in the output video.

On GPU enabled systems, the detection per frame takes around 0.1–0.15 s while on CPU it takes around 5–10 s. Fifty images were randomly selected for testing the proposed system. The accuracy found to be 95% for the images closer to the camera. The system fails to detect the violators if the region lies far away from the camera position. Figure 6 shows the test images in which violators were not recognized. It can be seen that only if the region of recognition is too far from camera view point, then the proposed system fails to identify the violators.

Fig. 5 Frames of summarized video highlighting the violators

Fig. 6 Non recognized images

5 Conclusion

The proposed method successfully summarizes the traffic surveillance videos to include the details about only rules violators. The method has been adopted to detect the two-wheeler riders without helmet. The automated system can be used in existing traffic monitoring setup without investing much on other costlier devices. Only training the system with annotations need to be performed, testing with actual videos can be done in real time as the time taken to detect violators is very less on GPU systems.

6 Future Scope

The system can be easily adopted to detect other violators like triple riders, wrong side drivers, etc., by just including more classes and updating the output layers. Recently, many works have been done to recognize the vehicle license plates [3, 10, 12]. The proposed system can be extended to detect and recognize the vehicle license numbers of the violators. The proposed system can be also extended to monitor other traffic related activities such as accident detection, traffic estimation, security threats, etc.

Acknowledgements We are grateful to Mr. Ashok Rao, Chief Security Officer, MIT Manipal for providing us the access to CCTV footage that helped us to carry out the research work.

References

1. Baran, R., Ruść, T., Rychlik, M.: A Smart Camera for Traffic Surveillance, pp. 1–15. Springer International Publishing, Cham (2014). http://dx.doi.org/10.1007/978-3-319-07569-3_1
2. Bradski, G.: The OpenCV Library. Dr. Dobb's J. Softw. Tools (2000)
3. Cheang, T.K., Chong, Y.S., Tay, Y.H.: Segmentation-free vehicle license plate recognition using convnet-rnn. CoRR abs/1701.06439 (2017). arXiv:1701.06439
4. Chiverton, J.: Helmet presence classification with motorcycle detection and tracking. IET Intell. Transp. Syst. **6**(3), 259–269 (2012)
5. Desai, M., Shubham Khandelwal, L.S.: Automatic helmet detection on public roads. Int. J. Eng. Trends Technol. (IJETT) **35**(5) (2016)
6. Dahiya, K., Singh, D., Mohan, C.K.: Automatic detection of bike-riders without helmet using surveillance videos in real-time. In: 2016 International Joint Conference on Neural Networks (IJCNN). pp. 3046–3051, July 2016
7. Dahl, R., Norouzi, M., Shlens, J.: Pixel recursive super resolutionc (2017). arXiv:1702.00783
8. Everingham, M., Eslami, S.M.A., Van Gool, L., Williams, C.K.I., Winn, J., Zisserman, A.: The pascal visual object classes challenge: a retrospective. Int. J. Comput. Vis. **111**(1), 98–136 (2015)
9. Iizuka, S., Simo-Serra, E., Ishikawa, H.: Let there be Color!: joint end-to-end learning of global and local image priors for automatic image colorization with simultaneous classification. In: ACM Transactions on Graphics (Proc. of SIGGRAPH 2016), vol. 35(4) (2016)
10. Jain, V., Sasindran, Z., Rajagopal, A., Biswas, S., Bharadwaj, H.S., Ramakrishnan, K.R.: Deep automatic license plate recognition system. In: Proceedings of the Tenth Indian Conference on Computer Vision, Graphics and Image Processing. pp. 6:1–6:8. ICVGIP '16, ACM, New York, NY, USA (2016). http://dx.doi.org/10.1145/3009977.3010052
11. Krizhevsky, A., Sutskever, I., Hinton, G.E.: Imagenet classification with deep convolutional neural networks. In: Proceedings of the 25th International Conference on Neural Information Processing Systems. pp. 1097–1105. NIPS'12, Curran Associates Inc., USA (2012). http://dl.acm.org/citation.cfm?id=2999134.2999257
12. Liu, X., Liu, W., Mei, T., Ma, H.: A Deep Learning-Based Approach to Progressive Vehicle Re-identification for Urban Surveillance, pp. 869–884. Springer International Publishing, Cham (2016). http://dox.doi.org/10.1007/978-3-319-46475-6_53
13. Otani, M., Nakashima, Y., Rahtu, E., Heikkilä, J., Yokoya, N.: Video summarization using deep semantic features (2016). arXiv:1609.08758
14. Qiu, S.: BBox-Label-Tool. https://github.com/puzzledqs/BBox-Label-Tool (2017). Accessed 11 Sept 2017

15. Ren, S., He, K., Girshick, R., Sun, J.: Faster R-CNN: Towards real-time object detection with region proposal networks. In: Advances in Neural Information Processing Systems (NIPS) (2015)
16. Silva, R., Aires, K., Veras, R., Santos, T., Lima, K., Soares, A.A.: Automatic motorcycle detection on public roads. CLEI Electron. J. **16**, 4–4 (2013). http://www.scielo.edu.uy/scielo.php?script=sci_arttext&pid=S0717-50002013000300004&nrm=iso
17. Sutikno, S., Indra Waspada, N.B.: Classification of motorcyclists not wear helmet on digital image with backpropagation neural network. TELKOMNIKA **14** (2016). http://dx.doi.org/10.12928/telkomnika.v14i3.3486
18. Vishnu, C., Singh, D., Mohan, C.K., Babu, S.: Detection of motorcyclists without helmet in videos using convolutional neural network. In: 2017 International Joint Conference on Neural Networks (IJCNN). pp. 3036–3041, May 2017
19. Yim, J., Ju, J., Jung, H., Kim, J.: Image Classification Using Convolutional Neural Networks With Multi-stage Feature, pp. 587–594. Springer International Publishing, Cham (2015). http://dx.doi.org/10.1007/978-3-319-16841-8_52
20. Zeiler, M.D., Fergus, R.: Visualizing and understanding convolutional networks (2013). arXiv:1311.2901

GAE: A Genetic-Based Approach for Software Workflow Improvement by Unhiding Hidden Transactions of a Legacy Application

Shashank Sharma and Sumit Srivastava

Abstract In organization numbers are increasing day by day with a drastic pace which prefers the extraction of the workflow of processes to interpret the operational processes. For a viably and sorted out approach to drive the development in the realm of digitization is utilized by the approach of work process extraction. The work process extraction/mining is otherwise called process mining. The goal of workflow mining is to get the extraction of data of an association's method of business by changing over the logs of occasion information recorded in association's frameworks. This impact to the enhance conformation of processes to organization regulation where workflow mining approach for analysis is actualized. Work process mining strategies absolutely rely upon the nearness of framework occasion log information. We accept to involve setting various endeavors on building our strategies or frameworks to record the greater part of the old information. The urge to comprehend and expand their procedures of businesses entails the process exploration practices. This paper displays a philosophy how programming occasion log information is analyzed to grasp and advance the product work process by utilizing arrangement which best in class utilized as a part of the product code clone streamlining for the human services area application.

1 Introduction

Workflow mining a.k.a. Process Mining is a considerably new and emerging area of academic research within data analytics. The key objective here is to deploy workflow-related data in directive to obtain pertinent info and knowledge by employing data analytic algorithms and determining a workflow model. This segment discusses the idea of how a log from events is the foundation of exploration along with other main building block of process mining. Work process mining articles to channel the hole between enormous information investigation and conventional business

S. Sharma (✉) · S. Srivastava
MU Jaipur, Jaipur, India
e-mail: shashanksharmaaa@gmail.com

© Springer Nature Singapore Pte Ltd. 2019
S. K. Bhatia et al. (eds.), *Advances in Computer Communication and Computational Sciences*, Advances in Intelligent Systems and Computing 759,
https://doi.org/10.1007/978-981-13-0341-8_12

Table 1 Typical event log example

Case id	Activity name	Event type	Originator	Timestamp	Extra data
011	Make order form	Start	Employee A1	19-10-1955 15:15:02	–
012	Make order form	Complete	Employee A2	19-10-1955 05:14:01	–

work process/process administration. This field can essentially be ordered into (1) Workflow revelation, (2) conformance checking and (3) improvement [1]. This permits the extraction of bits of knowledge about the by and large and inward conduct contained in any given procedure. Work process disclosure procedures underscore on utilizing the occasion information in order to decide work process models. Conformance checking strategies emphasize on supporting the occasion information on a work process model to confirm how well the model fits the information and the other way around [2]. In spite of the fact that enlargement methods utilize occasion information and work process models to repair or increase the work process show. Hence, workflow mining provides the conduit the gap between data mining and machine learning practices and the business process management discipline.

1.1 The Event Log as the Linchpin of Reasoning

Data is a crucial building block in various discovery domains. Process Mining uses data that is accounted by event logs. A sample log is shown in Table 1. An event log in terms of this situation can be defined as the process of recording of an action instance on the system. Action or activity instances are units of work that are registered by the system when work is piloted in the situation of an assured process. Statuses of activity or action are specified to a set of languages that are fixed to the workflow modeling hypothesis. Various workflow modeling hypothesis have unique implementation standards that are governed by their state evolution diagrams. A simple state evolution diagram has been explained in Fig. 1. More elaborate diagrams have been described in process modeling literature. Consider a case where the life phase action of Business process object notation is characterized by the Object Management Group (OMG) [3]. Adding to this, different process displaying ideal models, for example, YAWL (Yet Another Workflow Language), case taking care of [4] and explanatory methodologies (EM—BrA2CE [5], Declare [6]) which proposes unequivocal however undifferentiated from base outlines for their comparing delineation semantics.

It can get fascinating to experience different establishment semantics and state progress outlines, for work process mining and work process investigation; in any case, the real information is what is vital. Process Data is generally pooled from various vaults that are gotten from CRM, ERP, WFM, and other assorted data

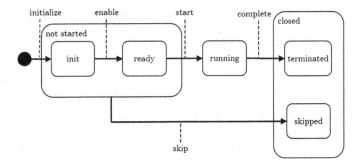

Fig. 1 Depiction of state transformation of a process

frameworks. This outcome in trouble while recognizing hypothetical establishment of an occasion as far as state progress outlines and genuine information found practically speaking. In different divisions, for example, CRM (customer relationship administration), item improvement, monetary administrations and so forth business work process depend intensely on heritage data frameworks or low work process arranged data frameworks. Further, the enrolled business process information is very much characterized with the end goal that exclusive a specific kind of state progress (e.g., Achievement of a movement event) can be mined by means of real information pieces. Work process mining can be viewed as most productive in measured conditions where business data frameworks have more extensive alternatives of conduct, here; the change of open information into an occasion log is regularly a critical undertaking.

1.2 How Petri Net Is Different from Workflow Net?

Petri net basically deals with mathematical modeling where as work flow net deals with XML modeling. Petri net basically deals with low level work flow where as work flow net deals with high level work flow. Petri net basically used to represent theoretical concepts of computer science whereas work flow net basically used for business process/management modeling. Filtration and extraction of event logs through business information systems is usually drifted out by text-based data scripts. In actuality, process data is more often distributed over various data sources and it is painstaking to define the precise scope of process that is studied. Further, a comprehensive ETL-phase is required before a concrete analysis is initiated. Adding to which, the data must be in an event log storage format.

One of the initial approaches to storing log based on event performed/executed is in the MXML format. From the date of its inception, i.e., 2003, this fact based standard is used, i.e., MXML format because it is highly integrated with the ProM-framework which is a framework used for academic purpose for workflow Process improvement.

It was not as of not long ago that IEEE team was favored rather than MXML for process/workflow mining. Then after the new standard has come into existence which enhances the efficiency and make this process easy and effective which is known as XES (eXtensible Event Stream) [7]. Logs behaves, as a foundation of the workflow mining, it is significant to express the variable necessities to which an occasion log must approve. The taking after three suppositions is mandatory and basic

- Activity instance must be well defined for the workflow instance of an event which specified by a unique activity name
- Unique process instances or ID which is used by cases must be referred by events.
- Ordered Timestamp must be recorded by the events.

2 Literature Review

As Dehnert et al. [8] and Pinter et al. [9] are the founders who originally able to extract the chances of getting to explore the split/join connections on the extracted models. The main idea behind to focus on model to extract work flow pattern [10] and each point in this system has an OR-split/join semantics. Without a doubt, each planned roundabout segment in the model has a boolean limit that evaluates to certified or false after an errand is executed. Grecco et al. [11] did not extract any of the loops through his research. The validation behind states that their approach is not able to extract meaningful understanding from the event logs after applying proposed algorithm. They tried to visualize the differentiating behavior of the mined models through various fundamental algorithms and tried to analyze the different behavior. Cycles or loops made this approach ineffective and efforts go in the vain. Some different procedures cannot mine self-assertive loops in light of the fact that their model documentation (or portrayal) does not bolster this sort of circle. The key inspiration driving why most methodologies can't mine non-neighborhood sans non-choice is that the larger part of their mining computations rely upon neighborhood information in the logs. The systems that do not mine nearby without non decision cannot do as such on the grounds that their portrayal does not bolster such a develop. Generally the system depends on a piece organized documentation, as Herbst et al. also, Schimm. Skip errands are not mined because of portrayal restrictions too. Split/join imperceptible undertakings are not mined by numerous methods, with the exception of for Schimm and Herbst et al. As a matter of fact, we likewise do not focus at finding such sort of assignments. Nevertheless, it is routinely the case that it is possible to fabricate an exhibit with no split/join imperceptible errands that impart a comparable lead to the approach with false or wrong assignments for the split/join constructs. Copy undertakings are definitely not mined in light of the fact that numerous methods expect that the mapping between the undertakings what's more, their marks is injective. By the day's end, the names are exceptional per task. The principle systems that mine duplicate endeavors are Cook et al. [12] for progressive structures just, and Herbst et al. [13] for both back to back and parallel methodology. We do not consider Schimm [14] to mine strategy models with duplicate errands since his approach expect that the revelation of the duplicate endeavors is done in a pre-getting ready stride. This progression identities all the copies and ensures that they have one of a kind identifiers when the occasion log is given as contribution to the mining calculation. As a matter of fact, every one of the procedures that we survey here would handle copy undertakings if this same pre-preparing step would be done before the sign in given as contribution to them.

3 Proposed Algorithm with Procedure and a Sample Case Study

The proposed process efficient GAE (Genetic Algorithm for Events) is based on GA and process mining of event logs. In this, function of fitness work is registered by the nature of a person. The nature of an individual is fundamentally set by its replaying of the log follow. The entire flow of GAE is shown in Flowchart 1. The steps of the proposed algorithm are written below

In order to apply an innate/hereditary count, we need to address people. Each individual thinks about to a possible methodology model and its depiction should be anything other than hard to manage. Our underlying thought was to speak to

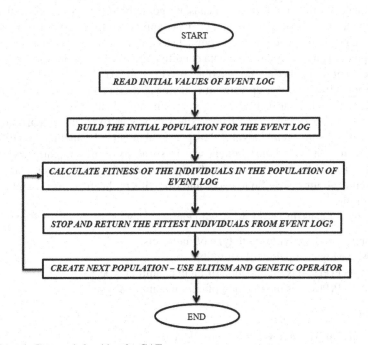

Flowchart 1 Proposed algorithm for GAE

forms straightforwardly by petri nets. Lamentably, petri nets end up being a less helpful way to speak to forms in this unique circumstance. The principal reason is that in petri nets there are places whose nearness cannot be gotten from the log, i.e., events simply imply the dynamic parts of the net (changes). Because of this it ends up being harder to make a hidden masses, describe inherited directors (half breed and change), and depict mixes of AND/OR-parts/joins. Note that given a log, it is definitely not hard to discover the activities and along these lines, the progressions that exist in the petri net. Be that as it may, implementing certain routings by just associating changes through spots is intricate (if certainly feasible). In this manner, we will utilize an alternate interior portrayal.

Table 2 demonstrates the inside portrayal of a specific individual which utilizes by mentioned hereditary mining approach. This purported causal network character-izes the relations of causal type between the exercises and if there should arise an occurrence of various info or yield exercises, the rationale is delineated. Consider for instance the line beginning with A. This column demonstrates that there is a not a causal connection among A and A (take note of the initial 0 in the column), in any case, there is a causal association continues among A and B (observe the underlying 1 in this segment). The following two sections in the segment show that there are also causal relations amongst A and C and A and D. The last component in the line demonstrates the directing rationale, i.e., $B \vee C \vee D$ demonstrates that A is trailed by B, C, or D. The section named "Yield" demonstrates the rationale relating an action to causally following exercises. The primary line underneath "Info" demonstrates the rationale relating a movement to causally going before exercises. Note that the data province of A is substantial, i.e., no data required. Activity G has $E \wedge F$ as data condition, i.e., both E and F need to complete a particular true objective to enable G. Development H has $B \vee C \vee G$ as data condition, i.e., B, C, or G needs to complete the process of remembering the true objective to enable H.

Subsequently after windup with introduction and groundwork we come back to the objective of this paper: hereditary process mining. With a specific end goal to apply a hereditary calculation we have to speak to people. Every individual compares to a conceivable procedure model and its portrayal ought to be anything but difficult to deal with. Our underlying thought was to speak to forms straightforwardly by petri nets. Tragically, petri nets end up being a less helpful way to speak to forms in this specific circumstance. So to overcome from this problem we use causal matrix [8, 15] (Table 3).

A matrix of causality is a row, X with elements named as (M, N, O, P) where

- M comprises of activities with type of finite sets,
- $N \subseteq M \times M$ is the relation of causality,
- $O \in A \rightarrow P(P(M))$ is the function of input condition type,3
- $O \in A \rightarrow P(P(M))$ is the function of output condition type,

such that

- $N = \{(m1, m2) \in M \times M \mid m1 \in O(m2)\}$,4
- $N = \{(m1, m2) \in M \times M \mid m2 \in P(m1)\}$,

Table 2 Causal matrix made and utilized for portrayal

→	Input								Output
	True	A	A	A	D	D	E∧F	B∨C∨G	
	A	B	C	D	E	F	G	H	Output
A	0	1	1	1	0	0	0	0	B∨C∨D
B	0	0	0	0	0	0	0	1	H
C	0	0	0	0	0	0	0	1	H
D	0	0	0	0	1	1	0	0	E∧F
E	0	0	0	0	0	0	1	0	G
F	0	0	0	0	0	0	1	0	G
G	0	0	0	0	0	0	0	1	H
H	0	0	0	0	0	0	0	0	true

Activity	Input	Output
Table 3 Concise and concrete encoding format of an individual in Table 2		
A	{}	{{B,C,D}}
B	{{A}}	{{H}}
C	{{A}}	{{H}}
D	{{A}}	{{E}},{{F}}
E	{{D}}	{{G}}
F	{{D}}	{{G}}
G	{{E}},{{F}}	{{H}}
H	{{B,C,G}}	{}

- $\forall m \in M \ \forall \ QQsR \in O(m) \ Q \cap QL = \emptyset \Rightarrow Q = QL$,
- $\forall m \in M \ \forall \ Q,QR \in P(m) \ Q \cap QL = \emptyset \Rightarrow Q = QL$,
- $N \cup \{(mo, mi) \in M \times M \mid mo\ N\bullet = \emptyset \wedge N\bullet mi = \emptyset\}$ is a connected graph of strong type.

3.1 Binding of Petri net and Matrix of Causality

Binding of a petri net through the framework of causality network included in light of the fact that it requires to "explore locations" and along with this need to explore more on lattice that supposed to be more detailed than its previous counter parts [8, 15].

Let $X = (M, N, O, P)$ be a matrix of causality representation type. The binding of petri net is a row $\Pi SX \rightarrow TS (X) = (T, U, V)$, where

- $T = \{w, y\} \cup \{wU,Q \mid U \in M \wedge q \in O(U)\} \cup \{Pu,q \mid U \in M \wedge Q \in P(U)\}$,
- $U = M \cup \{ZU1, U2 \mid (U1, U2) \in N\}$,
- $V = \{(w, u) \mid u \in M \wedge N\bullet u = \emptyset\} \cup \{(u, y) \mid U \in M \wedge U\ N\bullet = \emptyset\} \cup \{(wu,q, U) \mid U \in M \wedge q \in W(U)\} \cup \{(U, Pu,w) \mid U \in M \wedge w \in P(u)\} \cup \{(Pu1,q,zu1,u2) \mid (u1, u2) \in N \wedge q \in P(U1) \wedge U2 \in Q\} \cup \{(Zu1,u2, Wu2,q) \mid (u1, u2) \in N \wedge q \in W(U2) \wedge U1 \in Q\}$.

In this first we take the event log of hospital case of healthcare information system. In this particular event log we have 42 events or events flow. Our aim is to classify the events on the basis of type of events. We start with event log then we select MXML legacy classifier and process discovery algorithm for the extraction of petri net. Along with this we apply some user specified constraints to get expected result (Fig. 2).

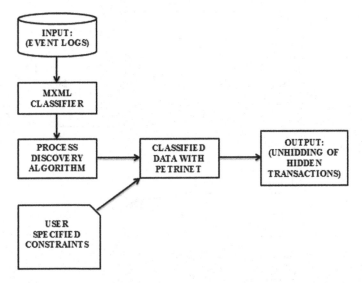

Fig. 2 Procedure for GAE

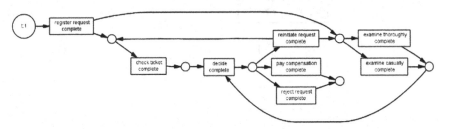

Fig. 3 Alpha algorithm

3.2 Comparison Analysis

For verifying and validating the effectiveness of GAE, we used following standard algorithm that are used for workflow management as well as analyze and compare it with proposed GAE. In figure number 6 you find black boxes (dark or complete black) which referred to hidden transactions, which shows that other or previous algorithms are unable to extract or locate hidden transactions although algorithms are able to classify the event log (Figs. 3, 4, 5, 6 and Table 4).

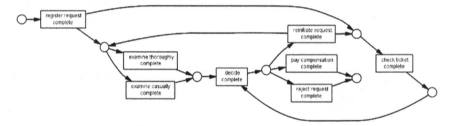

Fig. 4 Alpha ++ algorithm

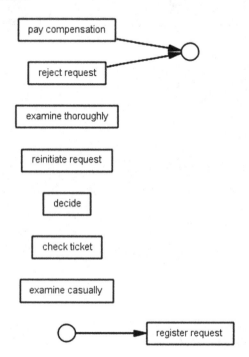

Fig. 5 Tsinghua-Alpha algorithm

4 Result and Outcomes

In this, we received classified result for that particular hospital case. We got eight classified event classes. For this classification, we used two criteria, i.e., on the basis of event functionality and other one on the basis of event type. Along with this, complete classification is further sub-divided into three parts, i.e., start event (count is 1), originators (count is 6) and end event (count is 2). After interpreting the petri net, we find that which work flow process model we need to work upon to improve the process of information system. Please find below the outcomes (Table 5).

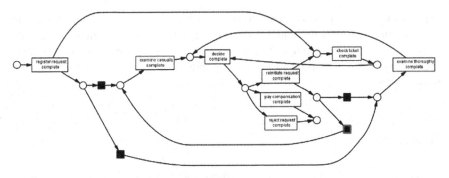

Fig. 6 Output petri net from proposed GAE algorithm

Table 4 Comparison of result for various algorithms

Name of algorithm	Input format	Output format	Intermediate output	Able to unhide hidden transactions
Alpha	**MXML Log File**	**Petri net**	**No**	**No**
Alpha++	MXML Log File	**Petri net**	**No**	**No**
Tsinghua-Alpha	MXML Log File	**Petri net**	**No**	**No**
GAE	MXML Log File	**Petri net**	**Heuristic Net**	**Yes**

Table 5 Classified event log data

Model element	Event type	Occurrences (absolute)	Occurrences (relative) (%)
Check ticket	Complete	9	21.429
Decide	Complete	9	21.429
Register request	Complete	6	14.286
Examine casually	Complete	6	14.286
Reinitiate request	Complete	3	7.143
Examine thoroughly	Complete	3	7.143
Pay compensation	Complete	3	7.143
Reject request	Complete	3	7.143

5   Conclusion and Future Work

"Workflow scientist" desires to possess particular/exact to initiate innovation in a progressively digitalized ecosphere. In this paper, we just only conceptualized our idea with a small case study. This enables us to investigate the operational procedure work process of healthcare information systems that frameworks under genuine situations, and utilize extraction systems for procedures to get particular and perceived programming change models. This paper oriented on associate the organized mining approach for the event classes from event data for the software process improvement by using the petri nets flow model approach. In the future, we aim at conducting additional experiments using different variety of event log data sets. A reasonable succeeding phase is to progress with tool support for domain-based information management systems.

References

1. van der Aalst, W.M.P.: Process Mining: Discovery Conformance and Enhancement of Business Processes. Springer, Berlin (2011)
2. Adriansyah, A., van Dongen, B.F., van der Aalst, W.M.P.: Towards robust conformance checking. In: Business Process Management Workshops, Lecture Notes in Business Information Processing, vol. 66, pp. 122–133. Springer, Berlin, Heidelberg (2011)
3. Object Management Group (OMG). Business Process Modeling Notation (BPMN)—Specification. OMG Document—formal/2011-01-03, Jan 2011
4. van der Aalst, W.M.P., Weske, M., Grünbauer, D.: Case handling: a new paradigm for business process support. Data Knowl. Eng. **53**(2), 129–162 (2005–2010)
5. Goedertier, S.: Declarative techniques for modeling and mining business processes. Ph.D. thesis, Katholieke Universiteit Leuven, Faculty of Business and Economics, Leuven, Sept 2008
6. Pesic, M., van der Aalst, W.M.P.: A declarative approach for flexible business processes management. In: Eder, J., Dustdar, S. (eds.) Business Process Management Workshops, 4103 of Lecture Notes in Computer Science, pp. 169–180. Springer (2006)
7. C. W. Günther. XES Standard Definition. www.xes-standard.org
8. Agrawal, R., Gunopulos, D., Leymann, F.: Mining process models from workflow logs. In: Sixth International Conference on Extending Database Technology, pp. 469–483 (1998)
9. Pinter, S.S., Golani, M.: Discovering workflow models from activities lifespans. Comput. Ind. **53**(3), 283–296 (2004)
10. IBM. IBM MQSeries Workow—Getting Started With Buildtime. IBM Deutschland Entwicklung GmbH, Boeblingen, Germany (1999)
11. Greco, G., Guzzo, A., Pontieri, L., Sacca, D.: Mining expressive process models by clustering workflow traces. In: Dai, H., Srikant, R., Zhang, C. (eds.) BIBLIOGRAPHY 365, PAKDD, volume 3056 of Lecture Notes in Computer Science, pp. 52–62. Springer (2004)
12. Cook, J.E., Du, Z., Liu, C., Wolf, A.L.: Discovering models of behavior for concurrent workflows. Comput. Ind. **53**(3), 297–319 (2004)
13. Herbst, J., Karagiannis, D.: Workow mining with InWoLvE. Comput. Ind. **53**(3), 245–264 (2004)
14. Schimm, G.: Mining exact models of concurrent workflows. Comput. Ind. **53**(3), 265–281 (2004)
15. Dehnert, J., van der Aalst, W.M.P.: Bridging the gap between business models and workflow specifications. Int. J. Coop. Inf. Syst. **13**(3), 289–332 (2004)
16. van der Aalst, W.M.P., Weijters, A.J.M.M., Maruster, L.: Workflow mining: discovering process models from event logs. IEEE Trans. Knowl. Data Eng. **16**(9), 1128–1142 (2004)
17. van der Aalst, W.M.P.: Data scientist: the engineer of the future. In: Mertins, K., Benaben, F., Poler, R., Bourrieres, J. (eds.) Proceedings of the I-ESA Conference, vol. 7 of Enterprise Interoperability, pp. 13–28. Springer, Berlin (2014)

18. Calders, T., Guenther, C., Pechenizkiy, M., Rozinat, A.: Using minimum description length for process mining. In: ACM Symposium on Applied Computing (SAC 2009), pp. 1451–1455. ACM Press (2009)
19. Dehnert, J., van der Aalst, W.M.P.: Bridging the gap between business models and workflow specifications. Int. J. Coopera. Inf. Syst. **13**(3), 289–332 (2004)

Research on Object Detection Algorithm Based on PVANet

Jianjun Lv, Bin Zhang and Xiaoqi Li

Abstract Based on the research and development of remote sensing image target extraction technology, the deep learning framework of cascade principal component analysis network is used to study the sea surface vessel detection algorithm. The visible image of the sea surface vessel is the input, the suspected target area is determined by the significance test, the PVANet model is extracted from the suspected target area, and the result is input into the support vector machine to obtain the final classification result. The experimental results show that the designed algorithm can successfully output the results of the detection of the sea area in the airspace, and verify the efficiency and accuracy of the PVANet model by comparing with the CNN algorithm. It proves the superiority of the PVANet model in feature extraction.

Keywords Vessel detection · Deep learning · Deep but lightweight neural networks · Significance detection

1 Introduction

As an emerging unmanned control system in the field of artificial intelligence, the unmanned surface of the water has played an important role in various industries with rapid momentum. The automatic target detection and identification technology is the core function module to ensure the selfless operation of the unmanned water. Therefore, this paper has carried on the related research on the sea vessel detection and recognition algorithm [1].

J. Lv · B. Zhang (✉) · X. Li
School of Information and Communication Engineering, BUPT, Beijing, China
e-mail: bluezb@bupt.edu.cn

J. Lv
e-mail: ljj19921026@bupt.edu.cn

X. Li
e-mail: clorislee@bupt.edu.cn

© Springer Nature Singapore Pte Ltd. 2019
S. K. Bhatia et al. (eds.), *Advances in Computer Communication and Computational Sciences*, Advances in Intelligent Systems and Computing 759,
https://doi.org/10.1007/978-981-13-0341-8_13

The difficulty of the sea vessel detection algorithm is the need for very high engineering practicality. In the complex dynamic context, it is necessary to quickly and efficiently detect sea targets, but also to use a priori as little as possible to ensure maximum utilization [2]. At the same time, it also requires to maintain the system detection rate and detection accuracy on the basis of improving the target detection distance as far as possible, and improve system early warning capability. At present, the detection of sea surface vessels mainly includes background modeling or background estimation [3], feature extraction and learning classification, image segmentation, significance detection [4], color space transformation, or color feature extraction [5] method.

2 Related work

As a result of the previous participation of the CCCV challenge—Remote Sensing Image Target Extraction Technology Challenge—there are 1000 sea ship image data for training. And all evaluations were done on Intel i7-6700 K CPU with a single core and NVIDIA GEFORCE GTX 1080 GPU.

In this paper, based on the analysis of the advantages and disadvantages of the above methods, use the "feature extraction + classifier design + search strategy" and the combination of significance detection algorithm to detect the sea ship.

In the process of feature extraction and classifier design, it is necessary to use the deep learning model to train and classify different scenes and datasets. The essence of deep learning is to build a nonlinear deep-level network structure to achieve the more complex function approximation and the purpose of the distribution of the characteristics of the target. Based on the deep but lightweight neural networks (PVANet) [6], this paper studies the sea surface vessel detection algorithm, learns the essential characteristics of the target from a small number of samples, and has high characteristic robustness.

3 Algorithm Framework

The objective of this paper is to target the detection and target identification of the ships in the input sea image, and to propose a solution to the difference between the vessel and the background in the image, using a simple deep learning model to train and classify in a number of different vessel samples and backgrounds, and ultimately output the results of the identification of the vessel, while providing a reference standard to achieve a wider range of target detection and target recognition.

The process chart of the sea vessel detection algorithm is shown in Fig. 1.

After entering a picture of a sea vessel, the first step is significant test [7]. The significance test is based on the human eye attention mechanism and imitates the physiological vision of the human eye, and achieved mainly through the construction

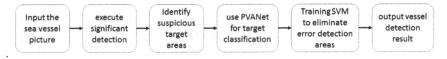

Fig. 1 The structure of the heterogeneous network

of regional complexity characteristics, regional differences, the characteristics of the surrounding differences, and other methods [8].

After obtaining the significance test result graph, the threshold is set to three times the average gray scale of the result graph, and binarizes the result graph: set 1 if the assignment greater than this threshold, and set 0 if the assignment less than this threshold. Second, in the binarized image obtained before, obtain the minimum circumscribed rectangle for each connected domain as a suspect region of the target.

After determining the target suspect region, use the PVANet model to classify each detected rectangular region. Each of the rectangular regions is input to PVANet and calculated the corresponding eigenvector. The eigenvector is input into the support vector machine (SVM) trained by the positive and negative samples of the vessel, and obtain the final classification result. For the different targets detected before the classification, through the identification of the PVANet model, exclude the wrong detection area and achieve the target detection and identification of sea vessels finally.

4 The PVANet Model

The network compression and decomposition of convolution show that the current network structure is highly redundant, so reducing redundancy has become the most direct way to enhance the network running. So Kye-Hyeon Kim proposes the PVANet. Elaborate adoption and combination of recent technical innovations on deep learning make us possible to redesign the feature extraction part of the faster R-CNN [9] framework to maximize the computational efficiency. Even though the proposed network is designed for object detection, we believe the design principle can be widely applicable to other tasks.

The network structure summary is as follows: Based on the basic principles of more layers with less channels, use C.ReLU (Concatenated rectified linear unit) [10] in the initial layer of the network and use inception [11] in the latter layers; Maximize the multi-scale nature of object detection tasks based on the application of multi-scale feature cascade; and Use plateau detection (reduce the learning rate several times when the loss is no longer decline in a certain number of iterations) [12] as the weight attenuation strategy. Experiments show that the PVANet can achieve efficient training through batch normalization [13] and residual connections.

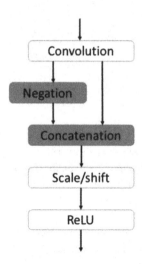

Fig. 2 C. ReLU building block

4.1 Feature Extraction Network

Feature extraction network consists of two building blocks—C.ReLU and Inception.

Figure 2 illustrates the C.ReLU implementation. Compared with the original C.ReLU, we add scaling and shift after the connection to allow the slope and activation thresholds of each channel to differ from the slope and activation thresholds of the opposite channels.

Figure 3 shows the inception building block—5 × 5 convolution which is replaced with two 3 × 3 convolutional layers [14] for efficiency. Figure 4 shows the inception for reducing feature map size by half.

Inception is one of the most efficient (computationally efficient) methods for capturing small targets in images. In order to capture the large target in the image, we need a large enough field of experience, which can be achieved by stacking a 3 × 3 filter. But in order to capture small targets, you need a little bit of experience, so 1 × 1 convolution of nuclear is suitable, which can avoid the problem of parameter redundancy caused by convolution. Based on the above two points, inception for the target detection is very appropriate.

4.2 Target Detection

In this part of the target detection network, Kye-Hyeon Kim basically follows the framework of the faster R-CNN, Fig. 5 shows the structure of the PVANet.

The feature map produces a 6 × 6 × 512 tensor after passing the ROI, and then output the result through the full connection layer. It is because we use the whole

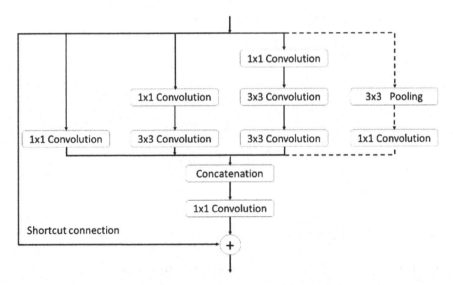

Fig. 3 Inception building block

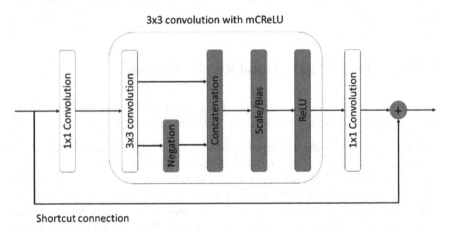

Fig. 4 Inception for reducing feature map size by half

connection layer, and we provide an opportunity to further compression in the back
of the model (and does not lose precision), which is getting PVANet + model through
bounding box vote [15].

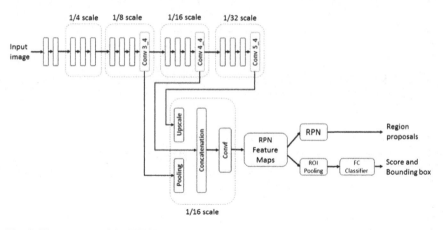

Fig. 5 The structure of the PVANet

Table 1 The correct rate of PVANet training for positive and negative sample

Positive sample			Negative sample		
Total	Correct number	Correct rate	Total	Correct number	Correct rate
300	281	93.7%	250	248	99.2%

5 Experimental Methods and Results Analysis

Using the PVANet algorithm to learn the characteristics of the vessel, judge whether the target of the significance is the vessel. The algorithm is as follows:

- **Phase 1**: After declaring and defining the function, read the image dataset, adjust the image size to 50×50 pixels, and perform the PVANet training;
- **Phase 2**: Perform SVM training. Set the SVM parameters and define SVM and the type of kernel function. At the same time, design termination criterion function, that is, terminating when the number of iterations reaches the maximum value. Training SVM needs to establish an instance of SVM class; the training model parameter is the previous set of parameters such as the input data, response data, etc.;
- **Phase 3**: Perform the target detection algorithm. Calculate the training error rate and execute the PVANet + SVM algorithm. When detecting the vessel, it executes the target detection algorithm.

5.1 Statistic Correct Rate of PVANet Training

The correct rate of PVANet training for positive and negative samples is shown in Table 1.

Table 2 Time comparison of PVANet and CNN algorithm

Algorithm	Computation cost (MAC)				Running time (ms)	mAP(%)
	Shared CNN	RPN	Classifier	Total		
PVANET	8.0	1.4	27.9	37.4	47	86.1
Faster R-CNN + ResNet-101	84.5	N/A	228.6	313.1	2310	87.6
Faster R-CNN + VGG-16	173.2	5.1	26.9	205.2	121	79.8
R-FCN + ResNet-101	131.9	0	0	131.9	143	85.6

The required image using the PVANet algorithm to train is a sea vessel, and we need to exclude the image of the non-vessel part of the sea scene. Therefore, it is desirable that the positive sample be identified as a vessel and the negative sample cannot be identified as a vessel. It can be seen from Table 2 that the correct rate of positive samples is 93.7%, the correct rate of negative samples is 99.2%, so the training results are good, PVANet algorithm can achieve the requirements of image classification and target ship identification.

5.2 PVANet + SVM Recognition Results

We output significant test result graph and binarization significance test result graph in the process of determining the suspicious area of the target by the significance test, and then classify each detected rectangular area by PVANet model and obtain the final result of the second classification by SVM.

Four sea surface vessel detection and recognition results are shown in Fig. 6.

5.3 Comparison Between PVANet and CNN Algorithms

In order to illustrate the accuracy and speed advantage of PVANet, we use CNN to classify and identify the same dataset. The results are shown in Table 2.

At the same time, after the significant test, we classify each detected suspect rectangular area by the CNN model and compare the faster R-CNN + VGG-16 + SVM recognition result with the PVANet + SVM recognition result. As shown in Fig. 6, the first line is PVANet + SVM recognition result, and the second line is CNN + SVM recognition results.

We can see it from Table 2, compared with CNN, PVANet has a faster training speed and higher accuracy. It can be seen from Fig. 7 that CNN algorithm sometimes detects the bow as a false detection area, and sometimes did not exclude the waves

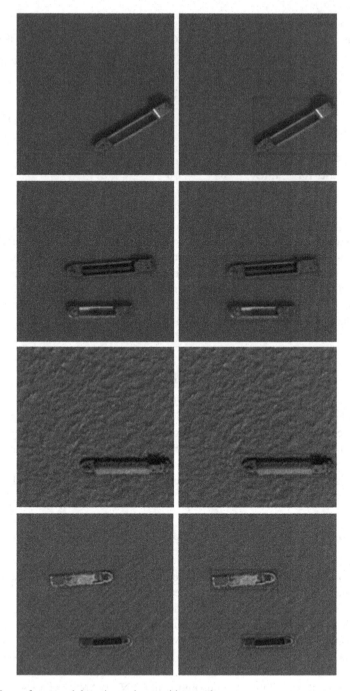

Fig. 6 Sea surface vessel detection and recognition results

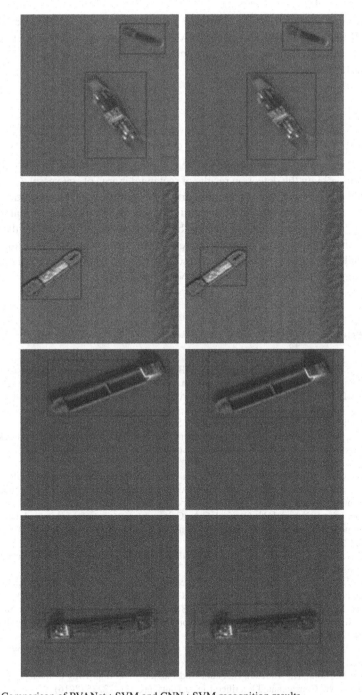

Fig. 7 Comparison of PVANet + SVM and CNN + SVM recognition results

and other non-vessel parts; the results of identify have some deviation compared to the PVANet algorithm. Therefore, PVANet algorithms in the training speed and training accuracy are more outstanding performance, suitable for the fast detection of sea surface targets.

6 Conclusion

In this paper, we design the sea vessel detection algorithm based on PVANet, and successfully obtain the detection results of rectangular area of sea vessel. Compared with CNN algorithm, PVANet method shortens the training time, improves the recognition accuracy, and has theoretical superiority. The sea vessel detection algorithm based on PVANet provides the theoretical basis and experimental basis for the future study of the deep learning network model. At the same time, the algorithm is of low complexity and easy to implement, so it is of great significance in the related engineering practice of water unmanned boat, and so on.

References

1. Kumar, M.P., Torr, P.H.S., Zisserman, A.: OBJ CUT. In: IEEE Computer Society Conference on Computer Vision and Pattern Recognition, pp. 18–25. IEEE Computer. Society (2005)
2. Socek, D., Culibrk, D., Marques, O., et al.: A hybrid color-based foreground object detection method for automated marine surveillance. In: International Conference on Advanced Concepts for Intelligent Vision Systems, pp. 340–347. Springer (2005)
3. Wu, Q., Cui, H., Du, X., et al.: Real-time moving maritime objects segmentation and tracking for video communication. In: 2006 10th International Conference on Communication Technology, pp. 1–4. IEEE (2006)
4. Albrecht, T., West, G.A.W., Tan, T., et al.: Visual maritime attention using multiple low-level features and naïve Bayes classification. In: International Conference on Digital Image Computing: Techniques and Applications, pp. 243–249. DBLP (2011)
5. Haarst, V.V., Leijen, A.V.V., Groen, F.C.A.: Colour as an attribute for automated detection in maritime environments. In: International Conference on Information Fusion, pp. 1679–1686. IEEE (2009)
6. Kim, K.H., Hong, S., Roh, B., et al.: PVANET: Deep but Lightweight Neural Networks for Real-time Object Detection (2016)
7. Hou, X., Zhang, L.: Saliency detection: a spectral residual approach. In: IEEE Conference on Computer Vision and Pattern Recognition, 2007. CVPR '07, pp. 1–8. IEEE (2007)
8. Cheng, M.M., Zhang, Z., Lin, W.Y., et al.: BING: binarized normed gradients for objectness estimation at 300fps. In: IEEE Conference on Computer Vision and Pattern Recognition, pp. 3286–3293. IEEE Computer Society (2014)
9. Ioffe, S., Szegedy, C.: Batch normalization: accelerating deep network training by reducing internal covariate shift. Comput. Sci. (2015)
10. Shang, W., Sohn, K., Almeida, D., et al.: Understanding and Improving Convolutional Neural Networks via Concatenated Rectified Linear Units, pp. 2217–2225 (2016)
11. Szegedy, C., Liu, W., Jia, Y., et al.: Going deeper with convolutions. In: Computer Vision and Pattern Recognition, pp. 1–9. IEEE (2015)

12. Mottaghi, R., Chen, X., Liu, X., et al.: The role of context for object detection and semantic segmentation in the wild. In: Computer Vision and Pattern Recognition, pp. 891–898. IEEE (2014)
13. Ren, S., He, K., Girshick, R., et al.: Faster R-CNN: towards real-time object detection with region proposal networks. IEEE Trans. Pattern Anal. Mach. Intell. **39**(6), 1137 (2016)
14. Lavin, A., Gray, S.: Fast algorithms for convolutional neural networks. Comput. Sci. (2015)
15. Gidaris, S., Komodakis, N.: Object Detection via a Multi-region and Semantic Segmentation-Aware CNN Model, pp. 1134–1142 (2015)

Characterization of Groundwater Contaminant Sources by Utilizing MARS Based Surrogate Model Linked to Optimization Model

Shahrbanoo Hazrati-Yadkoori and Bithin Datta

Abstract Unknown groundwater contaminant source characterization is the first necessary step in the contamination remediation process. Although the remediation of a contaminated aquifer needs precise information of contaminant sources, usually only sparse and limited data are available. Therefore, often the process of remediation of contaminated groundwater is difficult and inefficient. This study utilizes Multivariate Adaptive Regression Splines (MARS) algorithm to develop an efficient Surrogate Models based Optimization (SMO) for source characterizing. Genetic Algorithm (GA) is also applied as the optimization algorithm in this methodology. This study addresses groundwater source characterizations with respect to the contaminant locations, magnitudes, and time release in a heterogeneous multilayered contaminated aquifer site. In this study, it is specified that only limited concentration measurement values are available. Also, the contaminant concentration data were collected a long time after the start of first potential contaminant source(s) activities. The hydraulic conductivity values are available at limited locations. The performance evaluation solution results of the developed MARS based SMO for source characterizing in a heterogeneous aquifer site with limited concentration measurement, parameter values, and under hydraulic conductivity uncertainties are shown to be satisfactory in terms of source characterization accuracy.

Keywords Surrogate models based optimization · Groundwater contamination Source characterization

S. Hazrati-Yadkoori (✉) · B. Datta
Discipline of Civil Engineering, College of Science and Engineering, James Cook University, Townsville, Australia
e-mail: shahrbanoo.hazratiyadkoori@my.jcu.edu.au

B. Datta
e-mail: bithin.datta@jcu.edu.au

B. Datta
CRC for Contamination Assessment and Remediation of the Environment,
CRC CARE, ATC Building University of Newcastle, Callaghan, Australia

© Springer Nature Singapore Pte Ltd. 2019
S. K. Bhatia et al. (eds.), *Advances in Computer Communication and Computational Sciences*, Advances in Intelligent Systems and Computing 759,
https://doi.org/10.1007/978-981-13-0341-8_14

1 Introduction

The remediation of contaminated aquifers is one of the most important challenges encountered in groundwater management. This challenge arises due to insufficient information at the contaminated aquifers. This limited information availability and the uncertainties intrinsic in numerical groundwater flow and transport simulation models make contaminant source characterization a problem with uncertainties. Consequently, effective remediation process remains a difficult task. Therefore, developing an efficient methodology for source characterization has a crucial role in the remediation process. The methodologies proposed earlier are generally highly sensitive to concentration measurement errors and need a very large amount of data and computation time. For example, the linked simulation–optimization methodology which is the most frequently utilized methodology to tackle this problem is computationally intensive. This methodology requires an enormous amount of computational time when numerical simulation models are utilized in conjunction with optimization models. As a result, Surrogate Models linked to Optimization (SMO) have been suggested to solve these computational problems using much smaller computation time. In this proposed method, a simpler and faster model approximates groundwater numerical flow and contaminant transport models. Multivariate Adaptive Regression Splines (MARS) which is a robust tool in data mining and modeling is utilized to develop the SMO for source characterization.

A few of the prominent methodologies that were applied for source characterization are listed below. Least square regression method combined with linear programming [1], an embedded optimal source identification model [2], Genetic Algorithm (GA) [3], the Artificial Neural Network (ANN) [4], simulated annealing [5], and adaptive simulated annealing [6]. The methodologies which are applicable to real-world cases usually are computationally intensive and are sensitive to measurements errors, as any other method. Linking with a trained surrogate model can remove the computational limitations especially for large study areas.

The information related to an illustrative heterogeneous multilayered contaminated aquifer site are used in this study to assess the developed MARS based SMO for source characterization. For this study area, limited numbers of hydraulic conductivity and contaminant concentration measurement values are available. The performance evaluation results with uncertainties in hydraulic conductivity values and limited concentration measurements data demonstrate acceptability.

2 Methodology

2.1 Surrogate Models

Surrogate models are approximate models constructed to mimic the behavior of more complex systems in less time. These models are constructed by using limited data sets

Fig. 1 Schematic chart for
developing the SMO for
unknown groundwater
contaminant source
characteristics

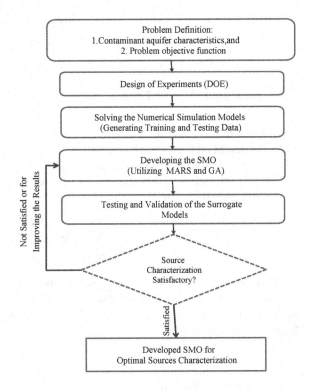

of computationally expensive simulation models [7]. The main steps of constructing a SMO for source characterization are explained in the following paragraphs [8]. Figure 1 presents the schematic chart for developing a SMO for characterization of unknown groundwater contaminant sources.

1. Problem definition: The problem to be solved needs to be completely defined.
2. Design of Experiments: The important variables of the system are selected. In this study, Latin Hypercube Sampling (LHS) technique as suggested in earlier studies is utilized for generating adequate numbers of possible contaminant source magnitudes [9].
3. Solving the numerical simulation models: Numerical groundwater flow and transport simulation models (MODFLOW [10] and MT3DMS [11]) are solved utilizing the data generated in step 2. The solutions provide the collection of training data in terms of concentration values at observation points corresponding to the randomly generated initial data sets.
4. Developing the surrogate models: in this step, the type of surrogate model and the architecture of it need to be addressed.
5. Testing the developed model(s) for performance evaluation: In this step, new sample data sets are used to evaluate the developed surrogate model(s). The testing results may be used to modify and improve the developed model(s) in the development stage.

6. Optimal solution: if the performance evaluation results are satisfactory, stop. Otherwise, go to step four for changing the surrogate model's type or design.

2.2 Surrogate Model Type

MARS is a stepwise linear technique [12] which is used in this study to develop MARS-based SOM for source characterizations. This algorithm is a nonparametric statistical tool which discovers the relationships between a set of input variables that known as predictors and a dependent target variable. This algorithm defines a model for a target variable by Eq. (1) [12].

$$y = \sum_{m=1}^{M} a_m B_m(x) \tag{1}$$

where $\mathbf{a_m}$ is an estimated constant coefficient by utilizing least-squares technique; and $\mathbf{B_m}$ is a basic function or a spline function. In this study, Salford Predictive Modeler 8.0 is used as the software to utilize the MARS algorithm for developing the surrogate models [13].

2.3 Objective Function

Equation (2) represents the objective function of this problem [2]. This equation is defined to minimize the difference between the estimated and the observed contaminant concentration values at possible observation points at specified times [2].

$$Minimize\ E = \sum_{t=1}^{T} \sum_{l=1}^{L} \left(Cest_l^t - Cobs_l^t \right)^2. \tag{2}$$

where $Cest_l^t$ and $Cobs_l^t$ are estimated and observed concentration values at observation point l and at time t, respectively. The whole set of possible observations times and locations are represented by T and L, respectively. Equation (3) represents w_l^t which is a weight related to the possible observation point l and time t [2].

$$w_l^t = \frac{1}{\left(Cobs_l^t + \eta \right)^2} \tag{3}$$

where η is defined as a constant coefficient. This coefficient needs to be large enough to prevent the solution being dominated by errors corresponding to very small measured concentrations [2]. MARS-based surrogate model linked to optimization model

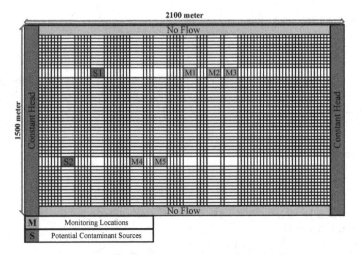

Fig. 2 Monitoring locations and contaminant sources in the illustrative contaminated aquifer

is the main constraint defining the approximate description of the flow and transport processes in the optimization model.

3 Study Area

The information of an illustrative study area is utilized (Fig. 2) in this study. The authors also used the information of this study area in [14]. This illustrative contaminated aquifer is heterogeneous and multilayered (three unconfined layers). The grid spacing of the illustrative aquifer site in X, Y, and Z directions are 30, 30, and 10 m, respectively. It is assumed that the hydraulic gradient, vertical anisotropy, and porosity are 0.00238, 5, and 0.3, respectively [14]. The longitudinal and transverse dispersivities are also assumed to be 15 and 3 m, respectively. Two possible potential contaminant sources and only one conservative contaminant are considered. The source1 is located in layer one. The source2 is located in layer two. Table 1 shows the contaminant source fluxes and their locations. It is assumed that the whole simulation times consist of five different stress periods. The contaminant sources are active only in the first four stress periods in which each of them lasts 2 years [14]. The stress period five's duration is 12 years. It is also specified that the groundwater contamination was observed for the first time at five monitoring locations in layer one just 2 years after the contaminant sources had stopped their activities. The breakthrough curves at the monitoring locations which are used for source characterization are presented in Fig. 3.

Table 1 Contaminant source fluxes and locations

Locations (row, column, layer)	Contaminant source fluxes (g/s)				
	Stress period 1	Stress period 2	Stress period 3	Stress period 4	Stress period 5
Contaminant source 1 (12, 15, 1)	6.3	4.6	9.0	5.6	0.0
Contaminant source 2 (38, 9, 2)	6.7	9.3	6.1	7.3	0.0

Fig. 3 Breakthrough curves at monitoring locations used for source characterization

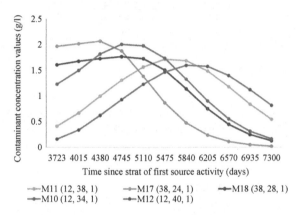

4   Evaluating Performance of the Developed SMO

The developed MARS-based SMO is utilized for performance evaluation. Two different scenarios are considered. In the first scenario, all the hydrogeologic parameters of the model are known and certain. In the second scenario, hydraulic conductivity values are available only at limited and sparse locations [14].

In the first scenario, in order to generate hydraulic conductivity throughout the entire study area, it is assumed that the hydraulic conductivity (HC) data fits a defined lognormal distribution [15]. Therefore, a new variable such as $Z = \log HC$ which follows normal distribution can be incorporated. The LHS method is also utilized to randomly generate the hydraulic conductivity field throughout the study area [9]. For the second scenario, Inverse Distance Weighting (IDW) methodology is utilized to generate hydraulic conductivity values at locations where these values are unknown.

Moreover, Root Mean Square Error (RMSE) criterion is utilized to quantify the accuracy of the developed procedure in an evaluation mode. The RMSE can be defined by Eq. (4).

$$RMSE = \sqrt{\frac{1}{(M \times N)}(\sum_{m=1}^{M} \sum_{n=1}^{N} ((S_m^n)_{est} - (S_m^n)_{obs}))^2} \qquad (4)$$

where M is the total numbers of potential contaminant sources and N is the numbers of specified times. $(S_m^n)_{obs}$ and $(S_m^n)_{est}$ are observed and estimated contaminant source fluxes at source number m, in stress period n, respectively. The actual source fluxes are assumed to be known only for the performance evaluation purpose.

5   Results and Discussions

After defining the problem and its objective function (Eq. 2), the following steps are followed to develop MARS-based SMO for source characterization.

- Design of Experiments (DOE): The LHS technique is used to provide two initial data set groups (500 and 1000 sets) at two potential contaminant sources. The possible range for contamination is 0–10 g/s.
- Solving the numerical simulation models: the explained numerical simulation models MODFLOW and MT3DMS are implemented to obtain corresponding concentration values for the generated data sets in the previous step.
- Developing the surrogate models based on SMO: MARS is used to construct MARS-based surrogate models and SMO. GA algorithm is also utilized as the optimization algorithm to develop MARS-based SMO.

The MARS algorithm for extracting the basis functions of a target variable needs the information of related predictor variables. In this study, the randomly generated source fluxes are considered as the predictor variables. The corresponding obtained contaminant concentration values are considered as the target variables. Therefore, 55 MARS models need to be developed. These models represented the observed values of contaminant concentration at five observation points at 11 specific times. Table 2 presents possible input data sets that can be used in an MARS model. After developing all these 55 MARS models, they are integrated to construct an integrated surrogate model. Then, this surrogate model is linked to the GA algorithm in the optimization model.

- Testing and validation of the surrogate model: 100 randomly generated new sample sets are used for testing the constructed model for source characterization. The solution result in terms of RMSE is equal to 0.7 for testing data.

The solution results of MARS-based SMO for source characterization are presented in Fig. 4. The source characterization result in terms of RMSE is equal to 0.5.

The constructed MARS-based SMO is also evaluated for source characterization under hydraulic conductivity uncertainty. In this second scenario, it is assumed that the measured hydraulic conductivity values are available at 20 locations. The IDW

Table 2 A typical input for developing a MARS model

ID	Predictors–source fluxes (g/s)								Target variable-contaminant concentration values (g/l)
	Source 1 (12, 15, 1)				Source 2 (38, 9, 2)				Monitoring location 1 (12, 34,1)
	Stress periods								3723 days after start of first source Activity
	1	2	3	4	1	2	3	4	
1	6.1	6.0	5.0	3.3	3.6	0.5	8.0	6.8	1.3
2	0.3	4.9	5.8	5.7	9.4	9.7	8.4	9.0	0.5
3	3.9	5.9	2.9	5.2	9.7	6.9	1.5	5.3	0.8
4	0.5	1.0	8.9	9.6	3.8	3.6	0.2	8.7	0.3
5	0.1	9.0	6.7	0.8	6.3	9.2	9.0	4.4	0.9

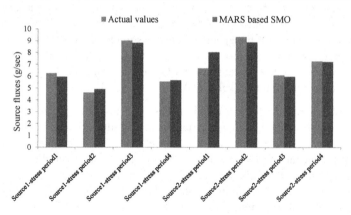

Fig. 4 Obtained source characterization solution results of the MARS-based SMO

methodology is utilized as the interpolation method to generate hydraulic conductivity values at the rest of required points. Next, MODFLOW and MT3DMS within GMS7 are solved by using new set of hydraulic conductivity values. Then, the new MARS-based SMO is formulated and applied for source characterization. The evaluation results are illustrated in Fig. 5. The source characterization solutions result in an RMSE value which equals to 1.7.

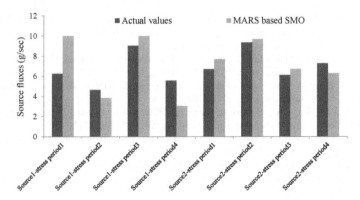

Fig. 5 Obtained source characterization solution results of the MARS-based SMO under hydraulic conductivity uncertainty

6   Conclusion

In this study, MARS-based SMO is developed to characterize unknown contaminant sources. For evaluation purpose only, the constructed MARS-based SMO is used in a heterogeneous, multilayered aquifer site. The contaminant concentration values of this contaminated aquifer are available at limited locations and times. The developed model is also utilized for source characterization under hydraulic conductivity uncertainty. In the first scenario tested, the hydraulic conductivity values and other model parameters are considered to be known with certainty. Then, in the second scenario, the hydraulic conductivity values are considered to be uncertain and its values are known only at limited numbers of locations (20 locations). The performance evaluation results obtained as solution for the two scenarios indicate that the constructed models are capable of adequately mimicking the MODFLOW and MT3DMS. The performance evaluation solution results also demonstrate the potential applicability of the MARS-based SMO to characterize contaminant sources. The MARS-based SMO solutions under hydraulic conductivity uncertainty also appear to be acceptable.

References

1. Gorelick, S.M., Evans, B., Remson, I.: Identifying sources of groundwater pollution—an optimization approach. Water Resour. Res. **19**(3), 779–790 (1983)
2. Mahar, P.S., Datta, B.: Optimal monitoring network and ground-water-pollution sources identification. J. Water Resour. Plan. Manag. **123**(4), 199–207 (1997)
3. Aral, M.M., Guan, J.B., Maslia, M.L.: Identification of contaminant source location and release history in aquifers. J. Hydrol. Eng. **6**(3), 225–234 (2001)
4. Singh, R.M., Datta, B., Jain, A.: Identification of unknown groundwater pollution sources using artificial neural networks. J. Water Resour. Plan. Manag.-Asce **130**(6), 506–514 (2004)

5. Prakash, O., Datta, B.: Optimal characterization of pollutant sources in contaminated aquifers by integrating sequential-monitoring-network design and source identification: methodology and an application in Australia. Hydrogeol. J. **23**(6), 1089–1107 (2015)
6. Jha, M., Datta, B.: Three-dimensional groundwater contamination source identification using adaptive simulated annealing. J. Hydrol. Eng. **18**(3), 307–317 (2013)
7. Gorissen, D., et al.: A surrogate modeling and adaptive sampling toolbox for computer based design. J. Mach. Learn. Res. **11**, 2051–2055 (2010)
8. Gong, W., Duan, Q.: An adaptive surrogate modeling-based sampling strategy for parameter optimization and distribution estimation. Environ. Model Softw. **95**, 16 (2017)
9. Dokou, Z., Pinder, G.F.: Optimal search strategy for the definition of a DNAPL source. J. Hydrol. **376**(3–4), 542–556 (2009)
10. Harbaugh, A.W.: MODFLOW-2005, The U.S. Geological Survey Modular Ground-Water Model-the Ground-Water Flow Process. U.S. Geological Survey Techniques and Methods 6–A16 (2005)
11. Zheng, C., Wang, P.P.: MT3DMS: a modular three-dimensional multispecies transport model for simulation of advection, dispersion, and chemical reactions of contaminants in groundwater systems; documentation and user's guide. 1999: US Army Corps of Engineers-Engineer Research and Development Center, Contract Report SERDP-99-1, p. 220 (1999)
12. Friedman, H.J.: Multivariate adaptive regression splines. Ann. Stat. **19**, 67 (1991)
13. Salford Predictive Modeller 8 (2017)
14. Hazrati, Y.S., Datta, B.: Self-organizing map based surrogate models for contaminant source identification under parameter uncertainty. Int. J. GEOMATE **13**(36), 8 (2017)
15. Freeze, R.A.: A stochastic-conceptual analysis of one-dimensional groundwater flow in nonuniform homogeneous media. Water Resour. Res. **11**, 17 (1975)

Performance Comparisons of Socially Inspired Metaheuristic Algorithms on Unconstrained Global Optimization

Elif Varol Altay and Bilal Alatas

Abstract In recent years, many efficient metaheuristic algorithms have been proposed for complex, multimodal, high-dimensional, and nonlinear search and optimization problems. Physical, chemical, or biological laws and rules have been utilized as source of inspiration for these algorithms. Studies on social behaviors of humans in recent years have shown that social processes, concepts, rules, and events can be considered and modeled as novel efficient metaheuristic algorithm. These novel and interesting socially inspired algorithms have shown to be more effective and robust than existing classical and metaheuristic algorithms in a large number of applications. In this work, performance comparisons of social-based optimization algorithms, namely brainstorm optimization algorithm, cultural algorithm, duelist algorithm, imperialist competitive algorithm, and teaching learning based optimization Algorithms have been demonstrated within unconstrained global optimization problems for the first time. These algorithms are relatively interesting and popular, and many versions of them seem to be efficiently used within many different complex search and optimization problems.

Keywords Optimization · Metaheuristics · Social-based algorithms

1 Introduction

Optimization is the process of searching for the optimal solution. Analytical, enumeration, and heuristic methods can be used for optimization task. Heuristic refers to experience-based techniques for problem solving and learning. Heuristics are problem-dependent and designed only for the solution of a specific problem. A metaheuristic is a higher level heuristic that may be used to obtain a sufficiently

E. V. Altay (✉) · B. Alatas
Department of Software Engineering, Firat University, Elazığ, Turkey
e-mail: evarol@firat.edu.tr

B. Alatas
e-mail: balatas@firat.edu.tr

© Springer Nature Singapore Pte Ltd. 2019
S. K. Bhatia et al. (eds.), *Advances in Computer Communication and Computational Sciences*, Advances in Intelligent Systems and Computing 759,
https://doi.org/10.1007/978-981-13-0341-8_15

Fig. 1 Computational intelligence search and optimization methods

good solution from search and optimization problems [1]. Over the past few decades, many novel metaheuristic algorithms inspired from nature have been proposed and efficiently utilized to deal with a wide range of complex search and optimization problems in many different fields.

General-purpose metaheuristic methods can be categorized according to ten different inspiration fields such as biology, physics, swarm, music, chemistry, sociology, sports, water, plant, and mathematics. Furthermore, there are hybrid methods which are combination of these [2].

Categorization is depicted in Fig. 1. Nevertheless, the majority of metaheuristic algorithms are based on either biological processes or social behaviors of mainly ants, termites, birds, insects, and fishes [3]. Most of the social phenomena that have been used for optimization are referred to as swarm intelligence [4]. Human beings are more intelligent than insects and animals. There are few works that utilize social phenomena of human being as source of inspiration to form optimization algorithm. Processes, concepts, rules, and events on social behaviors of humans have also been considered and modeled as novel efficient search and optimization methods with extremely effective exploration capabilities in many cases.

In this work, performance comparisons of socially inspired optimization algorithms, namely brainstorm optimization algorithm, cultural algorithm, duelist algorithm, imperialist competitive algorithm, and teaching learning based optimization algorithm, have been demonstrated within unconstrained global optimization problems for the first time.

Organization of this paper is as follows. Section two describes the socially inspired algorithms used for performance comparisons. Section three describes the unconstrained optimization problems and reports the experimental results obtained from the algorithms, and section four concludes the paper along with thoughts on future works and research directions.

2   Socially Inspired Metaheuristic Algorithms

Most of the metaheuristic algorithms proposed for optimization are based on biological phenomena or social behaviors of insects and animals. In fact, humans have been optimizing every aspect of daily life. Recent studies have demonstrated that many human activities and behaviors such as leadership, society status, teaching, learning, dueling, categorization of individuals in the society, brainstorming, etc.

may be utilized as the source of inspiration for search and optimization algorithms. Some researchers have inspired from these social phenomena in human societies and proposed social-based algorithms for complex problems.

Combination of the complexity and richness of the social behaviors with interactions among individuals enriches the computational power of the socially inspired search and optimization algorithms. Interactions among humans cannot be only explained biologically, and they demonstrate a high level of diversity. Even if there is not an optimization behind social behaviors, they also result in robust social structures that may demonstrate a high level of stability. Human society is more complex and effective than other animal categories. Hence, if one algorithm simulates the human society, the efficiency and effectiveness may be more robust than other swarm intelligence based algorithms. These characteristics enable social phenomena to become valuable sources of inspiration for search algorithms [3]. In this section, social-based algorithms selected for performance comparisons have been explained.

2.1 Brainstorm Optimization Algorithm

Brainstorm Optimization Algorithm (BOA) is an interesting socially inspired algorithm based on the human being idea generation process [5]. This algorithm is based on the notion of brainstorming that was developed by Osborn for the first time in 1939. Brainstorming process is defined as a group creativity technique in which a group of people with different backgrounds aim to obtain a conclusion for a specific problem [6, 7]. Flowchart of BOA is depicted in Fig. 2.

2.2 Cultural Algorithm

Cultural Algorithm (CA) is a population-based metaheuristic search and optimization method that uses evolutionary computation principles to model social behaviors and cultural evolution [8]. Working process of CA depends on principles of human social evolution taken from the social science literature. CA is a dual inheritance system that characterizes evolution in human culture at both the micro-evolutionary level in the population space and the macro-evolutionary level, in the belief space. Flowchart of CA is depicted in Fig. 3.

2.3 Duelist Algorithm

Duelist Algorithm (DA) is also an interesting and relatively new social-based search and optimization algorithm. DA is inspired by how duelist improves their skill in duel [9]. Duels among duelists form the core of DA. Outcome of a duel is win or

Fig. 2 Flowchart of BOA

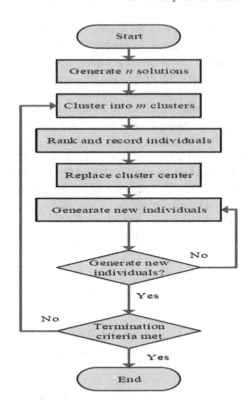

lose. Each loser learns from the winner, while the winner keeps innovating skills or techniques that may upgrade fighting capacities. A few duelists with highest fighting capacities are called as champion. Champions train new duelists that are as good as themselves to join the competition as a representative of each champion. All duelists are re-evaluated, and the worst fighting duelists are removed. Flowchart of DA is depicted in Fig. 4.

2.4 Imperialist Competitive Algorithm

Imperialist Competitive Algorithm (ICA) derived from the field of human social evolution is inspired by sociopolitical process of imperialistic competition of human beings [10]. ICA operators execute an imperialistic competition and the weak empires lose their power and their colonies. After some iterations of ICA, all the empires except the most powerful one will collapse and this unique powerful empire takes the control of all the colonies. Flowchart of ICA is depicted in Fig. 5.

Fig. 3 Flowchart of CA

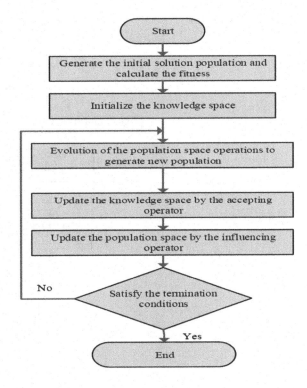

2.5 *Teaching Learning Based Optimization Algorithm*

Teaching Learning Based Optimization (TLBO) is inspired by the effect of influence of a teacher on the output of learners in a class. The algorithm uses the operators explaining the basic modes of the learning through teacher in teacher phase and through interaction with the other learners in learner phase [11]. Flowchart of TLBO is depicted in Fig. 6.

3 Experimental Results

The nature, complexity, and other characteristics of the benchmark functions can be easily derived from their definitions and have the nature and complexity of most engineering problems. The selected benchmark functions and their properties have been demonstrated in Table 1. Sphere function and Problem G1 function are unimodal with less complexity, and they can be used to evaluate the converging behaviors of algorithms. Expanded Schaffer function and Rastrigin function are multimodal problems with many local optima, and many algorithms are run on these problems for testing the global search ability in avoiding premature convergence. Shifted

Fig. 4 Flowchart of DA

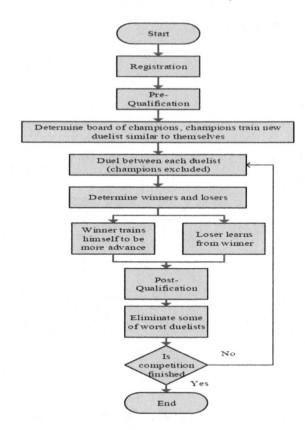

sphere and shifted Schwefel function are unimodal, shifted, and scalable. Global optimum $x^*=o$, $F_1(x^*)=f\_bias1=-450$. While shifted sphere function is separable, shifted Schwefel function is non-separable. These two functions are also used in CEC 2015.

In this section, performance comparisons of social-based optimization algorithms, namely BSO, CA, DA, ICA, and TLBO, have been demonstrated within unconstrained global optimization problems and experimentally investigated. In these studies, the tendency to converge to the optimum and the resultant values obtained are used as measure of performance. The results of the reviews are presented and interpreted through comparative tables and graphs.

All of the algorithms considered in this work are simulated 30 times, and the results have been recorded. For each method, the best, the worst, mean, median, and standard deviation of the five algorithms (A) for the functions (F) have been demonstrated in Table 2.

Fig. 5 Flowchart of ICA

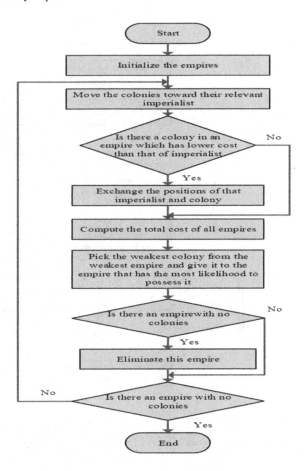

When Table 2 is examined, it is seen that the best result in terms of mean values in sphere function is obtained from TLBO. Dimension is 30, the maximum iteration number is 1000, and the number of population is 100 for all of the algorithms. When evaluated in terms of minimum values, it has been seen that the best result is obtained from DA. The mean convergence graphs of the algorithms with 30 runs for sphere function are shown in Fig. 7. Looking at Table 2, when the expanded Schaffer function is evaluated with 100 population numbers and 1000 iteration numbers in 2 dimensions, it is seen that all final values obtained are equal. According to this result, we can conclude that no algorithm for this function can supersede each other.

When Table 2 is examined, it is seen that the best in terms of mean values in Problem G1 function is obtained from TLBO. Dimension is 2, the maximum iteration number is 1000, and the number of population is 100 for all of the algorithms. When evaluated in terms of minimum values, it has been seen that DA has given the best result. The mean convergence graphs of the algorithms with 30 runs for Problem G1 function are demonstrated in Fig. 8.

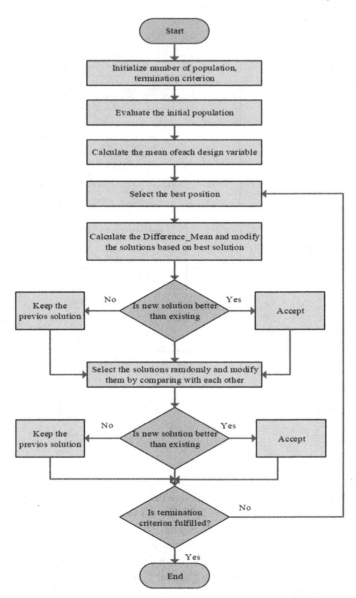

Fig. 6 Flowchart of TLBO

Table 1 Benchmark functions

Functions	Lower and upper limits	Min	D	Type	Equation
f_1–Sphere	$[-10, 10]$	0	30	UM	$\sum_{i=1}^{D} x_i^2$
f_2–Expanded Schaffer	$[-10, 10]$	0	2	MM	$0.5 + \frac{(\sin^2(\sqrt{x^2+y^2})-0.5)}{(1+0.001(x^2+y^2))^2}$
f_3–Problem $G1$	$[-10, 10]$	0	2	UM	$(x^2 + y^2)^{0.25} \times \sin\{30[(x + 0.5)^2 + y^2]^{0.1}\} + \|x\| + \|y\|$
f_4–Rastrigin	$[-5.12, 5.12]$	0	30	MM	$\sum_{i=1}^{D}[x_i^2 - 10\cos(2\pi x_i) + 10]$
f_5–Shifted Sphere	$[-5, 5]$	0	10	UM	$\sum_{i=1}^{D} z_i^2 + f\_bias, z = x - o, x = [x_1, x_2, \ldots x_D]$
f_6–Shifted Schwefel	$[-5, 5]$	0	10	UM	$\sum_{i=1}^{D}(\sum_{j=1}^{i} z_j)^2 + f\_bias\ z = x - o, x = [x_1, x_2, \ldots x_D]$

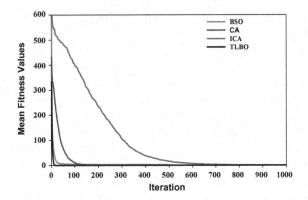

Fig. 7 Mean convergence graphs of algorithms for f1 function

When Table 2 is examined, it is seen that the best results in terms of both mean values and minimum values in Rastrigin function are obtained from DA. Dimension is 30, the maximum iteration number is 1000, and the number of population is 100 for all of the algorithms. The mean convergence graphs of the algorithms with 30 runs for Rastrigin function are shown in Fig. 9.

Looking at Table 2, DA's performance is best seen in functions f_5 and f_6. However, convergence has not been included in the results obtained. No steadily increasing or decreasing values have been found. The results of the other four algorithms seem to be very close to each other. Dimension is 10, the maximum iteration number is 5000, and the number of population is 100 for all of the algorithms. The mean

Table 2 Comparison of algorithms

F	A	Best value	Worst value	Mean	Median	St. Dev.
f1	BSO	3.12E-21	7.57E-21	4.84E-21	4.68E-21	1.06E-21
	CA	3.51E-16	5.63E-13	2.95E-14	7.54E-15	1.02E-13
	DA	0	4.90E-04	3.30E-03	0	1.26E-04
	ICA	7.86E-02	7.04E-01	3.07E-01	2.97E-01	1.54E-01
	TLBO	5.97E-178	7.39E-176	1.5E-176	8.44E-177	0
f2	BSO	−0.9359	−0.9359	−0.9359	−0.9359	0
	CA	−0.9359	−0.9359	−0.9359	−0.9359	0
	DA	−0.9359	−0.9359	−0.9359	−0.9359	0
	ICA	−0.9359	−0.9359	−0.9359	−0.9359	0
	TLBO	−0.9359	−0.9359	−0.9359	−0.9359	0
f3	BSO	5.08E-08	4.79E-07	2.86E-07	2.86E-07	8.86E-08
	CA	1.04E-35	2.47E-29	1.02E-30	4.71E-33	4.61E-30
	DA	0	2.87E-01	5.12E-02	0	8.95E-02
	ICA	4.89E-26	8.59E-23	1.31E-23	5.24E-24	1.91E-23
	TLBO	9.27E-191	2.07E-186	1.7E-187	3.72E-188	0
f4	BSO	1.69E+01	6.37E+01	4.03E+01	3.88E+01	1.08E+01
	CA	2.28E+02	3.61E+02	2.86E+02	2.85E+02	2.58E+01
	DA	0	0	0	0	0
	ICA	2.10E+01	3.74E+02	7.01E+01	4.19E+01	7.01E+01
	TLBO	0	1.99E+01	6.99E+00	7.96E+00	4.79E+00
f5	BSO	2.51E+00	7.81E+00	5.23E+00	5.24E+00	1.32E+00
	CA	2.86E+00	2.07E+01	8.59E+00	8.00E+00	3.51E+00
	DA	1.09E-06	3.56E-02	7.0E-03	1.08E-03	1.13E-02
	ICA	5.66E+00	1.99E+01	1.34E+01	1.29E+01	3.27E+00
	TLBO	3.30E+00	7.03E+00	5.28E+00	5.40E+00	9.97E-01
f6	BSO	2.95E+00	7.69E+00	5.53E+00	5.59E+00	1.38E+00
	CA	7.46E+00	2.58E+01	1.45E+01	1.40E+01	4.18E+00
	DA	4.39E-06	7.48E-03	2.75E-03	1.81E-03	2.87E-03
	ICA	5.36E+00	2.31E+01	1.19E+01	1.18E+01	3.57E+00
	TLBO	3.01E+00	7.48E+00	5.45E+00	5.53E+00	9.93E-01

convergence graphs of the algorithms with 30 runs for f_5 function and f_6 function are demonstrated in Fig. 10 and Fig. 11, respectively.

Fig. 8 Mean convergence graphs of algorithms for f3 function

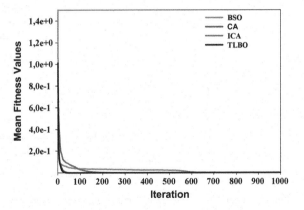

Fig. 9 Mean convergence graphs of algorithms for f4 function

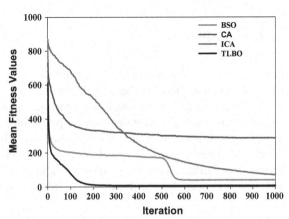

Fig. 10 Mean convergence graphs of algorithms for f5 function

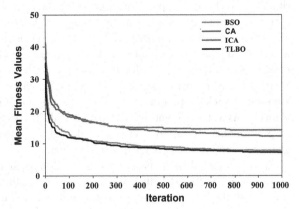

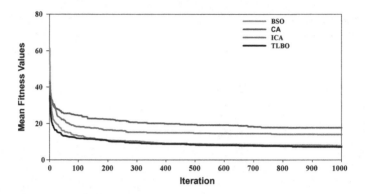

Fig. 11 Mean convergence graphs of algorithms for f6 function

4 Conclusions

Behaviors, interactions, and processes on human society are more complex, effective, and intelligent than those of other animals, insects, and plants. Many researchers have inspired from social phenomena in many human activities and behaviors such as society status, learning, teaching, leadership, dueling, categorization of individuals in the society, brainstorming, etc. and proposed social-based algorithms for complex search and optimization problems. Processes, concepts, rules, and events on social behaviors of humans have been considered and modeled as novel efficient search and optimization methods with extremely effective exploration capabilities in many cases.

In this work, performance comparisons of socially inspired optimization algorithms that utilize social phenomena of human being as source of inspiration to form optimization, namely brainstorm optimization algorithm, cultural algorithm, duelist algorithm, imperialist competitive algorithm, and teaching learning based optimization algorithm have been compared within unimodal and multimodal unconstrained global optimization problems for the first time. The basic versions of these algorithms have been used for performance comparisons. Teaching learning based optimization algorithm seems the best in terms of mean values and it is also more robust. Duelist algorithm is very new and many works should be performed in order to increase its performance. Standard deviation of duelist algorithm is high and it means the gathered solutions are very far from each other.

Some of these algorithms are very new and they can be improved in many ways. They can be combined with heuristic methods or other global optimization algorithms, and their hybrid versions may be proposed for efficient results for different complex search and optimization problems. More validation studies should be performed to discover the capabilities of these algorithms in dealing with the search and optimization problems. There is not a unique best algorithm for all types of problems. On the other hand, design, development, and implementation of novel algorithms are an important task for searching and obtaining better results. There are positive challenges in terms of efficiency and best possible use of the algorithms inspired by social phenomena of human being.

References

1. Du, K.L., Swamy, M.N.S.: Search and Optimization by Metaheuristics. Springer (2016)
2. Akyol, S., Alatas, B.: Güncel Sürü Zekası Optimizasyon Algoritmaları. Nevşehir Bilim ve Teknoloji Dergisi **1**(1), 36–50 (2012)
3. Khuat, T.T., Le, M.H.: A Survey on Human Social Phenomena inspired Algorithms. Int. J. Comput. Sci. Inf. Secur. **14**(6), 76 (2016)
4. Neme, A., Hernández, S.: Algorithms inspired in social phenomena. Nat.-Inspired Algorithm. Optim. 369–387 (2009)
5. Shi, Y.: Brain storm optimization algorithm. In: International Conference in Swarm Intelligence, pp. 303–309. Springer, Berlin, Heidelberg (2011)
6. Sun, C., Duan, H., Shi, Y.: Optimal satellite formation reconfiguration based on closed-loop brain storm optimization. IEEE Comput. Intell. Mag. **8**, 39–51 (2013)
7. Jordehi, A.R.: Brainstorm optimisation algorithm (BSOA): an efficient algorithm for finding optimal location and setting of FACTS devices in electric power systems. Int. J. Electr. Power Energy Syst. **69**, 48–57 (2015)
8. Reynolds, R.G.: An introduction to cultural algorithms. In: Proceedings of the Third Annual Conference on Evolutionary Programming, vol. 131–139. Singapore (1994)
9. Biyanto, T.R., Fibrianto, H.Y., Santoso, H.H.: Duelist algorithm: an algorithm in stochastic optimization method. In: Seventh International Conference on Swarm Intelligence Advances in Swarm Intelligence, pp. 25–30 (2016)
10. Atashpaz-Gargari, E., & Lucas, C.: Imperialist competitive algorithm: an algorithm for optimization inspired by imperialistic competition. In: CEC 2007, pp. 4661–4667 (2007)
11. Kumar, K.S., Samuel, R.H., Kumar, K.S., Samuel, R.H.: Teaching learning based optimization. Int. J. Innov. Res Sci. Technol. **1**(11), 413–419 (2015)

Comparative Analysis of Prediction Algorithms for Diabetes

Shweta Karun, Aishwarya Raj and Girija Attigeri

Abstract Machine learning is a widely growing field which helps in better learning from data and its analysis without any human intervention. It is being popularly used in the field of healthcare for analyzing and detecting serious and complex conditions. Diabetes is one such condition that heavily affects the entire system. In this paper, application of intelligent machine learning algorithms like logistic regression, naïve Bayes, support vector machine, decision tree, k-nearest neighbors, neural network, and random decision forest are used along with feature extraction. The accuracy of each algorithm, with and without feature extraction, leads to a comparative study of these predictive models. Therefore, a list of algorithms that works better with feature extraction and another that works better without it is obtained. These results can be used further for better prediction and diagnosis of diabetes.

Keywords Machine learning · Predictive algorithm · Feature extraction
Diabetes prediction · Classification · Ensemble

S. Karun · A. Raj · G. Attigeri (✉)
Department of ICT, Manipal Institute of Technology, Manipal University, Manipal, India
e-mail: girija.attigeri@manipal.edu

S. Karun
e-mail: shwetakarun@gmail.com

A. Raj
e-mail: aishwaryaraj16@gmail.com

© Springer Nature Singapore Pte Ltd. 2019
S. K. Bhatia et al. (eds.), *Advances in Computer Communication and Computational Sciences*, Advances in Intelligent Systems and Computing 759,
https://doi.org/10.1007/978-981-13-0341-8_16

1 Introduction

Diabetes is the most prevalent endocrine disease spread over the entire population and across all age groups. This disease is the fourth major cause of death in developed countries and considerable evidence shows that it is conquering epidemic shares in many developing and newly industrialized countries. About forty-four lakh Indians in their most productive years (20–79 years) are not aware of being diabetic, a disease that leaves them unprotected from heart attack, stroke, amputations, nerve damage, blindness, and kidney disease [1]. Data mining when applied to medical diagnosis can help doctors to take major decisions. Diabetes is a disease which has to be regularly kept in check by the patient so as not to cause severe damage to the body. Therefore, to analyze and predict diabetes is an important task that is most important for the patient. The main purpose of this paper is to analyze and predict how likely the women of age groups above 21 are being affected by diabetes and to find out factors responsible for an individual to be diabetic.

This paper aims to concentrate upon comparative analysis of prediction algorithms for diabetes, helping in its treatment using data mining techniques. Different classification algorithms like logistic regression, Gaussian naive Bayes, support vector machine, decision tree, k-nearest neighbors, neural networks, and random forest were understood and implemented to predict and analyze the dataset and make a comparison to reach to a final conclusion using Python. The dataset used is Pima Indians diabetes dataset which is for women aged 21 and above. Data mining is the process of selecting, exploring, and modeling huge data. This process has become a progressively extensive activity in all areas of research on medical science. Data mining has led to the uncovering of helpful hidden patterns from enormous databases. This paper could help in identifying relevant patterns for early diagnosis and treatment of diabetes and help combat it better. The rest of the paper is organized as follows. Section 2 presents review of literature. Section 3 discusses methodology feature selection, classification, and prediction algorithms. Section 4 describes the implementation of classification algorithms using R, and results are discussed in Sect. 5. Finally, Sect. 6 summarizes the work done.

2 Review of Literature

Jakhmola and Pradhan in [2] have come up with a method for controlled smoothening to provide different binning levels for different loss percentages given by the user. It supervises the binning level and allows disparate smoothening of data. This allows the user to smoothen the data to higher level of accuracy and hence attain a better prediction accuracy. They have used a new model which allows the user to be involved in guiding the output. Kayaer and Yildirim in [3] have used general regression neural network to examine the data. The results are compared to the previous results that used complex neural network structures. Three different neural network structures, multilayer perceptron, radial basis function, and general regression neural network (GRNN), were applied to the Pima Indians diabetes dataset. GRNN gave the best result for their test set. Karegowda et. al. in [4] have presented the development of a hybrid model for classifying Pima Indian diabetic database. K-means clustering and k-nearest neighbor classifier were applied for better classification of the given dataset. The resulting outcome indicated that the performance of the cascaded model can be improved even more by appropriate feature selection. Performance analysis was carried out in 70–30 ratio (training–test) splitting method in which classification accuracy and kappa statistics measures were deployed.

Karegowda et. al. in [5] have focused on the application of a hybrid model that integrates genetic algorithm (GA) and backpropagation network (BPN). Here, GA is used to initialize and optimize the connection weights of BPN. The hybrid GABPN model shows substantial improvement in the classification accuracy of BPN. The results showed that significant features selected by decision tree and GA-CFS (correlation-based feature selection) enhanced classification accuracy of GABPN further. Scherf and Brauer in [6] have used a batch-oriented, inter/intra-class distance-based approach for feature selection and feature weighting. This approach concentrates on the goal to strengthen the similarities between instances of the same class and degenerate the similarities between instances of different classes.

Chotirat and Gunopulos in [7] have focused on improving the accuracy of naive Bayes algorithm by using feature selection. Since, in C4.5, the highly correlated features cannot be separated, here only those features are considered that C4.5 use in building of a decision tree. The result obtained was faster and more accurate compared to both naive Bayes and C4.5. Campbell and Cristianini in [8] have suggested a simple training algorithm. It is used for training the classifier and getting optimized results with regard to accuracy and computational cost. Rudy Setiono and Huan Liu in [9] have offered a new algorithm which has an extended error function for neural network training and also an efficient quasi-Newton algorithm that minimizes the error function.

O. Hall et. al. in [10] have focused on dividing a large dataset into parts and then applying decision tree on each part since large datasets take a lot of time to process. The outcome of each decision tree is converted into a rule and then these rules are applied to the dataset to get same or better results compared to C4.5. Rajesh and Sangeetha in [11] applied various classification techniques on the diabetes dataset. They compared these on the basis of error rate and C4.5 emerged to be the one with the least error rate.

Authors in [12–14] have discussed application of diabetes prediction using data mining and machine learning algorithms. Diabetic complications, such as respiratory infections, diabetic ketoacidosis, hypoglycemia, periodontal disease, etc., and chronic complications due to diabetes, such as neuropathy, diabetic foot, heart failure, detection of these complications through data mining, and machine learning algorithms, are discussed in [15–18].

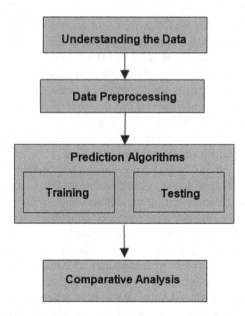

Fig. 1 Methodology used for diabetes prediction

3 Methodology

Methodology used for the work is shown in Fig. 1. Initially, the properties of the data are understood by computing mean, standard deviation, percentile scores, count, minimum, and maximum values which provide data description, and hence helps in understanding of the dataset. Histogram analysis of each attribute gives the frequency distribution of the data, leading to outlier detection. Heat maps are diagrams to plot features of data using colors. This is an interesting way of plotting the strength of correlation between the attributes to know if feature extraction is advisable.

Data preprocessing is a crucial step which includes handling missing values. There are many methods of doing this, one of them being substitution by mean value of the attribute. Feature extraction helps in eliminating the redundant attributes, making the dataset more informative and facilitating better results. Out of the many methods of performing feature extraction, chi-square analysis is one method to find similarities and dependencies between attributes, therefore removing the attributes that are irrelevant for classification. To bring in the data to a common format, we rescale it. Standardization rescales the data with mean as zero and standard deviation as one. This completes the preprocessing stage. In order to perform and test the different classification algorithms using cross-validation, the dataset is divided into testing and training sets. The models are trained based on the training dataset and tested to find the accuracy. The accuracy outcomes of different algorithms logistic regression (LR), naïve Bayes (NB), support vector machine (SVM), decision tree (DT), k-nearest neighbors (KNN), neural network (NN), and random decision forest

(RDF), with and without feature extraction lead to a comparative study and results in offering a set of algorithms that work better with feature extraction and the others that work better without it. Using the training set, new inputs in the dataset can also be classified using these algorithms and the final outcome can be generated using one of the ensemble approaches. One of these approaches is finding the modal value of the predicted outcomes of all the different algorithms. Therefore, this results in a final prediction.

4 Experimentation

Pima Indian diabetes dataset with eight attributes and a class variable have been used. For understanding the data, the count, mean, standard deviation, twenty-fifth, fiftieth, and seventy-fifth percentile, minimum and maximum values were calculated for each attribute. Histogram analysis was done to analyze the range of values for each attribute, after which a heat map showing the correlation between every pair of attributes was made. For data preprocessing, the missing values are handled by replacing by means of those particular attributes. Feature extraction using chi-square analysis was performed to find the four best variables for prediction, namely, glucose, insulin, body mass index, and age. Standardization of data was done to rescale it to a common format (with mean = 0 and standard deviation = 1). After this, the dataset was divided into training (80%) and testing sets (20%), and classification evaluation was performed using cross-validation score. The accuracy of the training sets was calculated using LR, NB, SVM, DT, KNN, NN, and RDF [19], once with feature extraction and again without it. Analysis was done on which algorithm performed better with feature extraction and which one without. Later, users were asked to input their values for the attributes for live prediction. Prediction result was declared on the basis of majority vote from all algorithms using mode.

5 Results and Discussions

Figure 2 indicates the description of the dataset with respect to mean and other statistical parameters. Figure 3 indicates the distributions of the variables across the dataset. Variable BMI, blood sugar, and glucose show normal distribution, whereas other variables are bit skewed in nature. Figure 4 indicates the heat map for depicting the correlation of class label (has diabetes or not) with other variables. The heat map shows that variable glucose and BMI are highly correlated with class variable. As the next step when feature extraction algorithm is applied to the dataset as shown in Fig. 5, BMI, glucose, age, and insulin are selected as important features for prediction of diabetes. Performance of various classification algorithms is shown in Fig. 6. It can be observed that logistic regression has highest true positives with and without feature extraction. SVM has best true positive rate when algorithm is applied to

DATASET DESCRIPTION

	Pregnancies	Glucose	BloodPressure	SkinThickness	Insulin
count	768.000000	768.000000	768.000000	768.000000	768.000000
mean	3.845052	121.681641	72.254557	26.595703	118.660417
std	3.369578	30.436015	12.115997	9.638053	93.080252
min	0.000000	44.000000	24.000000	7.000000	14.000000
25%	1.000000	99.750000	64.000000	20.500000	79.800000
50%	3.000000	117.000000	72.000000	23.000000	79.800000
75%	6.000000	140.250000	80.000000	32.000000	127.250000
max	17.000000	199.000000	122.000000	99.000000	846.000000

	BMI	DiabetesPedigreeFunction	Age
count	768.000000	768.000000	768.000000
mean	32.450911	0.471876	33.240885
std	6.875366	0.331329	11.760232
min	18.200000	0.078000	21.000000
25%	27.500000	0.243750	24.000000
50%	32.000000	0.372500	29.000000
75%	36.600000	0.626250	41.000000
max	67.100000	2.420000	81.000000

	Class variable (0 or 1)
count	768.000000
mean	0.348958
std	0.476951
min	0.000000
25%	0.000000
50%	0.000000
75%	1.000000
max	1.000000

Fig. 2 Dataset description

selected features. Comparison of the algorithms used is depicted in Figs. 7 and 8. The observed outcome of the experiment has been plotted as a graph. KNN gives the best prediction accuracy in the study with feature extraction, while LR stands the best for computation of prediction accuracy without feature extraction. Along with this, it can be observed that SVM, NN, and RDF also show better prediction accuracy with feature extraction and on the other hand, NB and DT perform better without it.

The output of individual algorithms and ensemble method when applied to a new user input is shown in Fig. 9. The user input is number of pregnancies: 0, glucose: 118, diastolic BP: 84, skin thickness: 47, insulin: 230, BMI: 45.8, DP function: 551, and age: 31. Except logistic regression, all the algorithms predict the outcome as 1. Ensemble method using the mode operation for combining results is predicted as the outcome 1.

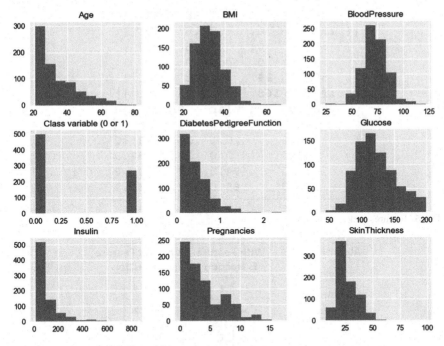

Fig. 3 Histograms for all the variables

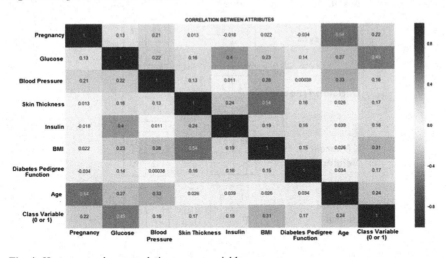

Fig. 4 Heat map to show correlation among variables

```
---------------FEATURE EXTRACTION--
[[ 148.      0.      33.6    50. ]
 [  85.      0.      26.6    31. ]
 [ 183.      0.      23.3    32. ]
 [  89.     94.      28.1    21. ]
 [ 137.    168.      43.1    33. ]]
     Glucose   Insulin    BMI    Age
0      148.0       0.0   33.6   50.0
1       85.0       0.0   26.6   31.0
2      183.0       0.0   23.3   32.0
3       89.0      94.0   28.1   21.0
4      137.0     168.0   43.1   33.0
```

Fig. 5 Feature extraction output

Algorithms	With Feature Extraction		Without Feature Extraction	
Logistic Regression	91	9	92	8
	34	20	30	24
Support Vector Machine	93	7	88	12
	34	20	33	21
Naïve Bayes	87	13	88	12
	36	18	29	25
Decision Tree	74	26	76	24
	31	23	24	30
K Nearest Neighbour	89	11	87	13
	27	27	32	22
Neural Network	91	9	91	9
	31	23	32	22
Random Forest	88	12	89	11
	29	25	35	19

Fig. 6 Output of classification algorithms

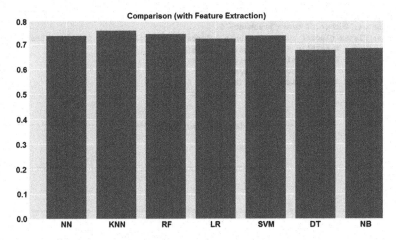

Fig. 7 Comparison of algorithms with all the features

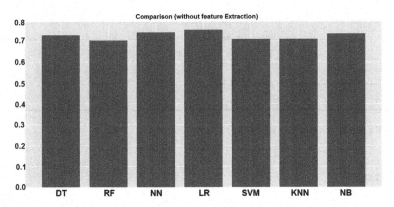

Fig. 8 Comparison of algorithms with selected features

6 Conclusion

To curb diabetes is becoming a concern of utmost importance. With various data mining techniques being used extensively, finding a solution to better prediction and early treatment has become an easier task. On applying the different prediction algorithms, we concluded that with feature extraction, SVM, KNN, NN, and RDF show better prediction accuracy, while LR, NB, and DT show better prediction accuracy without. This work can be carried forward by implementing different feature extraction algorithms to see how the results vary. More and more algorithms can be added for an enhanced comparison.

```
enter the details
Pregnancies: Number of times pregnant- 0
Glucose: Plasma glucose concentration a 2 hours in an oral glucose tolerance test- 118
BloodPressure: Diastolic blood pressure (mm Hg) 84
SkinThickness: Triceps skin fold thickness (mm) 47
Insulin: 2-Hour serum insulin (mu U/ml) 230
BMI: Body mass index (weight in kg/(height in m)^2) 45.8
DiabetesPedigreeFunction: Diabetes pedigree function.551
Age: Age (years) 31


-------------------------------Logistic Regression-------------------------------
[0]

-------------------------------Support Vector Machine----------------------------
[1]

-------------------------------Naives Bayes-------------------------------------
[1]

-------------------------------Decision Tree------------------------------------
[1]

-------------------------------K Nearest Neighbour------------------------------
[1]

-------------------------------RESULT-------------------------------------------
1
```

Fig. 9 Prediction for new user input

References

1. The Times of India, India—"44 lakh Indians don't know they are diabetic". http://timesofindia.indiatimes.com/india/44-lakh-Indians-dont-know-they-arediabetic/articleshow/17274366.cms

2. Jakhmola, S., Pradhan, T.: A computational approach of data smoothening and prediction of diabetes dataset. In: Proceedings of the Third International Symposium on Women in Computing and Informatics. ACM (2015)

3. Kayaer, K., Yıldırım, T.: Medical diagnosis on Pima Indian diabetes using general regression neural networks. In: Proceedings of the International Conference on Artificial Neural Networks and Neural Information Processing (ICANN/ICONIP) (2003)

4. Karegowda, A.G., Jayaram, M.A., Manjunath, A.S.: Cascading k-means clustering and k-nearest neighbor classifier for categorization of diabetic patients. Int. J. Eng. Adv. Technol. 1.3, 147–151 (2012)

5. Karegowda, A.G., Manjunath, A.S., Jayaram, M.A.: Application of genetic algorithm optimized neural network connection weights for medical diagnosis of pima Indians diabetes. Int. J. Soft Comput. 2.2: 15–23 (2011)

6. Scherf, M., Brauer, W.: Feature selection by means of a feature weighting approach. Inst. für Informatik (1997)

7. Ratanamahatana, C.A., Dimitrios, G.: Scaling up the Naive Bayesian Classifier: Using Decision Trees for Feature Selection (2002)

8. Campbell, C., Cristianini, N.: Simple Learning Algorithms for Training Support Vector Machines. University of Bristol (1998)
9. Setiono, R., Liu, H.: Neural-network feature selector. IEEE Trans. Neural Netw. **8.3**, 654–662 (1997)
10. Hall, L.O., Chawla, N., Bowyer, K.W.: Combining decision trees learned in parallel. In: Working Notes of the KDD-97 Workshop on Distributed Data Mining (1998)
11. Rajesh, K., Sangeetha, V.: Application of data mining methods and techniques for diabetes diagnosis. Int. J. Eng. Innov. Technol. (IJEIT) **2.3** (2012)
12. Vrushali, R., Balpande, R., Wajgi, D.: Prediction and severity estimation of diabetes using data mining technique. In: 2017 International Conference on Innovative Mechanisms for Industry Applications (ICIMIA), pp. 576–580 (2017)
13. Veena Vijayan, V., Anjali, C.: Computerized information system using stacked generalization for diagnosis of diabetes mellitus. In: 2015 IEEE Recent Advances in Intelligent Computational Systems (RAICS), pp. 173–178 (2015)
14. Kavakiotis, I., Tsave, O., Salifoglou, A., Maglaveras, N., Vlahavas, I., Chouvarda, I.: Machine learning and data mining methods in diabetes research. Comput. Struct. Biotechnol. J. **15**, 104–116 (2017). ISSN 2001-0370,2016
15. Lagani, V., Chiarugi, F., Thomson, S., Fursse, J., Lakasing, E., Jones, R.W., et al.: Development and validation of risk assessment models for diabetes-related complications based on the DCCT/EDIC data. J. Diabetes Complicat. **29**(4), pp. 479–487 (2015)
16. Lagani, V., Chiarugi, F., Manousos, D., Verma, V., Fursse, J., Marias, K., et al.: Realization of a service for the long-term risk assessment of diabetes-related complications. J. Diabetes Complicat. **29**(5), 691–698 (2015)
17. Sacchi, L., Dagliati, A., Segagni, D., Leporati, P., Chiovato, L., Bellazzi, R.: Improving risk-stratification of diabetes complications using temporal data mining. Conf. Proc. IEEE Eng. Med. Biol. Soc. **2015**, 2131–2213 (2015)
18. Huang, G.-M., Huang, K.-Y., Lee, T.-Y., Weng, J.: An interpretable rule-based diagnostic classification of diabetic nephropathy among type 2 diabetes patients. BMC Bioinform. **16**(S-1), S5 (2015)
19. Breiman, L.: Random forests. Mach. Learn. **45**(1), 5–32 (2001)
20. Han, J., Pei, J., Kamber, M.: Data Mining: Concepts and Techniques. Elsevier (2011)
21. Prima Indians Diabetes Data Set (2017). https://archive.ics.uci.edu/ml/datasets/Pima+Indians+Diabetes

An Improved Discrete Grey Model Based on BP Neural Network for Traffic Flow Forecasting

Ziheng Wu, Zhongcheng Wu and Jun Zhang

Abstract The forecasting of traffic flow is an important part of intelligent transportation system; actual and accurate forecasting of traffic flow can give scientific support for urban traffic guidance and control. As there is big forecast error when modeling toward traffic flow data with discrete grey model DGM (1, 1), this paper amends the equal interval time sequence. According to the characteristic of time coefficient and backpropagation (BP) neural network, we propose an improved grey model by combining DGM (1, 1) model with BP neural network model. The experimental result indicates that the improved grey model is scientific and effective for the forecasting of traffic flow.

Keywords Intelligent transportation · Traffic flow forecasting · Discrete grey model · Time coefficient · Backpropagation neural network

1 Introduction

Accurate and real-time traffic forecasting can provide scientific support for urban traffic guidance. The forecasting of traffic flow estimates traffic flow at the next time $t + \Delta t$ by using the existing traffic flow data at time t [1].

In 1982, professor Deng put forward Grey system theory, the features of which are "poor information" and "small sample", it has been applied in a variety of application areas [2–10]. However, there still many limitations in DGM (1, 1), to improve the forecasting accuracy, many researchers have conducted optimization and exten-

Z. Wu (✉) · Z. Wu · J. Zhang
High Magnetic Field Laboratory, Chinese Academy of Sciences, Hefei, China
e-mail: ziheng88@mail.ustc.edu.cn

Z. Wu
University of Science and Technology of China, Hefei, China

© Springer Nature Singapore Pte Ltd. 2019
S. K. Bhatia et al. (eds.), *Advances in Computer Communication and Computational Sciences*, Advances in Intelligent Systems and Computing 759,
https://doi.org/10.1007/978-981-13-0341-8_17

sive studies, for instance, Huiru Zhao proposed an improved DGM (1, 1) optimized by MVO [11]. Lifeng Wu put forward a novel NDGM with the fractional-order accumulation [12]. Based on initial value optimizing, Tianxiang Yao put forward an improved forecasting model [13]. A novel time-delayed polynomial grey prediction model which is abbreviated as TDPGM(1,1) is proposed [14].

Short-term traffic flow forecasting is a complex nonlinear problem, grey model can quickly forecast nonlinear sequences, and there have been many researchers using grey forecasting model for traffic flow forecasting [15, 16]. Although the DGM (1, 1) models mentioned above have achieved remarkable results in improving the forecasting performance of DGM (1, 1), yet limited efforts have focused on improving DGM (1, 1) from the insight of time sequence. This paper proposed an improved DGM (1, 1). According to the characteristic of time coefficient and BP neural network, we build the improved model by combining DGM (1, 1) model with BP neural network model. Experiments forecast performance and accuracy significantly.

2   DGM (1, 1)

Assume the traffic flow sequence is $X^{(0)} = \{x^{(0)}(1), x^{(0)}(2), \ldots, x^{(0)}(n)\}, x^{(0)}(k) > 0$.

$$X^{(1)} = \{x^{(1)}(1), x^{(1)}(2), \ldots, x^{(1)}(n)\} \quad x^{(1)}(k) = \sum_{i=1}^{k} x^{(0)}(i).$$

Definition 2.1.1 $x^{(1)}(k + 1) = \beta_1 x^{(1)}(k) + \beta_2$ is DGM (1, 1) (discrete grey model).

Theorem 2.1.1   *If $\widehat{\beta} = [\beta_1, \beta_2]^T$ is a parameter sequence and*

$$B = \begin{bmatrix} x^{(1)}(1) & 1 \\ x^{(1)}(2) & 1 \\ \vdots & \vdots \\ x^{(1)}(n-1) & 1 \end{bmatrix} \quad Y = \begin{bmatrix} x^{(1)}(2) \\ x^{(1)}(3) \\ \vdots \\ x^{(1)}(n) \end{bmatrix}$$

$$[\beta_1, \beta_2]^T = (B^T B)^{-1} B^T Y$$

then the recurrence formula of DGM model is

$$\hat{x}^{(1)}(k + 1) = \beta_1^k x^{(0)}(1) + \frac{1 - \beta_1^k}{1 - \beta_1} \beta_2 \tag{2}$$

So the time response sequence is

$$\hat{x}^{(0)}(k+1) = \hat{x}^{(1)}(k+1) - \hat{x}^{(1)}(k) = (x^{(0)}(1) - \frac{\beta_2}{1-\beta_1})(1 - \frac{1}{\beta_1})\beta_1^k \qquad (3)$$

In order to improve the forecasting accuracy, metabolism discrete grey forecasting model [17], in which the oldest information $x^{(0)}(1)$ is discarded and replaced by the latest information $x^{(0)}(n)$ during each modeling, is applied in this paper.

2.1 Calculation of Time Coefficient

If we assume that there is no error between the original data and the simulated data by DGM model, a computing method of time coefficient [18] is introduced.
Let

$$\hat{x}^{(1)}(\xi_k) = x^{(1)}(k) \qquad (4)$$

ξ_k is the time coefficient at time k.
From Eqs. (2) and (4), we can get

$$\hat{x}^{(1)}(\xi_k) = x^{(1)}(k) = \beta_1^{\xi_k-1} x^{(0)}(1) + \frac{1 - \beta_1^{\xi_k-1}}{1 - \beta_1}\beta_2$$

Then

$$\xi_k = \log_{\beta_1} \frac{x^{(1)}(k) - \frac{\beta_2}{1-\beta_1}}{x^{(0)}(1) - \frac{\beta_2}{1-\beta_1}} + 1 \qquad (5)$$

$$\sigma_k = \xi_k - k \qquad (6)$$

σ_k is called time coefficient error at time k.

3 BP Neural Network

Under the condition of appropriate weights and reasonable structure, BP neural network can approximate any continuous nonlinear function in theory, so BP is employed for time coefficient error forecasting (Fig. 1).

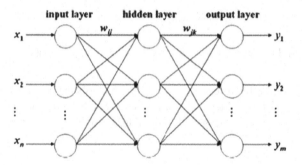

Fig. 1 The topology structure

If there are m outputs, n inputs, and s neurons in the hidden layer, s can be calculated by

$$s = \sqrt{m + n} + a$$

where a is the constant, and $1 < a < 10$.

The output of the hidden layer is

$$b_j = f_1(\sum_{i=1}^{n} w_{ij}x_i - \theta_j)$$

The output is

$$y_k = f_2(\sum_{j=1}^{s} w_{jk}b_j - \theta_k)$$

f_1, f_2 is the corresponding transfer function; w is the corresponding weight; and θ is the corresponding threshold.

Thus, we can get the error function:

$$e = \sum_{k=1}^{m} (y_k - o_k)^2$$

where o_k is the expectation output.

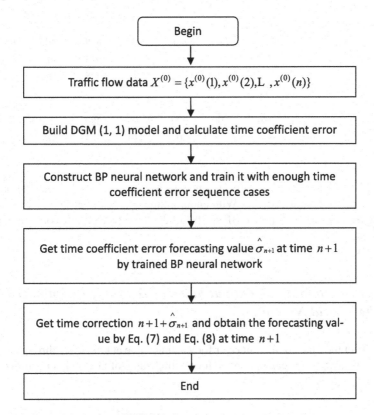

Fig. 2 Flowchart of the improved DGM (1, 1)

4 DGM (1, 1) Optimized by BP Neural Network

Figure 2 shows the flowchart of the improved DGM (1, 1), first, the DGM (1, 1) is built and the time coefficient error sequence is calculated; second, a BP neural network is constructed and trained using the training datasets; and finally the forecasting value of time coefficient error by the trained BP is adopted to correct the equal interval time sequence of DGM (1, 1).

4.1 Establishment of BP Neural Network Model Forecasting Time Coefficient Error

σ is the time coefficient error sequence, $\sigma = \{\sigma_1, \sigma_2, \ldots, \sigma_n\}$. In practice, if we assume the prediction step is R, the selected time coefficient error sequence $\sigma_{(k-R)}, \sigma_{(k-R+1)}, \ldots \sigma_{(k-2)}, \sigma_{(k-1)}$ is then considered as inputs to BP neural network,

Table 1 Part of the original traffic flow data

Date	7:00	7:15	7:30	7:45	8:00	8:15	8:30	8:45	9:00
1	213	273	266	243	312	284	315	363	350
2	372	356	374	429	417	392	428	417	432
$\vdots$					$\vdots$				$\vdots$
13	226	242	280	314	370	366	380	338	399
14	278	305	348	388	404	440	384	396	424

and the time coefficient error value at time k σ_k is considered as the expectation output value of BP neural network. With enough time coefficient error sequence cases training BP network, the trained BP can be used as an effective time coefficient error forecasting tool.

4.2 Calculation of the Output Value of DGM (1, 1) Optimized by Neural Network

After we get the time coefficient error forecasting value at time k: $\widehat{\sigma}_k$ through trained BP neural network, we can obtain the forecasting value of DGM (1, 1) optimized by neural network $\hat{x}_b^{(0)}(k)$:

$$\hat{x}_b^{(1)}(k) = \beta_1^{(k+\widehat{\sigma}_k-1)} x^{(0)}(1) + \frac{1 - \beta_1^{(k+\widehat{\sigma}_k-1)}}{1 - \beta_1} \beta_2 \tag{7}$$

$$\hat{x}_b^{(0)}(k) = \hat{x}_b^{(1)}(k) - \hat{x}_b^{(1)}(k - 1) \tag{8}$$

5 Experiments and Results

This paper selects the traffic flow data of one urban expressway, the data sampling period is 15 min, and part of the original traffic flow data [19] is shown in Table 1.

For the forecasting performance comparing, we adopt relative simulation error (RSE) and mean absolute percent error (MAPE) to validate the forecasting performance.

Respectively establish the improved model and the traditional DGM (1, 1) modeled by the traffic flow data at 7:00–8:45 to predict the traffic flow at 9:00. Table 2 shows the comparison of RSE and MAPE among two different models; from Table 2, we can see that the MAPE and RSE of the improved model are much lower than the traditional DGM (1, 1).

Table 2 Comparison of RSE and MAPE among different models

Date	Actual value	Traditional DGM (1, 1)		The improved model	
		Forecasting value	RSE (%)	Forecasting value	RSE (%)
1	350	358.35	2.38	351.21	0.34
2	432	444.14	2.81	427.05	1.14
3	431	443.79	2.96	428.77	0.51
4	420	433.42	3.19	427.69	1.83
5	425	434.47	2.23	434.29	2.18
6	434	428.95	1.16	426.30	1.77
7	368	359.02	2.43	364.47	0.95
8	404	375.53	7.04	379.59	6.04
9	354	387.63	9.50	374.12	5.68
10	431	451.11	4.66	429.41	0.36
11	428	446.16	4.24	443.28	3.57
12	372	381.31	2.50	367.03	1.33
13	399	404.96	1.29	397.76	0.30
14	424	436.21	2.88	432.84	2.08
		MAPE (%)	3.52	MAPE (%)	2.01

Fig. 3 Scatter diagram of forecasting value among two different models

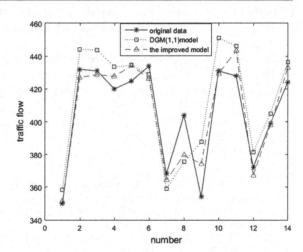

Figure 3 shows the forecasting value among two different models, and Fig. 4 shows the relative simulation error among two different models. From Figs. 3 and 4, we can see that compared with the traditional DGM (1, 1), the forecasting value of the improved model is more accurate, which demonstrates the effectiveness and accuracy of the proposed grey forecasting model.

Fig. 4 Scatter diagram of forecasting relative simulation error among two different models

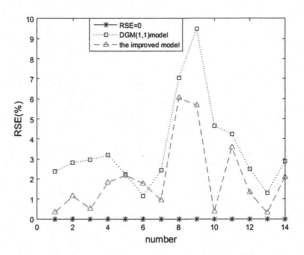

6 Conclusions

This paper introduced an effective DGM (1, 1) model optimized by BP neural network for traffic flow forecasting. In this model, the equal interval time sequence with time coefficient is amended by BP neural network and the computing method of time coefficient is given.

An experimental evaluation has been carried out to show that the improved grey model is scientific and, effective for the forecasting of traffic flow, can improve the forecasting performance greatly, and moreover has certain reliability.

References

1. Zhang, L., Wei, H., et al.: An improved k-nearest neighbor model for short-term traffic flow prediction. Procedia-Soc. Behav. Sci. **96**, 653–662 (2013)
2. Deng, J.L.: Grey Prediction and Grey Decision, 2nd edn. Press of Huazhong University of Science & Technology, Wuhan (2002)
3. Xie, N.-M., Liu, S.-F.: Grey System Theory and Its Application. Science Press, Beijing (2008)
4. Morita, H., Zhang, D.P., Tamura, Y.: Long-term load forecasting using grey system theory. Electr. Eng. Jpn. **115**(2), 11–20 (1995)
5. Xie, N.-M., Liu, S.-F.: Discrete grey forecasting model and its optimization. Appl. Math. Model. **33**(2), 1173–1186 (2009)
6. Yao, A.W.L., Chi, S.C., Chen, J.H.: An improved grey-based approach for electricity demands forecasting. Electr. Power Syst. Res. **67**(3), 217–224 (2003)
7. Wu, X.M., Chi, D.C., Pan, X., et al.: Waterlogging prediction based on the discrete gray DGM (1, 1) prediction model. J. Shenyang Agric. Univ. **01**, 104–107 (2013)
8. Xia, M., Wong, W.K.: A seasonal discrete grey forecasting model for fashion retailing. Knowl-Based Syst. **57**(2), 119–126 (2014)

9. Liu, J.F., Liu, S.F., Fang, Z.G.: Predict China's per capita GDP based on ending-point optimized discrete grey (1, 1) model. In: IEEE International Conference on Grey Systems and Intelligent Services, pp. 113–117 (2013)
10. Xie, N.M., Liu, S.F.: Research on discrete grey model and its mechanism. IEEE Int. Conf. Syst. Man Cybern. **01**, 606–610 (2005)
11. Zhao, H.R., Han, X.Y., Guo, S.: DGM (1, 1) model optimized by MVO (multi-verse optimizer) for annual peak load forecasting. Neural Comput. Appl. (2016). https://doi.org/10.1007/s00521-016-2799-1
12. Wu, L.F., Liu, S.F., Cui, W.: Non-homogenous discrete grey model with fractional-order accumulation. Neural Comput. Appl. **25**(5), 1215–1221 (2014)
13. Yao, T., Gong, Z., Zhu, X., et al.: The discrete grey prediction model based on optimized initial value. In: IEEE International Conference on Grey Systems and Intelligent Services, pp. 330–334 (2009)
14. Ma, X., Liu, Z.: Application of a novel time-delayed polynomial grey model to predict the natural gas consumption in China. J. Comput. Appl. Math. **324**, 17–24 (2017)
15. You, Z.S., He, L., Zhang, G.L.: Based on the improved GM(1,1) model of urban short-term traffic flow prediction research. J Chongqing Normal Univ. **1**, 168–172 (2016)
16. Wu, Z.Z., Fan, Y.Y., Ma, W.J.: Spot speed prediction model based on grey neural network. J. Southwest Jiaotong Univ. **47**(2), 285–290 (2012)
17. Luo, Y.X.: Non-equidistant step by step optimum new information GM(1,1) and its application. Syst. Eng.-Theory Pract. **30**(4), 2254–2258 (2010)
18. Wu, Z.H., Yang, A.P., Dai, W.Z.: The improvement method of DGM (1,1) based on time coefficient-amended 1. In: IEEE International Conference on Grey Systems and Intelligent Services, pp. 249–252. IEEE (2011)
19. Wang, F.Q., He, H.J.: Short tern traffic flow forecasting of grey system. J. Hubei Normal Univ. **35**(1), 20–24 (2015)

Influence of the System Parameters on the Final Accuracy for the User Identification by Free-Text Keystroke Dynamics

Evgeny Kostyuchenko and Ramil Kadyrov

Abstract The paper considers the selection of parameters for the user identification system based on free-text keystroke dynamics. The naive Bayes algorithm is used for identification. The dependence of the identification accuracy on the number of intervals of the classifier splitting is investigated. The effect of the size of the training set from each user on the accuracy of identification is considered. Graphic dependencies of the identification error probability on the volume of the training sample and the splitting parameter of the Bayesian classifier are obtained. Recommendations for the choice of the volume of the text training set and the splitting parameter of the Bayesian classifier are made. The obtained results allow the selection of these parameters for the construction of an identification system based on free-text keystroke dynamics.

Keywords Free-text keystroke dynamics · Bayesian classifier · Identification parameters · Parameter definition

1 Introduction

It is impossible to achieve complete security of the information system, but it is possible to minimize the risks of implementing threats and adequately strengthen the protection system. As measures to ensure information security, in addition to standard security measures, biometric authentication and identification tools can be used. Biometrics is the authentication of people on the basis of their physiological and behavioral characteristics. One of the most promising and actively researched at the moment areas of biometric identification of users is the identification of users on the free-text keystroke dynamics. It has some of advantages—simplicity and low cost of implementation, and a some of shortcomings—low accuracy, the fundamental

E. Kostyuchenko (✉) · R. Kadyrov
Tomsk State University of Control Systems and Radioelectronics, Lenina str. 40,
634050 Tomsk, Russia
e-mail: key@keva.tusur.ru
URL: http://www.tusur.ru

© Springer Nature Singapore Pte Ltd. 2019

199

S. K. Bhatia et al. (eds.), *Advances in Computer Communication and Computational Sciences*, Advances in Intelligent Systems and Computing 759,
https://doi.org/10.1007/978-981-13-0341-8_18

dependence of the parameters of the user model used from the typing language. Within the framework of this work, we investigate the influence of such parameters as the amount of text typed for identification and the number of intervals for the partitioning of the naive Bayesian classifier (used as the basis for building a user model) on the accuracy of the user identification procedure.

2 Current State of the Problem

The study of some works in foreign sources [1–7] suggests that at present biometric means of user identification and authentication are of particular interest among researchers, and interest in biometric authentication by keystroke dynamics is growing every year. Being one of the most promising and promising authentication methods based on behavioral biometrics, it is actively being investigated, and the number of articles and qualitatively new and effective algorithms is growing every year. If keystroke dynamics authentication based on a fixed passphrase is more or less studied and already implemented in practice, free-text keystroke dynamics and continuous authentication currently attracts researchers and this type of authentication is at the stage of active study.

The field for practical application of continuous authentication is wide, and there is a demand for such systems, because such systems will help to increase the security of computer systems from intrusions and increase the effectiveness of standard protection, while biometric systems based on keyboard handwriting remain one of the cheapest and easily realizable.

However, a number of shortcomings have been identified, such as low final authentication accuracy and lack of research on the dependence of authentication accuracy on the volume of typed text for the Russian language. In addition, the free-text keystroke dynamics user identification systems for Russian can be found only as separate independent modules, and final solutions on the market are not represented.

This allows us to talk about the relevance of the current study.

3 Mathematical Device for Free-Text Keystroke Dynamics Identification

The study is aimed at determining the parameters of the identification system by free-text keystroke dynamics for an approach based on

- using time characteristics for the pressing duration and the intervals between pressing;
- partition of a union array of time characteristics into intervals;
- using a naive Bayesian classifier to determine the degree of belonging of a sample of the keystroke dynamics to the reference sample.

As a mathematical apparatus used in the framework of this work, a naive Bayesian classifier is used to decide whether a set belongs to a particular user.

A naive Bayesian classifier is a simple probabilistic classifier based on the application of Bayes' theorem with strict assumptions of independence. More clearly, the principle of the operation of such a classifier can be described using the expression "work with independent characteristics" [8].

Despite the fact that the classifier is naive and works with simplified conditions, naive Bayesian classifiers have positively proved themselves in many real complex situations.

The big advantage of the naive Bayesian classifier is that it requires a relatively small amount of training data to work. Since it is assumed that the variables are independent, only the deviations of the characteristics within each class are to be taken into account.

A probabilistic model for a classifier is a conditional model over a dependent variable of class C with a small number of classes, depending on several variables from F_1 to F_n. The problem occurs with large numbers of properties n, or when properties can take a large number of values. Therefore, the model is reformulated to make it easy to process. From the Bayes theorem, it follows that

$$p(C|F_1, \ldots, F_n) = \frac{p(C)p(F_1, \ldots, F_n|C)}{p(F_1, \cdots, F_n)} \quad (1)$$

Since the denominator of the fraction does not depend on C and the values of the properties F_1 F_n are given, then the denominator is a constant, and only the numerator that is equivalent to the joint probability of the model, which can be written as follows using the conditional probability:

$$p'(C|F_1, \ldots, F_n) = p(C)p(F_1, \ldots, F_n|C) = p(C)p(F_1|c)p(F_2, \ldots, F_n|C, F_1) =$$
$$= p(C)p(F_1|c)p(F_2|c, F_1)p(F_3, \ldots, F_n|C, F_1, F_2) = \ldots$$
$$= p(C)p(F_1|c)p(F_2|c, F_1) \ldots p(F_n|C, F_1, F_2, \ldots, F_{n-1}) \quad (2)$$

We assume that each characteristic F_i is conditionally independent from any other characteristic F_j for $j \neq i$, then the joint model can be expressed as

$$p'(C|F_1, \ldots, F_n) = p(C)p(F_1|c)p(F_2|c, F_1) \ldots p(F_n|C, F_1, F_2, \ldots, F_{n-1}) =$$
$$= p(C)p(F_1|C)p(F_2|C) \ldots p(F_n|C) = p(C)\prod_i p(F_i|C) \quad (3)$$

Hence, from the assumption of independence, the conditional distribution over the class variable C can be expressed as follows:

$$p(C|F_1, \ldots, F_n) = \frac{1}{Z} p(C) \prod_i p(F_i|C) \qquad (4)$$

where Z is a scale factor that depends only on $F_1 \ldots F_n$, that is a constant if the values of the variables are known. A naive Bayesian classifier combines the model with the rule of solution. In accordance with one general rule, the most probable hypothesis should be chosen. This rule is known as a posteriori rule of decision (maximum a posteriori) [8]. The corresponding classifier is a function $classify$, defined as follows:

$$classify(f_1, \ldots f_n) = argmax \, p(C = c) \prod_i p(F_i = f_i|C = c) \qquad (5)$$

If this class and the value of the property never occur together in the training set, then the estimate based on the probabilities will be zero. This is a problem, since multiplying a zero score will result in loss of information about other probabilities. Therefore, it is preferable to make minor corrections to all probability estimates so that there are no probabilities equal to zero.

As an estimate of the degree of keystroke dynamics belonging to a certain user, based on the analysis of one combination, was selected a value determined by the formula:

$$I_i = \sum_i n_i lg \frac{p_{i,j}}{p_i} \qquad (6)$$

where n_i is the number of times an identifiable user hits to the ith interval of a common time array; $p_{i,j}$ is the probability of the jth user falling into the ith interval of a union array of times; s is the number of intervals divided the union array of times; and

$$p_i = \sum_j p_{i,j} \qquad (7)$$

4 Processing the Experiment

4.1 Planning of Experiment

The purpose of the experiment is to determine the values of the specified parameters of the identification system, which will allow to achieve the results of identification of legitimate users by keystroke dynamics on the available software product with a specified accuracy.

Objectives of the experiment are to determine the list of variable parameters of the identification system, the ranges of their changes in the experiment; receive user identification results for each value of the identification system parameters from the selected range; and make a choice of the system parameters in accordance with the purpose of the experiment and the given limitations.

The investigated parameters of the identification system based on the naive Bayesian classifier include the number of intervals for dividing the array of time characteristics of the user's keyboard handwriting (this parameter is associated with the chosen approach based on the naive Bayesian classifier) and the size of the identification data array (this parameter can refer to any system of dynamic biometric authentication). Previous experiments have shown that it makes no sense to increase the number of intervals of the partitioning more than six [9, 10].

In the course of an earlier study on an existing software product for a fixed number of intervals of four, the minimum size of the dataset for identification was obtained [9, 10]. In the course of the study, it was found out that when the size of the identification dataset is more than 0.29% of the total data array, the error probability will not exceed 3%. When the size of the identification data array is 0.13%, the number of identification errors does not exceed 10% for eight users. So the experiment will be carried out for the size of the identification data file from 0.01 to 0, 4% of the total data array in order to find on this interval the minimum value of the size of the identification data text, at which the error probability will not exceed 5%. The experiment for a wider range of the size of the array of identification data will not be conducted due to the inexpedient length of the experimental data. Thus, the experiment will consist of collecting information on the number of user identification errors on the keyboard for each interval within the specified limits (from two to six with subsequent expansion during the experiment), for each specific size of the identification data array (from 0.01% up to 0.4% of the total dataset), and for each of the eight users on a sample of one hundred unique identification sets. The choice of the parameters of the identification system will be carried out in such a way as to ensure a given identification accuracy and a minimum size of the identification data array.

4.2 Database for the Experiment

The database used to store and process data of the user's keystroke dynamics consists of the following tables: *users* (user information), *keyss* (information about keystrokes), *seans* (data about the session), *keyss_time* (duration of keystrokes), and *keyss_intervals* (intervals between keystrokes). The database schema is shown in Fig. 1.

At the moment, the database stores information about the time characteristics of eight users. The average number of keystrokes for all users is 83,245. The software that is available allows you to recreate text typed by users based on data stored in

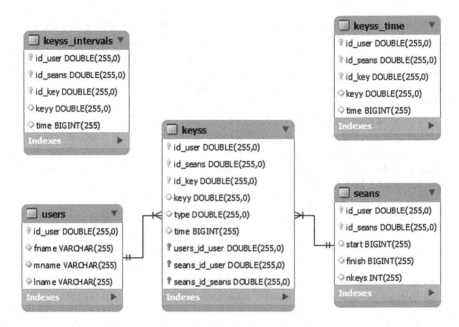

Fig. 1 The database schema

the database. So, for example, the recreated typed text of one of the users takes 50 pages and contains more than 10,000 words.

4.3   Results of the Experiment

In the framework of this work, it was necessary to investigate the influence of the identification system parameters on the accuracy of identification, that is, to show the dependence of the accuracy of user identification from the number of partition intervals and the size of the identification dataset or to prove that it does not exist. Based on the obtained results on the probabilities of user identification errors, graphs of the dependence of the probability of user identification errors from the size of the identification data array for each fixed interval of the array of time characteristics were constructed. A typical example of dependency for users 1 and 2 is shown in Fig. 2.

Based on the results of the experiment and the plotted dependencies, it was noted that with increasing the size of the data array for identification, a stable low value of the probability of identification error is achieved when the array of temporal characteristics is divided into two intervals. It is important to note that for different users, there is a different dependence of the probability of identification errors on the investigated parameters (the rate of decrease of the values described by the curves).

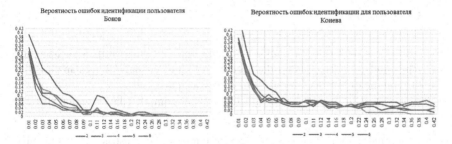

Fig. 2 Identification accuracy dependencies on the volume of identifying text for users 1 and 2

Table 1 The average values of the time intervals between keystrokes and their hold, as well as their variance, users 1 and 2

User	Mean between push	std between push	Mean of push	std of push
1	178,1	173,9	106,7	72,0
2	203,3	216,1	111,0	78,4

Table 2 The final average of the minimum amount of identifying data providing accuracy of identification is not worse than 0.95% for each of the users. The volume of the text—in fractions of the total

User	Text volume (%)	User	Text volume (%)
1	0,14	5	0,09
2	0,14	6	0,06
3	0,05	7	0,06
4	0,07	8	0,05

This dependence is due to the stability of the generated keystroke dynamics for users. So, for users 1 and 2, the average values of the time intervals between keystrokes and their hold, as well as their standard deviations are presented in Table 1.

It can be seen that the user's keyboard handwriting 1 is more established (greater typing speed, smaller time difference spread), which allows achieving greater identification accuracy with smaller text volumes, and also providing greater identification stability.

The final average of the minimum amount of identifying data providing accuracy of identification is not worse than 0.95% for each of the users, which are presented in Table 2.

It can be seen that to ensure the accuracy of 0.95 identification on the used database of examples, a maximum of 0.14% of the total text of the user text was sufficient, which in absolute values was about 120 characters of the text, which makes it possible to speak of the applicability of the approach under consideration at relatively small volumes of input text.

5 Conclusion

As a part of the work done, a study of the dependence of the identification accuracy from the number of intervals for the decomposition of the naive Bayesian classifier and the volume of text used for identification was made.

As a result of the experiment, conducted on a pre-compiled database of texts from eight users, it was found as follows:

- The optimal number of intervals for partitioning in the construction of the naive Bayesian classifier turned out to be 2. In the future, this fact allows us to speak of the applicability of the interval partition not only strictly in half but also on the basis of the probability distribution function, when as point for dividing the interval into two parts, observed value parameter will be used. The probability for the intervals on the left and right of this point will be calculated on the basis of the empirical probability distribution function;
- The dependence of the accuracy of identification from the keystroke dynamics is shown experimentally; it is shown that for users with unstable handwriting, the final accuracy estimates are also less stable;
- It has been shown experimentally that in order to achieve significant indicators of the accuracy of identification (in the present paper, 0.95), we must not use a colossal amount of text. So, for eight users, to achieve this accuracy, it turned out to be sufficient about 120 characters of text, which makes it possible to talk about the possibility of practical application of the method on relatively small texts, a decision can be made within a single paragraph.

In future, it is planned to conduct research on an extended database with a large number of users, and also to consider the practical applicability of constructing a naive Bayesian classifier on the basis of the distribution function for this task.

Acknowledgements The results were obtained as part of the implementation of the basic part of the state task of the Russian Federation Education and Science Ministry, project 8.9628.2017/8.9

References

1. Teh, P.S., Teoh, A.B.J, Yue, S.: A survey of keystroke dynamics biometrics. Sci. World J. **4082** (2013)
2. Morales, A., Fierrez, J., Ortega-Garcia, J.: Towards predicting good users for biometric recognition based on keystroke dynamics. In: Proceedigns of European Conference on Computer Vision Workshops, vol. 8926, pp. 711–724. Springer LNCS, Zurich, Switzerland (2014)
3. Mondal, S., Bours, P.: A computational approach to the continuous authentication biometric system. Inf. Sci. **304**(20), 2853 (2015)
4. Morales, A., Luna, E., Fierrez, J.: Score normalization for keystroke dynamics biometrics. In: Conference paper of 49th International Carnahan Conference on Security Technology, At Taipei (Taiwan), p. 7 (2015)
5. Bours, P.: Continuous authentication using biometric keystroke dynamics. In: The Norwegian Information Security Conference, pp. 1–12 (2015)

6. Brizan, D.G., Goodkind, A., Koch, P., Rosenberg, A.: Utilizing linguistically-enhanced keystroke dynamics to predict typist cognition and demographics. Int. J. Hum.-Comput. Stud. **82**, 5768 (2015)
7. Dora, R.A., Schalk, P.D., McCarthy, J.E., Young, S.A.: Remote Suspect Identification and the impact of demographic features on keystroke dynamics. In: Proceedings of the SPIE, Cyber Sensing, vol. 8757 (2013)
8. Naive Bayes classifier (2011). http://www.ic.unicamp.br/~rocha/teaching/2011s2/mc906/ aulas/naive-bayes-classifier.pdf
9. Kadyrov R., Kostyuchenko E.: Investigation of the influence of the parameters of the identification system on the final accuracy in the free-text keystroke dynamics identification. In: Scientific Session of TUSUR-2016: Materials of the International Scientific and Technical Conference of Students, Graduate Students and Young Scientists, Tomsk, 25–27 May 2016, vol. 5, pp. 49–51. B-Spectrum, Tomsk (2015)
10. Kadyrov, R., Kostyuchenko, E.: Determination of the minimum amount of input data for correct identification by free-text keystroke dynamics. In: Scientific Session of TUSUR-2016: Materials of the International Scientific and Technical Conference of Students, Graduate Students and Young Scientists, Tomsk, 25–27 May 2016, vol. 5, pp. 51–54. B-Spectrum, Tomsk (2015)

Correlation Criterion in Assessment of Speech Quality in Process of Oncological Patients Rehabilitation After Surgical Treatment of the Speech-Producing Tract

Evgeny Kostyuchenko, Roman Meshcheryakov, Dariya Ignatieva, Alexander Pyatkov, Evgeny Choynzonov and Lidiya Balatskaya

Abstract Within this work, the application of the criterion based on correlation coefficient for comparative assessment of speech quality in speech rehabilitation for patients after surgical treatment of speech-producing tract oncological diseases is considered. The sequence of the actions intended for receiving comparative assessment of the speech quality by comparison of the sound recordings made before and after operation is considered. As a material for the assessment, a set of syllables from Standard GOST 50840-95 Speech transmission over varies communication channels. Techniques for measurements of speech quality, intelligibility, and voice identification are used. Also, a set of syllables, compiled on the basis of the analysis of most prone to postoperative change phonemes, is used. The previously proposed criteria based on the comparison of the intensity of time-normalized syllable records spectra have a fundamental drawback. They need an additional normalization of signal power. The proposed approach, based on the use of the linear correlation coefficient as a measure of similarity, does not have this drawback. The comparability of the values received using new criterion with the values received using previous version of the criteria is shown. Results of comparison confirm the possibility of the new criterion practical use.

Keywords Assessment model · Speech quality · Cancer · Speech rehabilitation
Quality criteria · Correlation

E. Kostyuchenko (✉) · R. Meshcheryakov · D. Ignatieva
A. Pyatkov · E. Choynzonov · L. Balatskaya
Tomsk State University of Control Systems and Radioelectronics,
Lenina str. 40, 634050 Tomsk, Russia
e-mail: key@keva.tusur.ru
URL: http://www.tusur.ru

E. Choynzonov · L. Balatskaya
Tomsk Cancer Research Institute, Kooperativniy av. 5,
634050 Tomsk, Russia
URL: http://www.oncology.tomsk.ru/

© Springer Nature Singapore Pte Ltd. 2019
S. K. Bhatia et al. (eds.), *Advances in Computer Communication and Computational Sciences*, Advances in Intelligent Systems and Computing 759,
https://doi.org/10.1007/978-981-13-0341-8_19

1 Introduction

Currently, the number of people diagnosed with oral and pharyngeal cancer is constantly increasing. Each year about 13,000 new cases are diagnosed in the country. Total number of patients with cancer of oral and pharyngeal is currently more than 54,000 [1, 2]. Most cases require surgical intervention, after which there is a need for speech rehabilitation. Now, rehabilitation occurs by subjective assessment of speech quality by the treating doctor or speech therapist working with the patient. In view of the subjectivity of such assessment, it is difficult to talk about the accuracy of the evaluation and the description of the dynamics of speech restoration. However, it is the dynamics of quality that reflects the progress of rehabilitation in the best way. Therefore, there is a need to develop a criterion for assessing the quality of speech, which will later be used in an automated system for assessing the quality of speech in the rehabilitation of patients. One of the tasks is to minimize the participation of the speech therapist in the evaluation process. The ultimate practical goal of the work is to shorten the time of rehabilitation based on objective speech quality assessing using. This requires the development of a quantitative criterion for assessing the quality of pronunciation of various phonemes and syllables. There are no completed works in this direction at the world level, and localization of the Russian language would impose its specifics even in the case of their existence [3].

2 Current State of the Problem

As the first step toward a possibility of receiving quantitative assessment of quality of the speech was used the Standard GOST 50840-95 "Speech transmission over varies communication channels. Techniques for measurements of speech quality, intelligibility and voice identification" [4]. This approach, adapted to speech rehabilitation [5], allows to receive quantitative assessment, however, for correct use; it demands participation of several speech therapists (according to the recommendation of the Standard 5 people). In practice, as a rule, only one speech therapist participates in rehabilitation's process. Eventually, we have a subjectivity of the received results. Besides, this approach demands essential participation of the speech therapists in the process of assessment of quality of pronouncing a significant number of syllables (according to the Standard, 250 syllables are recommended) that also in practice is not convenient. According to this drawback, the task of formation of the objective automatically marked out criterion for evaluation of the speech quality is set.

Within earlier researches, a set of criteria has been offered [6], allowing to decide to achieve the objectives:

- the ratio of average distance between initial (before operation) and modified (after operation) signals to average distance between initial signals;
- the ratio of average distance between the initial and modified signals to average distance between signals of one type;
- the ratio of the minimum distinction between the initial and modified signals to the maximum distinction between initial signals;
- the proportion of pairs of initial signals, the distance between which exceeds the minimum distance between pairs of initial and modified signals, and the proportion of pairs of initial and modified signals, the distance between which is less than the maximum between pairs of initial signals.

The mathematical record of these criteria is presented in [6]. As a measure, the Minkowskian was used.

However, in the process of calculation of values, there are operations of a normalization of a signal, one of which (normalization on time) at this stage demands expert's participation and the following stage of the conducted researches will be devoted to automation of its carrying out. During process of assessment, there is a comparison of spoken syllables sets records before and after operation. The preoperative record was chosen as the reference, since the influence of the tumor on syllabic intelligibility of speech is insignificant most cases (at the studied diseases and stages), and the variability between several speakers is more essential. At this stage, the problem of elimination of the second normalization (normalization on power) is solved that will allow to simplify the procedure of spoken syllables quality assessment.

3 Model for Assessing the Quality of Speech

The mathematical model as a black box is presented in Fig. 1.
    System input:

- List of recorded syllables either according to GOST R 50840-95 or according to the compiled list based on speech modifications depending from surgery volume and localization [7];
- list of estimated phonemes;

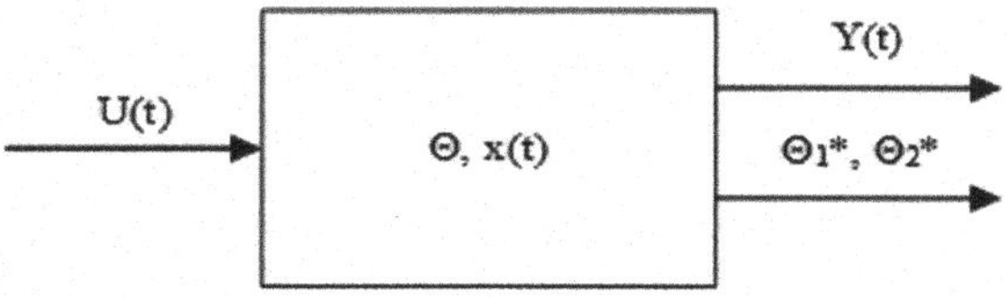

Fig. 1 Model of speech quality assessment as a black box

- Fourier spectrum parameters (the full or upper half of the Fourier spectrum, the length of the used window, the overlap value of adjacent signal segments, and the dimension of the discrete Fourier transform);
- distance metric parameter (if necessary, for example, a parameter for the Minkowskian metric); and
- values for the normalization of time and intensity, if it is necessary.

The output data is an array of numbers, depending on the input data and the calculation technique. Also, the output data is those parameters that were selected in the calculation of assessments, namely the parameters of Fourier spectrum and the parameter of metric. The designations for input and output parameters are as follows:

- $U(t) = \{u_1(t), u_2(t)\}$ the input data and the input influences:
- $u_1(t)$ preoperative records of syllables;
- $u_2(t)$ records of syllables after operation.
- $Y(t) = \{y_1(t)\}$ the output data:
- $y_1(t)$ quantitative estimates of quality of pronouncing phonemes.
- $x(t)$ internal variables that appear during the execution of the algorithm.
- $\theta = (\theta_1, \theta_2, \theta_3, \theta_4, \theta_5, \theta_6, \theta_7)$ system's parameters (do not change during time of modeling of system):
- θ_1 Fourier spectrum parameters;
- θ_2 distance metric parameter;
- θ_3 parameter selection criterion;
- θ_4 list of syllables;
- θ_5 list of estimated phonemes;
- θ_6 Enorm intensity normalization parameter;
- θ_7 Tnorm-time normalization parameter;
- $\theta_1{}^*$ selected Fourier spectrum parameters.
- $\theta_2{}^*$ selected distance metric parameter.

The model after decomposition, which essentially describes the sequence of actions in assessing the quality of speech within the developed approach, is presented in Fig. 2.

The model and the sequence of actions are described in more detail in [7]. Within this work, one of the stages of the model is intensity normalization. When this normalization occurs, the question arises of the choice of the reference intensity, which is not critical—we select and normalize by a preassigned value or by the intensity of the first reference signal. However, this problem can be avoided altogether if a criterion based on the correlation coefficient is used instead of normalization. In this case, the normalization by intensity is not required, the difference in the volume level is compensated for by the properties of the correlation coefficient.

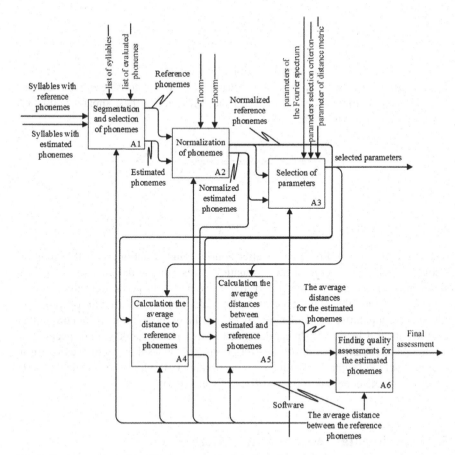

Fig. 2 Decomposition of model of assessment of quality of pronouncing phonemes

4 Correlation Criterion for Assessing the Quality of Spoken Phonemes

To get the value of the correlation criterion, the values of the time-normalized intensities of Fourier spectra before and after the operation are used. In this case, the value can be calculated both by individual phonemes and by whole syllables. Mathematically, the value of the correlation coefficient can be written as [8]

$$r_{XY} = \frac{cov_{XY}}{\sigma_X \sigma_Y} = \frac{\sum (X - \bar{X})(Y - \bar{Y})}{\sqrt{\sum (X - \bar{X})^2}\sqrt{\sum (Y - \bar{Y})^2}} \tag{1}$$

or the same, but is better for calculation

$$r_{XY} = \frac{n \sum XY - \sum X \sum Y}{\sqrt{n \sum X^2 - (\sum X)^2}\sqrt{n \sum Y^2 - (\sum Y)^2}} \tag{2}$$

where X, Y—arrays of data for correlation calculation; n—their length.

In relation to the two-dimensional spectrum of the signal considering the realization of its calculation, on the basis of this value, a criterion of the comparative quality of the speech can be formed, written as

$$K = \frac{n_t n_f \sum_t \sum_f S_0 S_1 - (\sum_t \sum_f S_0)(\sum_t \sum_f S_1)}{\sqrt{n_t n_f \sum_t \sum_f S_0^2 - (\sum_t \sum_f S_0)^2}} \cdot$$
$$\cdot \frac{1}{\sqrt{n_t n_f \sum_t \sum_f S_1^2 - (\sum_t \sum_f S_1)^2}} \tag{3}$$

where n_t is the count of time points after Fourier transformation; n_f is the count of frequency values after Fourier transformation; S_0 is a signal before surgery; S_1 is a signal after surgery; t is a time; f is a frequency; and K is a criterion value.

The closer the value of this criterion to 1, the closer the signal after the operation to reference preoperative signal. Obviously, due to the purpose of the linear correlation coefficient for detecting the linear dependence, its value is stable with respect to the multiplication operation. This leads to the absence of the need for normalization from the values of total signal power at the considered interval (phoneme or syllable—can be used scalable coders for different sizes of signal [9]).

5 Experimental Testing of the Criterion

By analogy with [6], testing was conducted in two stages: on a model signal simulating dysfunction when using a language by a healthy speaker and on real patient records before and after the operation. The need to use model signals is due to the fact that not all patients after the operation remain to the procedure of speech rehabilitation; they prefer to pass it at the place of residence. This leads to the fact that it turns out to make only records before the operation and at discharge (but before even basic exercises are set by the speech therapist).

In this case, the quality of the pronunciation after the operation is very low and obvious both by ear and by application of the considered criteria. This leads to the need to use records at least after one session with a speech therapist. Records allowing to evaluate the dynamics of rehabilitation are even less (for this, it is necessary to complete the full course of speech rehabilitation) and this study will be conducted after the accumulation of more patients records with the same stage of the disease and the full course of speech rehabilitation.

The obtained values of the syllable intelligibility by GOST 50840-95, previous criterion, and the proposed criterion (3) for model of patients 1, 2, and 3 are presented

Table 1 Mean criteria values for model patients before and after surgery (left column for speech before surgery, right column for speech after surgery)

Patient	Syl.Int. more = better		Prev.Crit more = worse		Prop.Crit more = better	
1	1	0.6	97.45	126.91	0.62	0.52
2	1	0.8	131.12	165.27	0.51	0.44
3	1	0.6	118.73	144.14	0.55	0.49

Table 2 Mean criteria values for real patients before and after surgery (left column for speech before surgery, right column for speech after surgery)

Patient	Syl.Int. more = better		Prev.Crit more = worse		Prop.Crit more = better	
1	1	0.6	127.41	141.54	0.56	0.48
2	1	0.6	115.21	139.28	0.56	0.51
3	1	0.6	141.81	165.53	0.43	0.35

in Table 1 (left column for speech before surgery, right column for speech after surgery).

The same information for real patients is presented in Table 2.

Both before and after the operation, five records were made.

6 Discussion

First, the estimates obtained for all three options for assessing the quality of speech do not contain contradictions, which confirms their consistency. Second, the evaluation of syllable intelligibility was obtained from the results of a comparison of only five signals, which reduces the confidence in the presented value. To increase accuracy, it is necessary either to increase the number of each syllable implementation or to supplement the sample with implementations of other syllables. However, this will lead to the availability of additional information in comparison with the other two criteria, which will affect the correctness of results comparison. The values for the other two criteria are also obtained on five signals, but on the basis of their pairwise processing, which leads to 10 values for comparison within one class (for example, for speech before the operation) and 25 values when comparing between classes (before and after operations). Third, the obtained value does not allow us to speak of the complete superiority of the criterion in question. The results are comparable, and the advantage of the proposed criterion is comparative simplicity. This is due to the lack of signal normalization by power. In addition, this value can be used as one of a variety of criteria values when implementing an integrated approach to assessing the quality of phoneme pronunciation.

7 Conclusion

In this paper, we considered the application of the correlation criterion for assessing the phoneme pronunciation quality. This task is actual in the development of methods for assessing the patient speech quality after surgery on the organs of the speech-forming tract in the treatment of cancer. This technique, in turn, will reduce the time to speech rehabilitation through the use of objective estimates obtained in automatic mode. The obtained results make it possible to speak about the consistency of the proposed criterion in comparison with the criteria developed earlier. The proposed criterion is simpler due to the lack of the power normalization need. The received criterion can be used at creation of a complex criterion for estimation of phonemes and syllables pronunciation quality, based on the analysis of several different parameters. As the next step of the study, automation of the temporal normalization of compared signals and the development of a complex comparative criterion for assessing the quality of phonemes and syllables will be considered.

Acknowledgements The study was performed by a grant from the Russian Science Foundation (project 16-15-00038). Special acknowlededgments to Professor Mikhail B. Stolbov (ITMO) for idea of using correlation values in our task.

References

1. Kaprin, A.D., Starinskiy, V.V., Petrova, G.V.: Status of Cancer Care the Population of Russia in 2014. MNIOI name of P.A. Herzen, Moscow (2015)
2. Kaprin, A.D., Starinskiy, V.V., Petrova, G.V.: Malignancies in Russia in 2014 (Morbidity and mortality). MNIOI name of P.A. Herzen, Moscow (2015)
3. Ronzhin, A.L., Karpov, A.A.: Russian voice interface. Pattern Recogn. Image Anal. 17(2), 321–336 (2007)
4. Standard GOST 50840-95. Speech transmission over varies communication channels. Techniques for Measurements of Speech Quality, Intelligibility and Voice Identification. Publishing Standards, Moscow (1995)
5. Balatskaya, L.N., Choinzonov, E.L., Chizevskaya, T.Y., Kostyuchenko, E.Y., Meshcheryakov, R.V.: Software for Assessing Voice Quality in Rehabilitation of patients after surgical treatment of cancer of Oral Cavity, Oropharynx and Upper Jaw. Speech and computer. SPECOM 2013. Lect. Notes Comput. Sci. 8113, 294–301 (2013)
6. Kostyuchenko, E.Y., Meshcheryakov, R.V., Ignatieva, D.I., Pyatkov, A.V., Choinzonov, E.L., Balatskaya, L.N.: Evaluation of the speech quality during rehabilitation after surgical treatment of the cancer of oral cavity and oropharynx based on a comparison of the Fourier spectra. Speech and computer. SPECOM 2016. Lect. Notes Comput. Sci. 9811, 287–295 (2016)
7. Kostyuchenko E.Y., Ignatieva D.I, Meshcheryakov R.V., Pyatkov A.V., Choinzonov E.L., Balatskaya L.N.: Model of system quality assessment pronouncing phonemes. Dyn. Syst. Mech. Mach. (Dynamics) 2016 (2016)
8. Pearson, K.: Notes on regression and inheritance in the case of two parents. Proc. R. Soc. Lond. 58, 240–242 (1895)
9. Petrovsky, A.A., Petrovsky, A.A.: A scalable speech and audio coders based on adaptive time-frequency signal analysis. SPIIRAS Proc. 50(1), 55–92 (2017)

Compressive Sensing Classifier Based on K-SVD

Xiaohua Xu, Baichuan Fan, Ping He, Yali Liang, Zheng Liao and Tianyu Jing

Abstract In recent years, compressive sensing has attracted considerable attention in many fields including medical diagnosis. However, due to the small number of medical samples, traditional methods might not be able to classify the medical data very well. In this paper, we propose a novel classification model based on K-SVD named Cluster-KSVD, which can obtain orthogonal dictionary from over-complete dictionary. With the reduced dictionary, we apply the sparsified features for classification, such as medical diagnosis. The sufficient experimental results demonstrate that our model has superior classification performance than the state-of-the-art classifiers.

Keywords Classification · K-SVD · Compressive sensing

1 Introduction

Recent research on the classification of medical data has attracted the attention of a large number of researchers. Traditional diagnosis is mainly made by doctor's subjective judgment but the diagnosis made by different doctors could be different, especially in a difficult case.

By far, various traditional machine learning methods, such as neural networks [1], decision trees [2], and support vector machines [3], are used to deal with medical data classification problems. Due to the inadequate medical samples and strong noise, these methods might hardly classify the medical data very well.

To overcome this problem, compressive sensing theory [4, 5] and several other related methods are proposed in recent years. Farouk et al. [6] proposed a method based on K-SVD using improved OMP algorithm to optimize adaptive dictionary, so that medical images can be denoised. Yang et al. [7] proposed a new method to enhance medical image resolution. Besides, Salman et al. [8] developed an automatic

X. Xu · B. Fan · P. He (✉) · Y. Liang · Z. Liao · T. Jing
Department of Computer Science, Yangzhou University, Yangzhou, China
e-mail: angeletx@gmail.com; arterx@gmail.com

© Springer Nature Singapore Pte Ltd. 2019
S. K. Bhatia et al. (eds.), *Advances in Computer Communication and Computational Sciences*, Advances in Intelligent Systems and Computing 759,
https://doi.org/10.1007/978-981-13-0341-8_20

framework for brain tumor classification and segmentation. However, insufficient number of in medical samples is still a great problem for over-complete dictionary in those K-SVD methods.

In this paper, we propose a novel Cluster-KSVD model on the basis of K-SVD [9–11]. It solves the small sample problem by replacing over-complete dictionary with orthogonal dictionary. In addition, we improve the alternating direction method of multipliers (ADMM) [12] for the objective function optimization. After the dictionary learning, we use the extracted features to classify medical data with some similarity-based methods.

The benefits of our methods lie in that the redundancy of over-complete dictionary could be removed and the learned dictionary is orthogonal. When dealing with insufficient medical samples, our proposed algorithm is capable of improving the classification accuracy of medical datasets and assisting medical diagnosing.

The rest of this paper is organized as follows: we review the related works on K-SVD in Sect. 2. Section 3 presents the compressive sensing model based on K-SVD. Section 4 describes the Cluster-KSVD algorithm, and Cluster-ADMM algorithm we propose. In Sect. 5, we discuss the experiments of our algorithm in comparison with other state-of-the-art methods on medical datasets. Section 5 concludes the whole paper.

2 Compressive Sensing Theory

Compressive sensing theory was proposed in the early years. The premise of compressive sensing theory is that the sparse representation of signal exists. Suppose there is an n-dimensional signal f, it could be formed by an orthogonal dictionary Ψ as

$$f = \Psi x \tag{1}$$

where x is an n-dimensional vector and the sparse representation of f.

For a signal f that can be sparse represented, given a observed matrix Θ that is independent of Ψ, we can obtain the m-dimensional $(m \ll n)$ compressed linear projection $y = \Theta f$. When we set $\Phi = \Theta \Psi$, it could be rewritten as

$$y = \Theta \Psi x = \Phi x \tag{2}$$

where $\Phi \in \mathbf{R}^{m \times n}$ is a sensing matrix and y could be regarded as the observed vector of sparse signal x to Φ. If Φ satisfies restricted isometry property, we can reconstruct the x by observed vector y using known algorithm and then restore y.

The core of compressive sensing theory is signal reconstruction method. Many scholars have proposed reconstruction algorithms that could find a suboptimal solution, including ADMM algorithm.

3   Cluster-KSVD

3.1   Compressive Sensing Model Based on K-SVD

In the classification problem based on compressive sensing, the optimization problem is $\min\|x\|_1$ s. t. $y = \Phi x$, where Φ is an over-complete dictionary consists of n training samples. We could find an optimal dictionary $D \in \mathbf{R}^{m \times k}$ to sparsify Φ using K-SVD and the optimization problem is as follows:

$$\min_{D,X} \|\Phi - DX\|_F^2 \text{ s. t. } \|\chi_i\|_0 \leq T_0 \tag{3}$$

where the number of nonzero entries T_0 has to be small, and $X = [\chi_1, \chi_2, \ldots, \chi_n]$ is the sparse representation of Φ.

K-SVD updates the dictionary by column and uses two-step iteration to solve the optimization problem:

At sparse code update step, suppose that dictionary D is known and obtain X by pursuit algorithm. Then, minimization problem can be decomposed as

$$\min_{\chi_i} \|\phi_i - D\chi_i\|_2^2 \text{ s.t. } \|\chi_i\|_0 \leq T_0 \tag{4}$$

At dictionary update step, suppose that coefficient X and dictionary D are fixed, only update the kth column d_k of the new dictionary D, and the kth row in sparse matrix X, that d_k is multiplied by, is χ_T^k. Then, the objective function can be rewritten as

$$\|\Phi - DX\|_F^2 = \left\| \Phi - \sum_{i=1}^{K} d_i \chi_T^i \right\|_F^2 = \left\| \left(\Phi - \sum_{i \neq k} d_i \chi_T^i \right) - d_k \chi_T^k \right\|_F^2 = \left\| E_k - d_k \chi_T^k \right\|_F^2 \tag{5}$$

where DX is decomposed as the sum of K rank $- 1$ matrices. Assuming that the $K - 1$ items of the K matrices are fixed, the other one is the kth column to be updated. The matrix E_k is the error for all the n samples without the kth atom d_k.

In order to keep the sparsity of χ_T^k, we only keep the nonzero position of E_k and obtain E_k^R. Afterward, we use SVD on E_k^R, and update d_k and χ_T^k.

Therefore, the compressive sensing model based on K-SVD dictionary learning is

$$\min\|x\|_1 \quad \text{s.t.} \quad y = Dx \tag{6}$$

In order to improve K-SVD, we propose Cluster-KSVD, which mainly improves the dictionary update step as follows:

(1) Cluster the sparse coefficient matrix X. The cluster rule is that the same sparse mode is shared in all the clusters. τ is the cluster row index after clustering,

C is the number of clusters, and τ_i is the number of the rows belong to the ith cluster;

(2) Use row transformation on sparse matrix X, and column transformation on dictionary D, i.e., $X = X^\tau$, $D = D^\tau$, $\Phi = D^\tau X^\tau$;

(3) Update the new dictionary D by cluster instead of updating a column at a time, which reduces the times of SVD:

$$
\begin{aligned}
\|\Phi - DX\|_F^2 &= \left\| \Phi - \sum_{i=1}^{C} d_{\tau_i} \chi_T^{\tau_i} \right\|_F^2 \\
&= \left\| \left(\Phi - \sum_{i \neq c} d_{\tau_i} \chi_T^{\tau_i} \right) - d_{\tau_c} \chi_T^{\tau_c} \right\|_F^2 \\
&= \left\| E_{\tau_c} - d_{\tau_c} \chi_T^{\tau_c} \right\|_F^2
\end{aligned}
\tag{7}
$$

where we only keep the entries in E_{τ_c} that share the same position with nonzero entries in the product of d_{τ_c} and $\chi_T^{\tau_c}$ when updating d_{τ_c} and $\chi_T^{\tau_c}$. In the end, we use SVD on $E_{\tau_c}^R$.

3.2 Cluster-ADMM Algorithm

We improve the ADMM algorithm to solve Eq. 6, and name it alternating direction method of multipliers based on Cluster-KSVD (Cluster-ADMM).

According to ADMM, Eq. 6 can be written as

$$
\min f(x) + \|z\|_1 \quad \text{s.t} \quad x - z = 0
\tag{8}
$$

where $f(x)$ is an indicator function on the closed convex set $\{x \in \mathbf{R}^K | y = Dx\}$.

So the augmented Lagrange is

$$
L_\rho(x, z, u) = f(x) + \|z\|_1 + u^T (x - z) + \frac{\rho}{2} \|x - z\|_2^2
\tag{9}
$$

where u is a dual variable, and $\rho > 0$ is the penalty parameter of the augmented item.

Accordingly, the update rules of $x, z, u,$ respectively, are

$$
x^{k+1} = \arg \min_x L_\rho(x, z^k, u^k)
\tag{10}
$$

$$
z^{k+1} = \arg \min_z L_\rho(x^{k+1}, z, u^k)
\tag{11}
$$

$$
u^{k+1} = u^k + \rho(x^{k+1} - z^{k+1})
\tag{12}
$$

For $f(x)$ being the indicator function of a nonempty closed convex set, based on [13], the update rule of x^{k+1} can be rewritten as

$$x^{k+1} = \Pi(z^k - u^k/\rho) \tag{13}$$

where Π is the projection on $\{x \in \mathbf{R}^K | y = Dx\}$. The Euclidean norm minimization problem is involved in the update of x^{k+1}, and the update rule of x^{k+1} can be rewritten as

$$x^{k+1} = (I - D^T(DD^T)^{-1}D)(z^k - u^k/\rho) + D^T(DD^T)^{-1} \cdot y \tag{14}$$

For z^{k+1}, it is common to compute the closed-form solution using subdifferential calculus and add a soft-threshold operator S because $\|z\|_1$ is not differentiable. The soft-threshold operator S is formed as

$$S_K(a) = \begin{cases} a - \text{sgn}(a) \cdot K, & |a| > K \\ 0, & |a| \le K \end{cases} = \begin{cases} a - K, & a > K \\ 0, & |a| \le K \\ a + K, & a < -K \end{cases} \tag{15}$$

So the update rule of z^{k+1} is

$$z^{k+1} = S_{1/\rho}(x^{k+1} + u^k/\rho) \tag{16}$$

In order to simplify the form, we set a scaled dual variable $v = (1/\rho)u$, which gives the new update rules of x, z, v:

$$x^{k+1} = (I - D^T(DD^T)^{-1}D)(z^k - v^k) + D^T(DD^T)^{-1} \cdot y \tag{17}$$
$$z^{k+1} = S_{1/\rho}(x^{k+1} + v^k) \tag{18}$$
$$v^{k+1} = v^k + x^{k+1} - z^{k+1} \tag{19}$$

3.3 Classification

The optimization dictionary D obtained through Cluster-KSVD cannot directly give the classification. However, sparse matrix X can be treated as the sparse representation of the training sample on dictionary D and x can be treated as the sparse representation of the test sample on D so that we can classify the sample according to the similarity analysis between X and x.

Suppose all the samples have c clusters and the number of the training sample of the ith cluster is n_i, $i \in \{1, 2, \ldots, c\}$, so the training sample of the ith cluster makes the matrix $\Phi_i = [y_{i,1} \ y_{i,2}, \ldots, y_{i,n_i}]$, the sparse representation of matrix

Table 1 Dataset information

Dataset	Total number	Gene dimensions	Cluster name	Sample number
DLBCL	77	5469	DLBCL	58
			FL	19
Brain	90	5920	Medulloblastoma	60
			Malignant glioma	10
			AT/RT	10
			Normal cerebellum	4
			PNET	6

Φ_i on D makes the matrix $X^{(i)} = [X_{i,1}, X_{i,2}, \ldots, X_{i,n_i}]$. Then, the average sparse representation of the training samples of the ith cluster is

$$\overline{X^{(i)}} = \sum_{j=1}^{n_i} \chi_{i,j}/n_i \tag{20}$$

And the similarity between x and $\overline{X^{(i)}}$ can be written as

$$sim(x, \overline{X^{(i)}}) = \frac{\left|x^T \overline{X^{(i)}}\right|}{\|x\| \left\|\overline{X^{(i)}}\right\|} \tag{21}$$

Due to the higher similarity between the sparse representations of the sample of the same cluster, the cluster k is the cluster of $\overline{X^{(i)}}$ that bears the highest similarity to x, i.e.,

$$k = \arg\max_{i, 1 \leq i \leq c} sim(x, \overline{X^{(i)}}). \tag{22}$$

4 Experiments

In the experiments, we compared the performance of the Cluster-ADMM algorithm with the other three state-of-the-art methods: SVM, decision tree and KNN, on the task of classification of medical data in two commonly used cancer gene expression datasets: DLBCL and brain. The sample information of datasets is shown in Table 1.

We chose one sample as test sample and the rest samples as training sample. Then, we changed test sample and training samples until all the samples had been chosen as test sample to compute the accuracy of the classification.

Table 2 The classification accuracy on DLBCL

Methods	Dimensions		
	8	16	32
Cluster-ADMM	0.8182	0.9091	**0.9351**
SVM	**0.9610**	**0.9740**	**0.9351**
D-tree	0.8312	0.8052	0.8961
KNN	0.7662	0.7662	0.7922

Table 3 The classification time on DLBCL

Methods	Dimensions		
	8	16	32
Cluster-ADMM	0.277	0.483	0.731
SVM	0.468	0.281	0.219
D-tree	11.627	23.003	44.241
KNN	0.234	0.219	0.140

Table 4 The classification accuracy on brain

Methods	Dimensions		
	8	16	32
Cluster-ADMM	**0.8111**	**0.8889**	**0.8667**
SVM	0.7889	0.8111	0.8000
D-tree	0.7111	0.6667	0.6889
KNN	0.7333	0.7667	0.7222

Because the dictionary obtained by Cluster-KSVD changes, we did the classification experiments 100 times, respectively, on every dimension of data when datasets were reduced to 8 dimensions, 16 dimensions, and 32 dimensions. Then, we evaluated the classification accuracy and the average time of 100 runs on each dataset.

Considering different dimensionality reduction methods may have an impact on the classification, we compared the accuracy of classification by different methods and found that the accuracy is higher when using PCA + SVD.

In Tables 2 and 3, our proposed method reaches a high accuracy comparable to SVM, which is implemented in LIBSVM [14]. With the increase of dimension, the classification accuracy of Cluster-ADMM becomes higher. The performance of SVM is better because the LIBSVM can adjust the classifier during classification and the classification time of SVM is short.

In Tables 4 and 5, it is obvious that the accuracy of our method is better than that of the other three methods. The non-balanced distribution of sample in brain greatly affects the performance of results. The accuracy of Cluster-ADMM is the highest and the time is relatively short.

In general, the method we propose could effectively classify the medical data and have a relatively high accuracy compared to the other classification methods.

Table 5 The classification time on brain

Methods	Dimensions		
	8	16	32
Cluster-ADMM	0.436	0.545	0.907
SVM	0.593	0.686	0.312
D-tree	19.380	37.365	74.763
KNN	0.202	0.234	0.172

5 Conclusion

In this paper, we develop a novel classification model based on compressive sensing theory, named Cluster-KSVD. It is able to optimize the over-complete dictionary to make it orthogonal. We also incorporate an iterative optimization method, i.e., Cluster-ADMM, with Cluster-KSVD. This algorithm could extract compact features and classify samples based on similarity analysis method.

We compare the classification performance of Cluster-ADMM with other state-of-the-art classification algorithms, including support vector machine, on medical datasets. We demonstrate that our model can produce higher classification accuracy than the other classification algorithms.

Acknowledgements This research was supported in part by the Chinese National Natural Science Foundation under Grant nos. 61402395, 61472343, and 61502412, Natural Science Foundation of Jiangsu Province under contracts BK20140492, BK20151314, and BK20150459, Jiangsu overseas research and training program for university prominent young and middle-aged teachers and presidents, Jiangsu government scholarship funding, and practice innovation plan for college graduates of Jiangsu Province under contracts SJLX16_0591.

References

1. Khan, J., Wei, J.S., Ringner, M., et al.: Classification and diagnostic prediction of cancers using gene expression profiling and artificial neural networks. Nat. Med. **7**, 673–680 (2001)
2. Lavanya, D., Rani, K.U.: Performance evaluation of decision tree classifiers on medical datasets. Int. J. Comput. Appl. **26**, 1–4 (2011)
3. Statnikov, A., Aliferis, C.F., et al.: A Comprehensive evaluation of multicategory classification methods for microarray gene expression cancer diagnosis. Booinformatics **21**, 631–643 (2004)
4. Donoho, D.: Compressed sensing. IEEE Trans. Inf. Theory **52**, 1289–1306 (2006)
5. Candès, E.J.: Compressive sampling. In: Proceedings of the International Congress of Mathematicians, pp. 1433–1452. IEEE Press, Madrid (2006)
6. Farouk, R.M., Elsayed, M., Aly, M.: Medical image denoising based on log-Gabor wavelet dictionary and K-SVD algorithm. Int. J. Comput. Appl. **141** (2016)
7. Yang, J., Zhang, X., Peng, W., et al.: A novel regularized K-SVD dictionary learning based medical image super-resolution algorithm. Multimedia Tools Appl. **75**, 13107–13120 (2016)
8. Salman Al-Shaikhli, S.D., Yang, M.Y., Rosenhahn, B.: Brain tumor classification and segmentation using sparse coding and dictionary learning. Biomedizinische Technik Biomed. Eng. **61**, 413 (2016)

9. Aharon, M., Elad, M., Bruckstein, A.M.: The K-SVD: an algorithm for designing of over-complete dictionaries for sparse representations. IEEE Trans. Image Process. **54**, 4311–4322 (2006)
10. Aharon, M., Elad, M., Bruckstein, A.: K-SVD: design of dictionaries for sparse representation. Proc. SPARS **5**, 9–12 (2005)
11. Bruckstein, A.M., Donoho, D.L., Elad, M.: From sparse solutions of systems of equations to sparse modeling of signals and images. SIAM Rev. **51**, 34–81 (2009)
12. Boyd, S., Parikh, N., Chu, E., Peleato, B., Eckstein, J.: Distributed optimization and statistical learning via the alternating direction method of multipliers. Found. Trends Mach. Learn. **3**, 1–122 (2011)
13. Combettes, P.L., Pesquet, J.C.: Proximal Splitting Methods in Signal Processing. Springer, New York (2009)
14. Chang, C.C., Lin, C.J.: LIBSVM: A library for support vector machines. ACM Trans. Intell. Syst. Technol. **2**, 1–27 (2012)

Nonlinear Manifold Classification Based on LLE

Ping He, Tianyu Jing, Xiaohua Xu, Lei Zhang, Zheng Liao and Baichuan Fan

Abstract Classification is one of the most fundamental problems in machine learning and data mining. In this paper, we propose a novel nonlinear manifold classification algorithm based on a well-known manifold leaning method called Locally Linear Embedding (LLE). LLE is a classical manifold learning algorithm, which preserves the local neighborhood structure in low-dimensional space. On the basis of LLE, our algorithm incorporates the label information of training data into the manifold mapping, so that the transformed manifold becomes more discriminative than the original manifold. The incorporated label information can help increase the similarity of homogeneous data from the same class and decrease the similarity of heterogeneous data from different classes. Sufficient experimental results demonstrate that our method exhibits better classification performance over other well-known manifold classification algorithms on seven real-world datasets.

Keywords Manifold · Classification · LLE

1 Introduction

Classification is an important problem in machine learning and data mining. Great endeavor has been paid to the research of classification problems [1]. However, with the increase of data complexity, the traditional classification algorithms can hardly handle the data with high dimensionality. For this reason, researchers tried to simplify the original data by employing dimension reduction techniques before classification. In recent years, dimension reduction algorithms have shown pretty performance on the treatment of high-dimensional data [2, 3]. It not only decreases the space and time costs but also reveals the underlying data structure.

Manifold learning is a promising branch of nonlinear dimension reduction which assumes that the data of interest lie on a low-dimensional embedded nonlinear man-

P. He · T. Jing · X. Xu (✉) · L. Zhang · Z. Liao · B. Fan
Department of Computer Science, Yangzhou University, Yangzhou, China
e-mail: arterx@gmail.com; angeletx@gmail.com

© Springer Nature Singapore Pte Ltd. 2019
S. K. Bhatia et al. (eds.), *Advances in Computer Communication and Computational Sciences*, Advances in Intelligent Systems and Computing 759,
https://doi.org/10.1007/978-981-13-0341-8_21

ifold within the original feature space [4]. Many approaches have been developed under such manifold assumption, including Locally Linear Embedding (LLE) [5]. In LLE algorithm, the main idea is preserving the locally linear structures of the nearest neighborhood when mapping into a low-dimensional subspace as much as possible based on the local linear assumption.

In this paper, we proposed a Nonlinear Manifold Classification algorithm based on LLE (NMCL). Compared with traditional LLE that only aims at dimension reduction, NMCL is developed for classification by adapting the local linear weight matrix so that not only the discriminative information is enforced but also the local linear structure of data can be preserved. By incorporating the supervised information, NMCL modifies the local weight matrix to preserve the similarity of homogeneous data and expands the difference of the heterogeneous data simultaneously. The supervised version of LLE is Supervised LLE (SLLE) [6]. It integrates the label information into the neighborhood determination by redefining a distance that increases the between-class scatter. Local Sensitive Discriminant Analysis (LSDA) [7] constructs both inter-class graph and intra-class graph to characterize the discriminant and geometrical structure of data manifold, and then determines the linear transformation matrix by preserving the combined local neighborhood information. The extensive experimental results on the real-world datasets demonstrate the superiority of our proposed approach in comparison with SLLE and LSDA.

The rest of this paper is organized as follows: In Sect. 2, the procedure of NMCL is described in details. The experimental results on the artificial and real-world datasets are discussed in Sect. 3. In the end, Sect. 4 concludes the whole paper.

2 Locally Linear Embedding

Locally linear embedding is a nonlinear dimensionality reduction algorithm which is proposed by ST Roweis et.al [5]. LLE can preserve the original manifold local structure well after dimension reduction [8]. As a local optimization method, it adopts the linear reconstruction of local neighborhood to remain the locally linear geometry structure in low-dimensional manifold. LLE assumes that each data can be represented by its several neighbors. After reducing the dimension by LLE, the corresponding linear relations are preserved. Suppose that a dataset $\chi = \{x_1, x_2, \ldots, x_N\}$ is sampling from a high-dimensional nonlinear manifold. For arbitrary data point x_i, we can find its k neighbors x_i^j by using KNN methods. Next, we can calculate the reconstruction weight by the following function:

$$\min \varepsilon(W) = \sum_{i=1}^{N} \left\| x_i - \sum_{j=1}^{k} w_i^j x_i^j \right\|^2 \tag{1}$$

where w_i^j is the element of the reconstruction weight matrix W.

According to the assumption of LLE, the local geometrical structure of manifold remains unchanged after dimension reduction. The objective function to obtain the low-dimensional embedding Y is minimizing the loss function with the same linear reconstruction weight W.

$$\min \Phi(Y) = \sum_{i=1}^{N} \left\| y_i - \sum_{j=1}^{k} w_i^j y_i^j \right\|^2 \tag{2}$$

where y_i and y_i^j are the corresponding data of x_i and x_i^j in the low-dimensional embedding.

3 Nonlinear Manifold Classification Based on LLE

Inspired by traditional LLE algorithm, NMCL changes the local weight matrix by incorporating the supervised information. In this way, it not only can preserve the similarity of homogeneous data but also can expand the difference of the heterogeneous data. The details of NMCL will be described as follows.

3.1 Computing Local Weight Matrix

Because of the assumption of LLE, each point in the local patch of manifold can be denoted by linear representation of its neighbors and the connecting information between neighborhoods can be denoted by overlapping part. Furthermore, the linear relation stays the same after reducing the dimension. The coefficient of linear relation is called as reconstruction weight. In NMCL, we adopt the traditional LLE algorithm to compute the local linear reconstruction for dimension reduction.

The steps of computing local reconstruction weight matrix are as follows:

1. Compute the neighbors of each data point. Suppose that $\chi = \{x_1, x_2, \ldots, x_N\}$ is a high-dimensional dataset, where $x_i \in R^D$. Next, select k neighbors of x_i, where k is a parameter. Based on the manifold assumption, LLE adopts Dijkstra method to obtain the geodesic distance.
2. Compute the reconstruction weight matrix of each sample point. The objective function is

$$\min \varepsilon(W) = \sum_{i=1}^{N} \left\| x_i - \sum_{j=1}^{N_i} w_i^j x_i^j \right\|^2 \tag{3}$$

where $x_i^j \, (j = 1, 2, \ldots, N_i)$ is the jth neighbor of x_i, w_i^j is the local weight of x_i^j for linear reconstruction of x_i, and it satisfies the condition $\sum_{j=1}^{N_i} w_i^j = 1$. To solve Eq. (4), we need to first construct local covariance matrix C_i of x_i:

$$C_i(j, k) = \left(x_i^j - x_i\right)^T \left(x_i^k - x_i\right) \tag{4}$$

With Eq. (4) and $\sum_{j=1}^{N_i} w_i^j = 1$, the local reconstruction weight matrix can be obtained by using the Lagrangian multiplier method:

$$w_i^j = \frac{\sum_{k=1}^{N_i} C_i^{-1}(j, k)}{\sum_{p=1}^{N_i} \sum_{q=1}^{N_i} C_i^{-1}(p, q)} \tag{5}$$

3.2 Adjusting Linear Weight

To direct at the better classification result, NMCL incorporates the supervised information so that it can preserve the similarity of homogeneous data and expand the difference of the heterogeneous data. To increase local reconstruction weight matrix of intra-class data points and decrease the local reconstruction weight matrix of inter-class points simultaneously, we adjust the linear weight by means of incorporating the supervised information. $\forall w_i^j > 0$, it can be formulated as follows:

$$w_j^i \leftarrow \begin{cases} w_j^i + \delta & \text{if } x_i, x_j \text{ are in the same class} \\ w_j^i - \delta & \text{otherwise} \end{cases} \tag{6}$$

where δ is a parameter which controls the amount of distance adjustment. We can obtain δ through grid search. Because of the constraint $\sum_{j=1}^{N_i} w_i^j = 1$, we still need to normalize $\sum_{j=1}^{N_i} w_i^j = 1$ after weight adjustment.

By this way, NMCL preserves the similarity of homogeneous data and expand the difference of the heterogeneous data.

3.3 Embedding of Low-Dimensional Discriminative Manifold

On the basis of the adjusted local reconstruction weight matrix, then we embed the training data onto a low-dimensional discriminative manifold. NMCL preserves the local linear structure in high-dimensional space by minimizing the following reconstruction error as objective function:

$$\min \varepsilon(Y) = \sum_{i=1}^{N} \left\| y_i - \sum_{j=1}^{N_i} w_i^j y_i^j \right\|^2 \tag{7}$$

$\varepsilon(Y)$ is the loss function and y_i is the low-dimensional embedding of x_i. $y_i^j (j = 1, 2, \ldots, N_i)$ is the neighbor of y_i. In addition, it needs to satisfy two conditions:

$$\sum_{i=1}^{N} y_i = 0, \quad \sum_{i=1}^{N} y_i y_i^T = I \tag{8}$$

where I is an identity matrix.

3.4 Extension to Out-of-Sample Test Data

Suppose that X_S is the training data and X_T is the test data. On account of the lack of supervised information, we cannot apply test data to the method like training data directly. We define the kernel function $K_{LLE}(X_S, x)$ as the computation of the local weight matrix of X_S in the linear reconstruct of X_T on the original manifold. Meanwhile, let $\eta = Z_S \Lambda z^T$ be the relationship between the training data mapping Z_S and an arbitrary test data z (row-wise vector) on the low-dimensional manifold. By minimizing their difference of the training–test data relationship before and after the out-of-sample manifold mapping, we can obtain the optimal mapping of test data in the low-dimensional space. Assume that the optimal mapping of an arbitrary test data x_i is z^*, it can be formulated as follows [9]:

$$\begin{aligned} z_i^* &= \arg\min_z \left\| K_{LLE}(X_S, x_i) - Z_S \Lambda z_i^T \right\|^2 \\ &= K_{LLE}(X_S, x_i)^T Z_S \left(Z_S^T Z_S \right)^\dagger \Lambda^{-1} \\ &= K_{LLE}(X_S, x_i)^T Z_S \Lambda^{-1} \end{aligned} \tag{9}$$

3.5 Classification

NMCL employs linear support vector machine to classify the embedded low-dimensional data. Assume that the mapped data are linearly separable. We simply solve the optimized decision boundary as quadratic programming problems:

$$\min_{w,b} \frac{1}{2} \|w\|^2 \ s.t. \ y_i \left(w^T x_i + b \right) \geq 1, \quad i = 1, 2, \ldots, N \tag{10}$$

With Lagrangian multiplier method, we can solve Eq. (7):

Table 1 Comparison of classification accuracy among SLLE, LSDA, and NMCL

Datasets	SLLE (%)	LSDA (%)	NMCL (%)
Liver	83.23	86.25	**88.3**
Diabetes	73.96	79.41	**82.73**
Meningitis	72.98	73.42	**75.87**
Breast cancer	71.55	74.98	**75.25**
Dystrophy	83.72	84.58	**90.25**
Mammographic mass	70.22	71.35	**74.54**
VLBW infant	74.67	76.52	**78.62**

$$\min_{w,b,\zeta} \frac{1}{2}\|w\|^2 + c\sum_{i=1}^{l}\zeta_i \tag{11}$$

where relaxation $\zeta \geq 0$ and C is penalty parameter.

4 Experiments

We validate our NMCL by using seven real-world medical datasets, namely liver disorder dataset, diabetes dataset, acute bacterial meningitis dataset, breast cancer dataset, muscular dystrophy dataset, mammographic mass dataset, and very low birth weight (VLBW) infant dataset. SLLE and LSDA are compared on these datasets.

Figure 1 shows the procedures of NMCL of dataset of liver disorder (Hot colors represent normal samples; cold colors represent disordered samples). About this dataset, there are seven attributes in total which can be used to diagnose the alcoholic liver disease (ALD). Figure 1a shows the training data after normalized mapped, which distribute chaotically on the coordinate. And we can find a satisfied result of recognition in Fig. 1c. In Fig. 1b, there are out-of-sample data without labels (the colors on the out-of-sample data are used to evaluate the experimental results). Finally, we can observe that the result of recognition of testing data also performs pretty well. In addition, it shows us visual procedure which is helpful to understand and make further analysis.

Similarly, we demonstrate the superiority of our approach over SLLE and LSDA in respect of the classification accuracy on the datasets given in Table 1, which proves the effectiveness of our method.

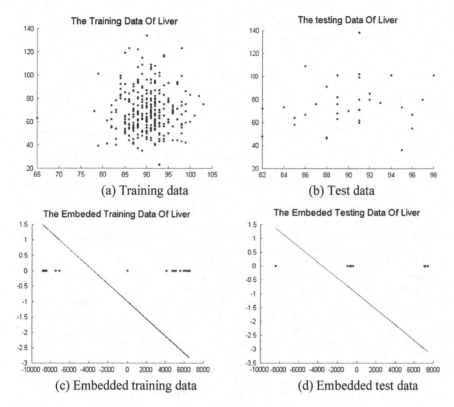

(a) Training data   (b) Test data

(c) Embedded training data   (d) Embedded test data

Fig. 1 The experimental results of NMCL on liver disorder dataset

5 Conclusions

In this paper, we propose a nonlinear manifold classification algorithm based on Local Linear Embedding (LLE). Traditional LLE manifold learning algorithm is unsupervised dimension reduction method, thus cannot take full advantage of the label information for high-dimensional and complex data. By taking leverage of training data label information, NMCL increases the within-class data similarity and enlarges the difference of the inter-class data. The discrimination structure of training data is also extended to test data by the our-of-sample extension strategy. To evaluate the performance of our method, we compare NMCL with other state-of-the-art manifold classification algorithms on multiple real-world medical datasets. Promising experimental results demonstrate the effectiveness of our proposed method.

Acknowledgements This research was supported in part by the Chinese National Natural Science Foundation under Grant nos. 61402395, 61472343, 61379066, and 61502412, Natural Science Foundation of Jiangsu Province under contracts BK20140492, BK20151314, and BK20150459, Jiangsu overseas research and training program for university prominent young and middle-aged teachers and presidents, Jiangsu government scholarship funding, and practice innovation plan for college graduates of Jiangsu Province under contracts SJLX16_0591.

References

1. Nasrabadi, N.M., Bishopr, C.M.: Pattern Recognition and Machine Learning. Academic Press (2006)
2. Abdi, H., Williams, L.J.: Principal component analysis. Wiley Interdiscip. Rev. Comput. Stat. **2**(4), 433–459 (2010)
3. Qu, T., Cai, Z.: A fast multidimensional scaling algorithm. In: IEEE International Conference on Robotics and Biomimetics, pp. 2569–2574 (2015)
4. Camastra, F., Vinciarelli, A.: Feature extraction methods and manifold learning methods. In: Machine Learning for Audio, Image and Video Analysis. Springer, London (2015)
5. Roweis, S.T., Saul, L.K.: Nonlinear dimensionality reduction by locally linear embedding. Science (New York, N.Y.) **290**(5500), 2323–2326 (2000)
6. Ridder, D.D., Kouropteva, O., Okun, O., Pietikäinen, M., Duin, R.P.W.: Supervised locally linear embedding. In: Artificial Neural Networks and Neural Information Processing—ICANN/ICONIP 2003. Springer, Berlin, Heidelberg (2003)
7. Liu, X.M., Deng, S.G., Yin, J.W., Li, C., Feng, Z.L., Dong, J.X.: Locality sensitive discriminant analysis based on matrix representation. J. Zhejiang Univ. (2009)
8. Daza-Santacoloma, G., Castellanos-Dominguez, G., Principe, J.C.: Locally linear embedding based on correntropy measure for visualization and classification. Neurocomputing **80**(2), 19–30 (2012)
9. Bengio, Y., Vincent, P., Delalleau, O., Roux, N.L., Ouimet, M.: Out-of-sample extensions for LLE, Isomap, MDS, Eigenmaps, and Spectral Clustering. In: International Conference on Neural Information Processing Systems, vol. 16, pp. 177–184. MIT Press (2003)

Application of Deep Autoencoders in Commerce Recommendation

Gali Bai and Zhiqiang Zhan

Abstract Collaborative filtering aims to provide personalized recommendations using user feedback. Such algorithms look for latent variables in a large sparse matrix of ratings. To solve the accurate problems of neighbor-based collaborative filtering algorithm and space problems of storage similarity matrix on sparse rating data, we propose a collaborative filtering algorithm based on the deep autoencoders. The results on training data provided by RecSys 2015 challenge show that the proposed algorithm not only significantly reduces the overhead of model training compared to the existing collaborative filtering algorithms in solving the TopN recommendation problem with implicit feedback behavior but also has substantial increase on coverage and novelty of result.

Keywords Collaborative filtering · Deep autoencoders · Recommendation

1 Introduction

Personalized recommendation technology has become a popular field of computer science research because it can help users find valuable information from their massive data and solve the problem of "information overload" effectively. In a variety of personalized recommendation technology, collaborative filtering technology has been widely concerned about and researched, because of it is simple and effective, and does not depend on the structured information of the recommended object itself. However, the existing collaborative filtering algorithm still has the following problems in practical applications.

G. Bai (✉) · Z. Zhan
School of Information and Communication Engineering, BUPT, Beijing, China
e-mail: baigali@bupt.edu.cn; baigali@bupt.com

Z. Zhan
e-mail: zqzhan@bupt.edu.cn

© Springer Nature Singapore Pte Ltd. 2019
S. K. Bhatia et al. (eds.), *Advances in Computer Communication and Computational Sciences*, Advances in Intelligent Systems and Computing 759,
https://doi.org/10.1007/978-981-13-0341-8_22

First, the sparseness of scoring data. In recent years, the number of users and items that recommended system should treat has been greatly increased, the scoring data becomes very sparse, and the similarity measure system adopted by the traditional cooperative filtering algorithm has became unstable end ineffective in this case. For example, the user-based collaborative filtering algorithm calculates the similarity of two users based on two users' common scores. In the case of sparsely populated data, the two users can score little or no. The similarity of the user is not reliable enough to be calculated. Secondly, the scalability of the algorithm. The neighbor-based collaborative filtering algorithms need to calculate and store the similarity between all users or items, and the time and space complexity is proportional to the square of the number of users or items. Therefore, neighbor-based collaborative filtering algorithms cannot be applied to real large level data sizes on internets. Finally, the applicability of the algorithm. Many collaborative filtering algorithms are proposed to solve the problem of scoring prediction under explicit feedback behavior. The problem of scoring prediction under explicit feedback behavior is of great theoretical significance, but the more practical value is the TopN recommendation problem of implicit feedback behavior. That is, the system analyzes the implicit user feedback behavior (usually only 0 and 1 Rating), generates a recommended list of N items for each user, and tries to make the list meet the user's interests and needs.

On the one hand, the recent study [1] shows that the algorithm that achieves the lowest error in the scoring prediction problem does not necessarily perform well in the TopN recommendation problem. On the other hand, the coverage and novelty of results that the existing cooperative filtering algorithm recommended applied on the implicit feedback data, which is not ideal. These algorithms tend to recommend popular items and greatly weakened the ability of the recommendation system to mine personalized items.

We proposed a collaborative filtering algorithm based on deep autoencoders to solve the TopN recommendation problem of implicit feedback behavior effectively. The results on training data provided by RecSys 2015 challenge show that the algorithm proposed in this paper can not only reduce the space cost but also increase the coverage and novelty of the proposed results compared with the existing collaborative filtering algorithm.

2 Related Work

The main idea of the collaborative filtering recommendation algorithm is to use the user community's past behavior to predict which items the current user is interested in. The mainstream collaborative filtering technology can be divided into two types: Neighbor-based algorithm and model-based algorithm.

The neighbor-based collaborative filtering algorithms mainly contain user-based and item-based. User-based algorithm found the nearest neighbor user similar to target user behavior, and then the score is predicted according to the score of the target object and the similarity with the target user. The user-based collaborative filtering is studied comprehensively in [2], and the influence of various similarity measure systems, weighting systems, and the choices of neighboring number on recommendation result is compared. In order to solve the problem of excessive number of users [3], noted that the number of items in the general system is much smaller than that of users, and proposed object-based collaborative filtering algorithm. The algorithm uses the historical score to calculate the similarity between items, and then predicts the score according to the similarity between the target user and the target object. The paper also compares the similarity of the cosine similarity and the improved cosine similarity. In [4], a collaborative filtering method on user-based and item-based methods is proposed for implicit feedback user data. The disadvantage of the neighbor-based algorithm is that it becomes increasingly difficult and impossible to calculate the similarity matrix as the number of users and items in the system increases, and when the score data is very sparse, the similarity between users or items becomes unreliable and difficult to calculate.

The most representative algorithm of the model-based collaborative filtering algorithm is the matrix decomposition (SVD) algorithm [5]. By decomposing the user–object score matrix into the product of two low-rank matrices, the matrix decomposition algorithm can extract a set of potential factors from the scoring data and describe the user and item characteristics with the vectors of these factors, and then generate the score prediction. In recent years, researchers have done a lot of work in the framework of matrix decomposition. For example, the matrix decomposition is extended from the Gaussian distribution to arbitrariness probability distribution [6]. In order to solve the problem of the scalability of the matrix decomposition algorithm, an algorithm based on multi-core processor parallel acceleration training is proposed in [7]. The disadvantage of matrix decomposition is that it is designed for score prediction problems, although it can achieve a high score prediction accuracy, but for TopN recommendation, the accuracy is not as good as other methods.

In recent years, the development of the deep learning theory makes it possible to extract features from large-scale non-tagging data using neural networks. Autoencoders have achieved very good results in the text processing, image classification task. These research works let the neural network be applied to the recommendation system, and use autoencoders to extract the user interest characteristic.

3   Autoencoders and CF

The main idea of this algorithm is to use autoencoders to obtain low-dimensional representation of user behavior, and then calculate the similarity between users in low-dimensional space, and then produce recommendations. The following will first introduce the principles of automatic encoder, and then elaborate the improved depth of the autoencoders.

Fig. 1 Neural network
structure for an autoencoders

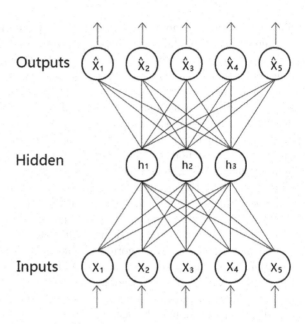

3.1 Autoencoders

Autoencoder is a kind of artificial neural network (ANN), which is often used to extract the characteristics of input data in deep learning. It is the same as the ordinary neural network, composed of input layer, hidden layer, and output layer. In particular, the target value of the autoencoders output layer is equal to the input value of the input layer, that is, the autoencoders attempts to learn an identity function so that the output is as close as possible to the input. Therefore, the automatic encoder is an unsupervised learning algorithm and does not need to manually annotate the training data.

An example of a neural network structure for autoencoders is shown in Fig. 1. It contains five input neurons, five output neurons (training sample dimension equal to 5), and three hidden layer neurons (the characteristic dimension after encoding is equal to 3).

3.2 Application of Deep Autoencoders

The autoencoders learn a constant function through the coding and decoding process, although it seems meaningless, but if the number of hidden neurons is limited, some special structures in the input data can be excavated. For example, assuming that the input of the autoencoders is a 32×32 image (the input layer contains 1024 neurons) and the hidden layer has only 100 neurons, such a neural network structure will force the autoencoders to learn the compressed representation of the input data. Otherwise, it will not be able to recover 1024 pixels of the original image from the 100-dimensional eigenvector of the hidden layer.

If the input data is random distribution, it will be very difficult to get this compression. However, if the input data implies some special structure, then the autoencoders can find these special structures in the input data, and thus obtain a compressed representation of the raw data.

The mining data inherent structure of autoencoders, and the characteristics that obtain the compression representation of original data, has been successfully applied in natural language processing, speech recognition, computer vision, and other fields.

In this paper, we consider to use the autoencoders into the collaborative filtering algorithm, try to use the autoencoders to mine the inherent special structure of the user's interest behavior, obtain the compression expression of the user behavior characteristic, and improve the quality of the recommendation result.

However, due to the particularity of the scoring data in collaborative filtering, the user behavior vector contains a large number of "missing" values, so the user behavior vector cannot be entered directly into the existing autoencoders model. Specifically, the user is set to U, the set of items is I, and the behavior of a user $u \in U$ can be expressed by an $|I|$-dimensional vector y^u as shown in equation:

$$y_j^u = \begin{cases} 1 & \text{user u like item j} \\ 0 & \text{user u dislike item j} \quad 1 \leq j \leq |I| \\ missing & \text{otherwise} \end{cases}$$

Since a user usually only expresses interest preferences for a small portion of all items, most of the user behavior vectors are "missing" values, indicating that the user's interest in the item is not yet clear. It should be noted that, unlike the vector space model (VSM) used in information retrieval, the "missing" value in collaborative filtering cannot be filled directly with 0. If all the "missing" values are filled with 0. All "missing" values will be treated as negative samples, making the number of positive and negative samples unbalanced, leading to the model tend to recommend popular items [8].

Therefore, in order to adapt to the scenario of the collaborative filtering problem, we have the following autoencoders model to do the following improvements:

Deep autoencoder is a stack of autoencoders, exactly multilayer autoencoders, and the output from the previous autoencoders (the middle hidden layer) is an input of the latter autoencoders. In fact, some encoding part is superimposed and some decoding part is superimposed. This is autoencoders with multiple hidden layers [9].

Training deep network can be used layer-by-layer greedy training methods, each time only train a hidden layer, and it can use supervised pretraining or unsupervised pretraining. There unsupervised sparse self-coding algorithm to learn the characteristics of hidden layer. Because it is a multilayer sparse self-coding neural network and is encoded layer by layer, we also can call it stacked autoencoders (Fig. 2).

Coding steps of deep autoencoders neural network:

Fig. 2 Architecture of the deep (Stacked) autoencoders

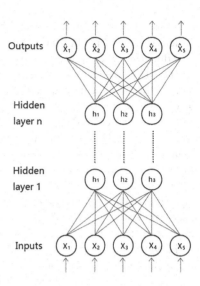

$$a^{(l)} = f\left(z^{(l)}\right)$$
$$z^{(l+1)} = W^{(l,1)}a^{(l)} + b^{(l,1)}$$

Decoding steps of deep autoencoders neural network:

$$a^{(n+1)} = f\left(z^{(n+l)}\right)$$
$$z^{(n+l+1)} = W^{(n-1,2)}a^{(n+1)} + b^{(n-l,2)}$$

Deep autoencoders have powerful expression and all the advantages of deep neural networks. Since the encoder tends to learn the characteristics of the data representation, then the first layer can learn the first-order features, the second layer can learn second-order features, and so on.

4 Experiments

4.1 Datasets and Evaluation Metric

We use the real-world consumption data that RecSys Challenge 2015 provides. The training data contain clicks and buy events that operated by users on an electronic business website, and test data only contains click events that are operated by users. Each click event consists of session id, timestamp, item id, and the category of the item. And each buy event consists of session id, timestamp, item id, price, and quantity.

Table 1 Performance of two autoencoder models	Model	Evaluation score
	Autoencoders	57832
	Deep (stacked) autoencoders	57935

Take the following score as the evaluation metric:

$$Score(Sl) = \sum_{s \in Sl} \begin{cases} \frac{|S_b|}{|S|} + \frac{|A_s \cap B_s|}{|A_s \cap B_s|} \ if \ s \in S_b \\ -\frac{|S_b|}{S} \qquad else \end{cases}$$

Sl sessions in submitted solution file
S all sessions in the test set
s sessions in the test set
S_b sessions in the test set which end with buy
A_s predicted bought items in session s
B_s actual bought items in session s.

The evaluation indicator takes into account how to accurately predict what the user will buy [10]. This is not only related to the whether the user will buy but also related to which items will be purchased. In this way, the provider can recommend certain items to the user based on the forecast data.

4.2   Results

In this section, we compare the performance of collaborative filtering algorithm based on deep autoencoders and collaborative filtering algorithm based on traditional autoencoders under the above evaluation index. We used the same magnitude of the RecSys Challenge 2015 data to test the performance of autoencoders and deep autoencoders. We tested 5 times in the same operating environment, and scored the average of the evaluation score.

Table 1 shows the result of two autoencoder models, autoencoders and deep autoencoders. Our deep autoencoders features perform slightly better than the traditional autoencoders. It states that stack of autoencoders can increase the accuracy of the recommendation, because the network is able to find a better parameter space by stacking several autoencoders in a greedy layer-wise manner for pretraining.

5 Conclusion

In this paper, a collaborative algorithm based on deep autoencoders is proposed. The algorithm uses the autoencoders to learn the user behavior and encodes the high dimension and sparse user behavior vector into low dimension, dense user characteristic vector.

The results on training data provided by RecSys 2015 challenge show that the method of this paper can greatly reduce the spatial complexity of the algorithm and greatly improve the coverage and novelty of the recommended results in the case of guaranteeing the recommended accuracy.

References

1. Cremonesi, P., Koren, Y., Turrin, R.: Performance of recommender algorithms on top-n recommendation tasks. In: Proceedings of the Fourth ACM Conference on Recommender systems, pp. 39–46. ACM (2010)
2. Herlocker, J.L., Konstan, J.A., Borchers, A., et al.: An algorithmic framework for performing collaborative filtering. In: Proceedings of the 22nd Annual International ACM SIGIR Conference on Research and Development in Information Retrieval, pp. 230–237. ACM (1999)
3. Sarwar, B., Karypis, G., Konstan, J., et al.: Item-based collaborative filtering recommendation algorithms. In: Proceedings of the 10th International Conference on World Wide Web, pp. 285–295. ACM (2001)
4. Verstrepen, K., Goethals, B.: Unifying nearest neighbors collaborative filtering. In: Proceedings of the 8th ACM Conference on Recommender Systems, pp. 177–184. ACM (2014)
5. Koren, Y., Bell, R., Volinsky, C.: Matrix factorization techniques for recommender systems. Computer **42**(8) (2009)
6. Bauer, J., Nanopoulos, A.: A framework for matrix factorization based on general distributions. In: Proceedings of the 8th ACM Conference on Recommender Systems, pp. 249–256. ACM (2014)
7. Chin, W.S., Zhuang, Y., Juan, Y.C., et al.: A fast parallel stochastic gradient method for matrix factorization in shared memory systems. ACM Trans. Intell. Syst. Technol. (TIST) **6**(1), 2 (2015)
8. Pradel, B., Usunier, N., Gallinari, P.: Ranking with non-random missing ratings: influence of popularity and positivity on evaluation metrics. In: Proceedings of the Sixth ACM Conference on Recommender Systems, pp. 147–154. ACM (2012)
9. Lore, K.G., Akintayo, A., Sarkar, S.: LLNet: a deep autoencoder approach to natural low-light image enhancement. Pattern Recogn. **61**, 650–662 (2017)
10. Yan, P., Zhou, X., Duan, Y.: E-Commerce item recommendation based on field-aware factorization machine. In: Proceedings of the 2015 International ACM Recommender Systems Challenge, p. 2. ACM (2015)

Part II
Intelligent Hardware and Software Design

An Efficient Pipelined Feedback Processor for Computing a 1024-Point FFT Using Distributed Logic

Hung Ngoc Nguyen, Cheol-Hong Kim and Jong-Myon Kim

Abstract This paper proposes an effective fast Fourier transform (FFT) processor for 1024-point computation based on the radix-2 of decimation-in-frequency (R2DIF) and uses the pipelined feedback (PF) technique via shift registers to efficiently share the same storage between the inputs and outputs during computation. The large memory footprint of the complex twiddle factor multipliers, and hence, area on a chip, of the proposed design is reduced by employing the coordinate rotation digital computer (CoRDiC), which replaces the complex multipliers and does not require memory blocks to store the twiddle factors. To enhance the efficient usage of the hardware resources, the proposed design only uses distributed logic. This can eliminate the use of dedicated functional blocks, which are usually limited to the target chip. The entire proposed system is mapped on a Virtex-7 field-programmable gate array (FPGA) for functional verification and synthesis. The achieved result is the proposed FFT processor more effective in terms of the speed, precision, and resource, as shown in experimental results.

Keywords Radix-2 FFT · Pipelined feedback · CoRDiC algorithm · FPGA

1 Introduction

The fast Fourier transform (FFT) is an algorithm for spectral analysis of digital signals, which is widely applied in general digital signal processing (DSP) systems

H. N. Nguyen · J.-M. Kim (✉)
School of Electrical Engineering, University of Ulsan, Ulsan, South Korea
e-mail: jongmyon.kim@gmail.com

H. N. Nguyen
e-mail: hungnguyenvldt@gmail.com

C.-H. Kim
School of Electronics and Computer Engineering, Chonnam National University,
Gwangju, South Korea
e-mail: cheolhong@gmail.com

© Springer Nature Singapore Pte Ltd. 2019   245
S. K. Bhatia et al. (eds.), *Advances in Computer Communication and Computational Sciences*, Advances in Intelligent Systems and Computing 759,
https://doi.org/10.1007/978-981-13-0341-8_23

and monitoring systems in particular. Spectrum analysis of signals in the frequency domain provides more useful information about the amplitude, phase, and frequency vibrations, providing support for the maintenance systems in order to make reliable and accurate decisions during classification and diagnosis [1]. Therefore, it is very critical to implement an accurate and efficient FFT processor that is suitable for the requirements of real-time processing, particularly focusing on its precision and low costs in many applications. The advantages of integrated circuit technologies have promoted more interest in the hardware implementation of signal processing algorithms. The FFT algorithms are mainly implemented on field-programmable gate arrays (FPGAs) [2, 3] that provide better performance and are more suitable for implementing the high-speed processing algorithms due to their parallel signal processing architecture. Thus, FPGA is a perfect solution for hardware implementation of FFT.

In this paper, the efficiency pipelined FFT processor is performed on hardware as an IP logic core, which can be integrated into DSP systems for high-speed spectrum analysis. Here, the FFT based on the radix-2 of decimation-in-frequency (R2DIF) is exploited to reduce the computational complexity of DFT [4, 5]. In addition, the proposed pipelined technique for FFT permits all the processing stages in architecture to carry out parallelism at the same time to enhance the processing speed of the system [6, 7]. Particularly, in this research, the pipelined feedback (PF) technique is proposed for implementation of FFT on FPGA because it has less resources requirement on the chip, and a rather simple in setting up the control logic. Furthermore, conventional FFT implementations need a large number of memory to store processing data and the twiddle factors for complex multipliers in architecture, which consume large areas on chip lead to increasing the cost of hardware designs. To address this issue, the proposed R2DIF, PF-based FFT architecture (R2PF) is more efficiently improved by employing the coordinate rotation for digital computer (CoRDiC) algorithm to replace the complex twiddle factor multiplication [8]. This will reduce the memory requirements and lead to improvements in performance. The R2PF is both very versatile and hardware efficient since it requires only the add and shift operations for the butterfly operations in FFT. Instead of storing the actual twiddle factors, the proposed CoRDiC-based R2PF design only needs to store the twiddle factor angles for the butterfly operation, thereby reducing the memory requirements of storage, saving resources and area on chip, while significantly enhancing its precision. Moreover, the proposed design only uses distributed logic, making the FFT processor more appropriate for further design as an ASIC.

The main contribution of the proposed FFT processor is exploiting the PF technique for hardware implementation of R2DIF, with improved resource utilization and computational time. The CoRDiC algorithm is proposed to replace the complex twiddle factor multipliers, reducing the memory requirements and resulting in improvements in the area and processing speed. The proposed design totally eliminates the need for dedicated functional blocks on chip, which are usually limitedly embedded on the target FPGA, while still guaranteeing a high processing speed and accuracy for the system. Moreover, the proposed architecture uses shared registers instead of slower block memories to store both the input data and intermediate results, making the computation process faster at each stage and saving more registers in the design.

The rest of the paper includes the follows. A summary of the radix-2 DIF FFT algorithm is given in Sect. 2. A brief description of the CoRDiC algorithm and its hardware implementation are discussed in Sect. 3. Section 4 presents the implementation of the proposed R2PF design for 1024-point FFT computation, whereas Sect. 5 discusses the results of the proposed design. Finally, Sect. 6 concludes this paper.

2   The Radix-2 DIF FFT Algorithm

The N-point DFT of $x(n)$ is defined as follows:

$$X(k) = \sum_{n=0}^{N-1} x(n)W_N^{kn}; \qquad 0 \le k \le N-1, \tag{1}$$

where W_N^{kn} is the twiddle factor, defined as the follows:

$$W_N^{kn} = e^{(-2\pi kn/N)} = \cos\left(2\pi kn/N\right) - j\,\sin\left(2\pi kn/N\right) \tag{2}$$

From the above equation, the N-point of $X(k)$ are decomposed into even samples $X(2k)$ and odd samples $X(2k+1)$ using the R2DIF algorithm, as given in Eqs. (3) and (4), respectively.

$$X(2k) = \sum_{n=0}^{N\frac{1}{2}-1} \left(x(n) + x\left(n+\frac{N}{2}\right)\right) W_{N/2}^{kn} \tag{3}$$

$$X(2k+1) = \sum_{n=0}^{\frac{N}{2}-1} \left(x(n) - x\left(n+\frac{N}{2}\right)\right) W_{N/2}^{kn} W_N^n \tag{4}$$

Here, $0 \le k \le N/2 - 1$. Computing an N-point FFT is accomplished via repeating the radix-2 butterfly operation in architecture, as illustrated in Fig. 1.

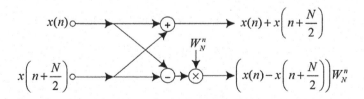

Fig. 1 The butterfly unit of radix-2 DIF algorithm

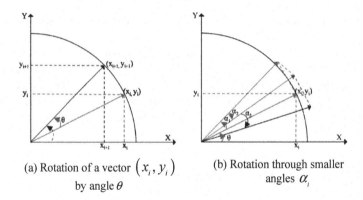

(a) Rotation of a vector $\left(x_i, y_i\right)$ by angle θ

(b) Rotation through smaller angles α_i

Fig. 2 The rotation of a vector on a two-dimensional plane by the CoRDiC algorithm

3 The Coordinate Rotation Digital Computer (CoRDiC) Algorithm

The CoRDiC is a simple and efficient iterative algorithm which can be used to carry out complex twiddle factor multiplication in an FFT processor without using any dedicated multipliers, thus decreasing the chip area and consequently reducing the system cost.

3.1 The CoRDiC Algorithm for Complex Multipliers

In the proposed FFT design, the rotation mode of the CoRDiC algorithm is used to carry out the twiddle factor multiplication, i.e., $x(n) \cdot w_N^n$, where w_N^n is equivalent to rotating $x(n)$ by an angle $\theta = 2\pi n/N$, as shown in Fig. 2a. This can be derived from the rotation transform, i.e., when a vector with coordinates (x_i, y_i) is rotated by an angle θ, the new coordinates (x_{i+1}, y_{i+1}) are given as follows:

$$
\begin{aligned}
x_{i+1} &= x_i \cos \theta - y_i \sin \theta \\
y_{i+1} &= y_i \cos \theta + x_i \sin \theta
\end{aligned} \tag{5}
$$

The terms in Eq. (5) can be rearranged to yield the following:

$$
\begin{aligned}
x_{i+1} &= \cos \theta (x_i - y_i \tan \theta) \\
y_{i+1} &= \cos \theta (y_i + x_i tan\theta)
\end{aligned} \tag{6}
$$

A scheme is proposed to split the rotation angle θ into a series of smaller angles α_i, as shown in Fig. 2b, where $\theta = \sum_{i=0}^{N-1} d_i \alpha_i$ and $d_i = \pm 1$. The resultant series can utilize the property of a tangent function, i.e., $\tan \theta = \pm 2^{-i}$, in which multipliers are realized by a sequence of simple shift operations through i bit locations. Thus, arbitrary angles can be obtained by performing a series of rotations iteratively. At each iteration, the direction of rotation is chosen by obtaining the difference between the actual angle and the reference angle obtained by rotation, which can be given as follows:

$$
\begin{aligned}
x_{i+1} &= k_i \left(x_i - d_i y_i 2^{-i} \right) \\
y_{i+1} &= k_i \left(y_i + d_i x_i 2^{-i} \right)
\end{aligned} ,
\tag{7}
$$

where k_i is the scale factor for each iteration and is given in Eq. (8):

$$
k_i = \cos\theta = \cos(\arctan(2^{-i})) = \frac{1}{\sqrt{1 + \tan^2(\theta)}} = \frac{1}{\sqrt{1 + 2^{-2i}}}.
\tag{8}
$$

If N iterations are performed, then the scale factor K is defined as follows:

$$
K = \prod_{i=0}^{N-1} k_i = \prod_{i=0}^{N-1} \frac{1}{\sqrt{1 + 2^{-2i}}}.
\tag{9}
$$

Here, if the number of iterations is sufficiently large, then $K \approx 0.6072523$. The direction for each rotation is defined by d_i, which is equal to the sign of differential angle z_i between the desired and reference angle for the i-th rotation:

$$
d_i = \mathrm{sgn}(z_i) = \begin{cases} -1, & z_i < 0 \\ +1, & z_i \geq 0 \end{cases} ,
\tag{10}
$$

where z_i is the angle accumulator for each iteration. For the subsequent rotation, it is given as

$$
z_{i+1} = z_i - d_i \arctan\left(2^{-i}\right).
\tag{11}
$$

The equations for the rotation mode of the CoRDiC algorithm are given in Eq. (12) after removing the scale factor for each iteration.

$$
\begin{aligned}
x_{i+1} &= x_i - d_i y_i 2^{-i} \\
y_{i+1} &= y_i + d_i x_i 2^{-i} \\
z_{i+1} &= z_i - d_i \alpha_i \\
\alpha_i &= \arctan\left(2^{-i}\right)
\end{aligned}
\tag{12}
$$

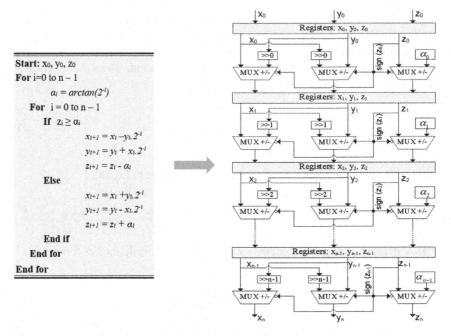

Start: x_0, y_0, z_0
For i=0 to n − 1
    $\alpha_i = arctan(2^{-i})$
    For i = 0 to n − 1
        If $z_i \geq \alpha_i$
            $x_{i+1} = x_i - y_i.2^{-i}$
            $y_{i+1} = y_i + x_i.2^{-i}$
            $z_{i+1} = z_i - \alpha_i$
        Else
            $x_{i+1} = x_i + y_i.2^{-i}$
            $y_{i+1} = y_i - x_i.2^{-i}$
            $z_{i+1} = z_i + \alpha_i$
        End if
    End for
End for

Fig. 3 The basic hardware structure of the N-stage pipelined CoRDiC algorithm

3.2 Implementation of the Pipelined-Based CoRDiC Algorithm

Hardware implementation of CoRDiC is realized using a high-performance pipelined technique, as shown in Fig. 3, which only consists of shifters, adders, and registers between different stages of the pipeline. The registers are inserted between the different stages of pipelined implementation. Therefore, the CoRDiC algorithm directly computes the input data that allows a substantial reduction in memory size and total removal of the complex multipliers in FFT rotation.

For each clock cycle, the controller determines the amount of shift and type of operation to execute. The operation type is determined by the direction of vector rotation, which is obtained from the sign bit of the previous angle accumulator value z_i. The initial angle z_0 equals the desired angle of vector rotation; after N iterations, z_{N-1} approaches zero, which also determines the convergence of the algorithm. The computation precision of the algorithm increases by increasing the number of stages in the architecture, where a pipeline stage corresponds to an iteration of the CoRDiC algorithm. Generally, N iterations of the CoRDiC algorithm are required to have N-bit output precision. As mentioned earlier, for a sufficiently large value of N, the constant scaling factor is $K \approx 0.6072523$. The final results need to be multiplied by this factor.

4 Hardware Implementations of the Pipelined FFT Processor

In this paper, the pipelined feedback (PF) technique via shared shift registers is chosen for the FFT implementation on hardware, achieving higher performance of speed and resources. There is always a balance of speed and resource in hardware implementation. The pipeline technique is a suitable solution for the real-time processing applications.

4.1 Implementation of the R2DIF Pipelined Feedback FFT Architecture

To achieve high-performance pipelined FFT architecture, the data buffer structure is designed based on the PF technique. Instead of using memory blocks, the PF architecture uses the shift registers, sharing the same storage between the inputs and outputs of the butterfly unit. The proposed PF-based R2DIF architecture (R2PF) is used for computing a 16-point FFT, as shown in Fig. 4. It requires four pipelined stages, using 15 shift registers, 8 adders, 3 multipliers, and some multiplexers to control data flow for computing a 16-point FFT. The controller blocks the control read and write process to the memory and creates the consistent controlling signals at each stage in the pipelined architecture.

We can describe the operation of the R2PF design in the continuous flow procedure. A single data stream goes through multiplier in each stage. A data buffer (shift registers) is designed in a feedback manner for storing immediate data at each stage. The dimension of the shift registers reduces by half according to butterfly operation in architecture when the number of stages increases by one. Each pipelined stage in FFT architecture is a PF module with various sizes. The structure of each stage is referred as the processing element that contains butterfly units for addition and subtraction between two input data of each stage, a block of the data buffer to store

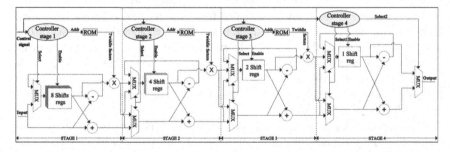

Fig. 4 The R2PF pipelined architecture for 16-point FFT on hardware

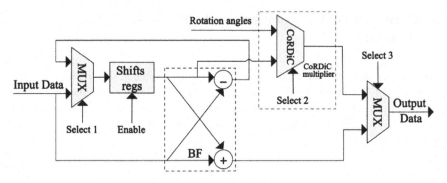

Fig. 5 The hardware architecture of a CoRDiC-based R2PF pipelined stage

the intermediate data for butterfly operation and a complex multiplier for the twiddle factor multiplication in FFT computation. A ROM is used to store and generate twiddle factors for multiplication, which can be designed based on resources, such as slice registers, look-up tables (LUTs), or embedded memory blocks. The pipelined architecture for N-point has the same PF module, which is repeated for $\log_2 N$ stages. It is very flexible for designing of any point FFTs.

4.2 The Proposed CoRDiC-Based R2PF FFT Processor

Implementation of the multiplication for the twiddle factors in FFT using the CoRDiC algorithm reduces the hardware complexity and chip area in two ways. First, instead of storing the complex twiddle factors, it only needs to store the real rotation angle values, reducing its memory requirements. Second, it disposes complex multiplication, which involves four real multiplications, two real additions, and one complex addition; in an FPGA complex, multiplication is realized with either an embedded DSP block or using real multipliers and adders.

The CoRDiC-based multiplication scheme replaces the complex multipliers in each of the $\log_2 N$ stages of the proposed R2PF design, as shown in Fig. 5. The memory required at each stage is reduced by using the shift registers for storing both the input and output values. In the first stage, the input data consists of real values only, and hence the CoRDiC algorithm of the first stage is also optimized accordingly to save hardware resources. Furthermore, in the butterfly operations, multiplication by the twiddle factor $W_N^{N/2} = -j$ is realized by simply interchanging the real and imaginary parts of the product. Therefore, multiplication by $-j$ requires only multiplexers and inverters.

The proposed design is evaluated by comparing its performance with the conventional R2PF pipelined FFT architecture, which uses memory blocks to store the precomputed complex twiddle factors, and its more efficient modified form that uses the distributed logic of configurable logic blocks (CLBs), slice registers, and LUTs,

Table 1 Evaluation of the obtained hardware results using three different designs

Resource usages	(A)	(B)	(C)
# of slice registers	2,066	1,611	1,413
# of slice LUTs	3,159	16,336	13,298
# of in/out buffers	92	92	92
# of block RAMs	4	0	0
# of block DSPs	30	0	0
# of clocking buffers	1	1	1
Total number of clock cycles for execution	2,477	2,065	1,239
Execution time (μs)	12.387	10.325	6.195

(A): The conventional R2PF design using functional blocks of BRAMs and DSPs
(B): An efficiently modified R2PF design using mainly distributed logic
(C): The proposed CoRDiC-based R2PF design

instead of memory blocks to store both twiddle factors and instant data. The performance of the designs in this study is analyzed for 1024-point FFT computation in the next section.

5 Experimental Results

The efficient R2PF processor with the aforementioned designs for computing a 1024-point FFT is performed on a Virtex 7 XC7VX485T FPGA, respectively. The data paths are flexibly designed in a signed fixed-point format with a 16-bit word length and 10-bit precision for the fractional part, respectively. The conventional design stores precomputed complex twiddle factors using block memories [9]. It uses slice registers and LUTs to store only the intermediate data during operation and realizes the complex multipliers using embedded DSP blocks in the FPGA.

The other efficiently modified architecture does not require any dedicated functional blocks on the FPGAs, such as memory blocks or DSP blocks. The block memories are replaced with slice registers and LUTs to store both the intermediate data and the complex twiddle factors. The pre-calculated twiddle factors are stored in a look-up table. The shift–add method is used for implementation of complex multipliers, only using adders and shifters. By this solution, the design can save 100% of the functional blocks on chips.

In contrast to these conventional designs, the proposed CoRDiC-based R2PF processor uses only the distributed logic of slice registers and LUTs, replacing the complex multipliers as well as the block memories required for their storage. The CoRDiC improves resource utilization and system performance while significantly reducing the hardware complexity. Table 1 presents a comparison of the three implementations.

Table 2 The percentage savings on hardware resource utilization of the proposed design

Logic utilization	Modified design	Proposed design	Saving %
# of slices registers	1,611	1,413	12.29
# of slices LUTs	16,336	13,298	18.6
Execution time (μs)	10.325	6.195	40

Table 3 The results for the 1024-point FFT using three different schemes

Usages	Designs							
	# of slices	DSP48		BRAM		Total # of slices	Frequency (MHz)	Execution time (μs)
		# of blocks	Equivalent slices	# of blocks	Equivalent slices			
(A)	5,225	30	15,000	4	6,800	27,025	200	12.387
(B)	17,947	0	0	0	0	17,947	200	10.325
(C)	14,711	0	0	0	0	14,711	200	6.195

The proposed design requires approximately 12.29% fewer slice registers, 18.6% fewer LUTs, and improves the processing speed by about 40%, as shown in Table 2.

In this paper, the slice is a metrics to evaluate the area of design, which is the principal element of the FPGAs. A comparison of the efficiency in terms of area and execution time of three implementation solutions for 1024-point FFT is shown in Table 3. The proposed design uses approximately 45.56% fewer slices of distributed logic.

Computing the average relative percentage error and the results obtained via Matlab as a baseline for evaluating the precision of design, the average relative percentage error of the design is 0.72%, which is far better in comparison to design in [10] (3.22%). This experimental result shows the high precision of the system. Table 4 presents the comparison of the proposed design with the existing designs in the efficiency of resources utilization and processing speed.

6 Conclusions

In this paper, an efficient R2PF processor for computing a 1024-point FFT is proposed. A comparison with other designs is evaluated via terms of speed, precision, and hardware complexity. The proposed FFT processor is successfully accomplished on the pipelined feedback technique and enhances its efficiency of speed and resources by employing the CoRDiC algorithm in computation. This proposed method operates faster and requires fewer hardware resources. Furthermore, the proposed design also optimizes storage of the twiddle coefficients, which are traditionally stored via the much slower memory blocks. Using efficiently distributed logic in architecture makes the proposed design more efficient and less costly compared to other designs.

Table 4 A comparison of the performance of the proposed design and the previous designs

Resource usages	Design schemes		
	Derafshi [4]	Xilinx [11]	The proposed design
# of FFT points	1,024	1,024	1,024
Operation clock frequency (MHz)	100	395	200
# of slice registers	2,472	2,264	1,413
# of slice LUTs	10,353	1,987	13,298
# of block RAMs	16	10	0
# of block DSPs	10	12	0
Total # of slices	29,625	27,251	14,711
Total number of clock cycles for execution	2,600	9,430	1,239
Execution time (μs)	26	23.87	6.195

Acknowledgements This work was supported by the Korea Institute of Energy Technology Evaluation and Planning (KETEP) and the Ministry of Trade, Industry & Energy (MOTIE) of the Republic of Korea (Nos. 20162220100050, 20161120100350, 20172510102130). It was also funded in part by The Leading Human Resource Training Program of Regional Neo Industry through the National Research Foundation of Korea (NRF) funded by the Ministry of Science, ICT and Future Planning (NRF-2016H1D5A1910564), and in part by the Basic Science Research Program through the National Research Foundation of Korea (NRF) funded by the Ministry of Education (2016R1D1A3B03931927).

References

1. Kang, M., Kim, J., Wills, L.M., Kim, J.-M.: Time-varying and multiresolution envelope analysis and discriminative feature analysis for bearing fault diagnosis. IEEE Trans. Ind. Electron. **62**, 7749–7761 (2015)
2. Sanchez, M.A., Garrido, M., Lopez-Vallejo, M., Grajal, J.: Implementing FFT-based digital channelized receivers on FPGA platforms. IEEE Trans. Aerosp. Electron. Syst. **44**, 1567–1585 (2008)
3. McKeown, S., Woods, R.: Power efficient, FPGA implementations of transform algorithms for radar-based digital receiver applications. IEEE Trans. Ind. Inf. **9**, 1591–1600 (2013)
4. Derafshi, Z.H., Frounchi, J., Taghipour, H.: A high speed FPGA implementation of a 1024-point complex FFT processor. In: 2010 Second International Conference on Computer and Network Technology (ICCNT), pp. 312–315. IEEE (2010)
5. Ma, Z.-G., Yin, X.-B., Yu, F.: A novel memory-based FFT architecture for real-valued signals based on a radix-2 decimation-in-frequency algorithm. IEEE Trans. Circ. Syst. II Express Briefs **62**, 876–880 (2015)
6. Garrido, M., Parhi, K.K., Grajal, J.: A pipelined FFT architecture for real-valued signals. IEEE Trans. Circ. Syst. I Regul. Pap. **56**, 2634–2643 (2009)
7. Wang, Z., Liu, X., He, B., Yu, F.: A combined SDC-SDF architecture for normal I/O pipelined radix-2 FFT. IEEE Trans. Very Large Scale Integr. (VLSI) Syst. **23**, 973–977 (2015)

256

8. Meher, P.K., Valls, J., Juang, T.-B., Sridharan, K., Maharatna, K.: 50 years of CORDIC: algorithms, architectures, and applications. IEEE Trans. Circ. Syst. I Regul. Pap. **56**, 1893–1907 (2009)
9. Nguyen, N.-H., Khan, S.A., Kim, C.-H., Kim, J.-M.: An FPGA-based implementation of a pipelined FFT processor for high-speed signal processing applications. In: International Symposium on Applied Reconfigurable Computing, pp. 81–89. Springer (2017)
10. Kumar, M., Selvakumar, A., Sobha, P.: Area and frequency optimized 1024 point radix-2 FFT processor on FPGA. In: 2015 International Conference on VLSI Systems, Architecture, Technology and Applications (VLSI-SATA), pp. 1–6. IEEE (2015)
11. Inc, X.: Logic core IP fast Fourier transform v8.0. Product specifications DS808 (2012)

Design of Low-Voltage CMOS Op-Amp Using Evolutionary Optimization Techniques

K. B. Maji, R. Kar, D. Mandal, B. Prasanthi and S. P. Ghoshal

Abstract The comparative optimizing efficiency of particle swarm optimization (PSO) and simplex particle swarm optimization (Simplex-PSO) method is explored in this work. A CMOS operational amplifier (Op-Amp) with low voltage has been optimized using this method. The concept of PSO is based on communal manner of bird flocking. PSO suffers from stagnation problem and premature convergence. Nelder–Mead simplex method (NMSM) is hybridized with PSO to produce simplex-PSO. Simplex-PSO is very fast and efficient optimization technique. Simplex-PSO gives high accuracy in terms of computational complexity. The main idea is to reduce the overall circuit's area of low-voltage CMOS op-amp. PSO and simplex-PSO based optimized results are verified by SPICE. SPICE-based results demonstrate that the design and essential specifications are approximately reached. Simplex-PSO shows the better optimizing efficiency than PSO for the designed circuit.

Keywords Analog IC · Two-stage CMOS op-amp · Evolutionary optimization techniques · PSO · Simplex-PSO

1 Introduction

Analog circuits are important part of the design of integrated circuits (IC). Obtaining aspect ratios (ratio of transistor's width and length) of MOS transistor is a tedious task. Evolutionary technique can be used for the automatic sizing of MOS transistor aspect

K. B. Maji · R. Kar (✉) · D. Mandal · B. Prasanthi
Department of Electronics and Communication Engineering, National Institute of Technology Durgapur, Durgapur, India
e-mail: rajibkarece@gmail.com

K. B. Maji
e-mail: kbmaji@gmail.com

S. P. Ghoshal
Department of Electrical Engineering, National Institute of Technology Durgapur, Durgapur, India
e-mail: spghoshal@gmail.com

© Springer Nature Singapore Pte Ltd. 2019
S. K. Bhatia et al. (eds.), *Advances in Computer Communication and Computational Sciences*, Advances in Intelligent Systems and Computing 759,
https://doi.org/10.1007/978-981-13-0341-8_24

ratios. Analog circuit sizing and design using evolutionary optimization techniques are proposed as follows. Eberhart et al. [1, 2] proposed the PSO technique. Area optimization of MOS transistors for the op-amp utilizing PSO is proposed in [3, 4]. Folded-cascode op-amp design is presented in [5]. Analog circuit optimization technique is proposed in [6]. Design of CMOS op-amp is reported [7]. Stagnation problem and premature convergence are the two main disadvantages of PSO [8, 9]. To eliminate these limitations, in this article, simplex-PSO [10–12] is applied for designing a low-voltage CMOS op-amp. This work is organized as follows: PSO and simplex-PSO are described concisely in Sect. 2. Design procedure and the cost function of a CMOS op-amp are explained in Sect. 3. Simulation plots and discussion for the designed circuit are given in Sect. 4. Conclusion part of this work is presented in Sect. 5.

2 Evolutionary Algorithm Applied for Low-Voltage Two-Stage Op-Amp

Detailed descriptions of PSO and simplex-PSO are found in [3, 4] and [11, 12], respectively. The different parameters utilized for PSO and simplex-PSO algorithm are tabulated in Table 1.

3 Specification and Design Criteria

The low-voltage CMOS operational amplifier is designed as follows: slew rate (SR), output capacitance (C_L), unity gain bandwidth (UGB), maximum ICMR ($V_{IC (max)}$), and minimum ICMR ($V_{IC (min)}$). The design parameters which are obtained from evolutionary techniques are listed as follows: I_{BIAS}, aspect ratios of MOS transistors

Table 1 Parameters for PSO and simplex-PSO algorithm

Parameters	PSO	Simplex-PSO
Size of population (m)	10	10
Size of the optimization problem (n)	5	5
Iteration cycle	100	100
C_1	2	–
C_2	2	–
W	0.99	–
c_0	–	0.8
c_1	–	0.6
c_2	–	0.08

Table 2 Inputs and technology utilized

Inputs and technology	Considered values
Positive supply voltage (V_{DD}) in volt	2
Negative supply voltage (V_{SS})	-2
Thresh hold voltage for PMOS (V_{tp})	-0.6513
Thresh hold voltage for NMOS (V_{tn})	0.4761
Trans-conductance of NMOS (K_n') in $\mu A/V^2$	181.2
Trans-conductance of PMOS (K_p') in $\mu A/V^2$	65.8
Technology	0.35 μm

Table 3 Design specifications of CMOS op-amp

Parameters	Specified values
SR (V/μs)	≥ 10
C_L (pF)	≥ 10
UGB (MHz)	≥ 10
Phase margin ($°$)	>60
$V_{IC\,(min)}$ (V)	≥ 0.5
$V_{IC\,(max)}$ (V)	≥ 1.5
$V_{out\,(max)}$ (V)	≥ 1.75
$V_{out\,(min)}$ (V)	≥ 0.1
Length (L_i) in μm	1

deployed in designing the CMOS Op-Amp are described in [13] (Table 2). The design specifications for low-voltage CMOS op-amp are presented in Table 3.

4 Results and Discussions

Both the PSO and simplex-PSO algorithms have been designed in MATLAB R2016a; Intel® core™ i5-6600 CPU@3.30 GHz processor with 8 GB RAM to design a CMOS op-amp circuit (Fig. 1). Input variables and technical specification are presented in Table 2.

For the low-voltage CMOS op-amp circuit, the evolutionary techniques are utilized individually to get the optimal value of design parameters I_{BIAS}, C_L C_c, R_1, and W_i ($i = 1, 2, ..., 16$). Both the PSO and simplex-PSO have been executed in 50 iterations to achieve the best sets of results for design. Transistor level simulations of the circuit were executed with Cadence software (IC 614) for the authentication purpose. PSO and simplex-PSO based design results are described elaborately in this section. Process technology parameter used is 0.35 μm. PSO results in a total circuit area of 173 μm^2 with optimal design parameters (I_{BIAS}, W_i, C_L, C_c, and R_1)

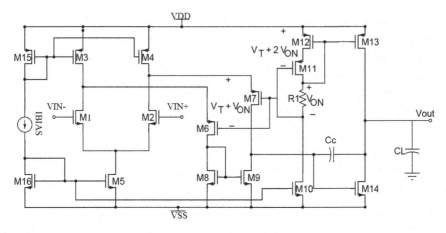

Fig. 1 Circuit diagram of the CMOS op-amp

Table 4 Optimal design parameter obtained using evolutionary techniques

Design variables	PSO	Simplex-PSO
I_{BIAS} (μA)	52	30
W_1/L_1; W_2/L_2 (μm/μm)	3.5/1	2/1
W_3/L_3; W_4/L_4 (μm/μm)	11.25/1	9.5/1
W_5/L_5; W_{10}/L_{10}; W_{16}/L_{16} (μm/μm)	8.5/1	9/1
W_6/L_6; W_7/L_7; W_{11}/L_{11}; W_{12}/L_{12}; W_{15}/L_{15} (μm/μm)	6.5/1	9/1
W_8/L_8; W_9/L_9 (μm/μm)	1/1	1/1
W_{13}/L_{13} (μm/μm)	62.5/1	3.5/1
W_{14}/L_{14} (μm/μm)	21/1	8/1
C_L (pF)	14.4	11.5
C_c (pF)	3.2	2.8
R_1 (k-Ω)	10.2	12.5

in 3.746 s. Simplex-PSO results in a total circuit area of 108.5 μm^2 with optimal design parameters (I_{BIAS}, W_i, C_L, C_c, and R_1) in 3.148 s. The design parameters obtained from evolutionary techniques for the low-voltage CMOS op-amp are listed in Table 4.

The optimally designed low-voltage CMOS op-amp circuit is reconstructed and simulated in CADENCE environment. SPICE simulation results achieved from PSO-based low-voltage CMOS op-amp circuit's design are presented in Figs. 2, 3, 4, and 5, respectively. Figures 6, 7, 8, and 9 show the SPICE simulation results obtained from simplex-PSO based optimal design of the circuit. Finally, Table 5 presents comparison results obtained using PSO and simplex-PSO techniques.

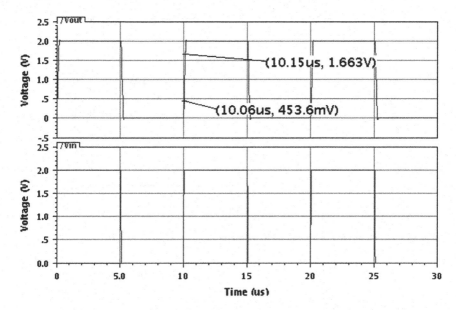

Fig. 2 Slew rate estimation of the CMOS op-amp based on PSO

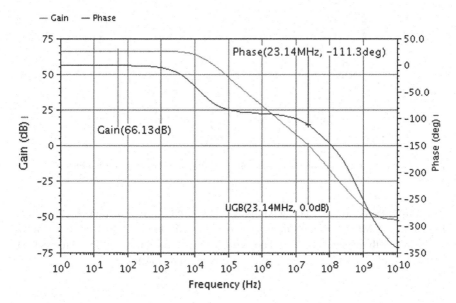

Fig. 3 Gain–phase plot of the PSO-based CMOS op-amp

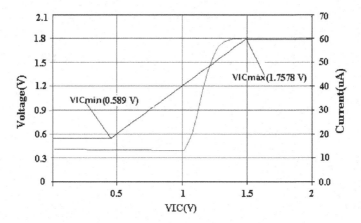

Fig. 4 ICMR plot of the PSO-based CMOS op-amp

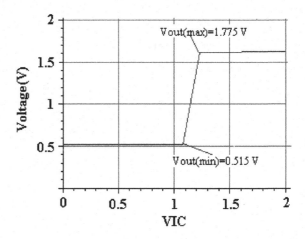

Fig. 5 Plot to estimate the swing in output voltage of PSO-based operational amplifier

It (Table 5) is proved that the simplex-PSO based optimal design technique for low-voltage CMOS op-amp circuit better efficient in terms of SR, gain, and total MOS area with respect PSO-based design. Therefore, the simplex-PSO exhibits best optimizing capability for the example considered in this article.

4.1 Statistical Results for PSO and Simplex-PSO

Figures 10 and 11 show whisker and box plots of PSO and simplex-PSO base design of CMOS op-amp, respectively. The median value of CF_{op-amp} is 173.60×10^{-12} obtained using PSO and CF_{op-amp} values varies from 173×10^{-12} to 174.20×10^{-12}.

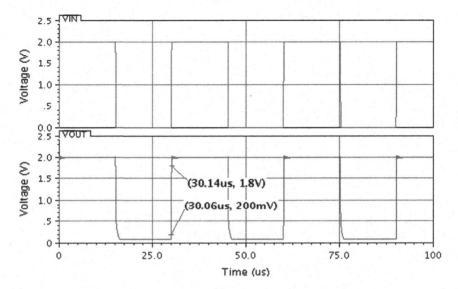

Fig. 6 Slew rate calculation of the CMOS op-amp based on simplex-PSO

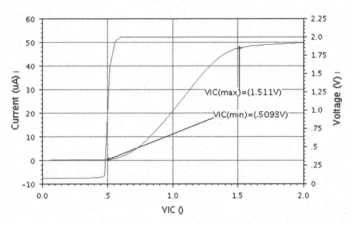

Fig. 7 Plot of gain phase of CMOS op-amp based on simplex-PSO

For simplex-PSO based design, the median value of CF_{op-amp} is 108.7×10^{-12} and CF_{op-amp} values vary from 108.5×10^{-12} to 108.95×10^{-12}. Simplex-PSO shows more stable than PSO in terms of CF values as it is small for the designed circuit.

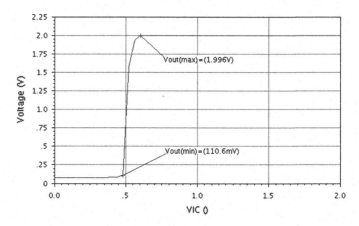

Fig. 8 ICMR plot for CMOS op-amp based on simplex-PSO

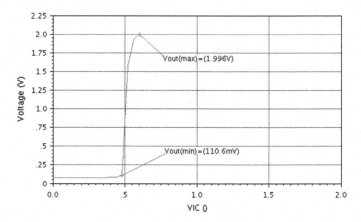

Fig. 9 Plot to calculate output swing of simplex-PSO based operational amplifier circuit

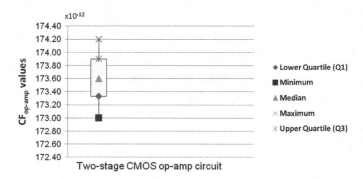

Fig. 10 Statistical plots for simplex-PSO based designed circuit

Table 5 Result of comparison between PSO and simplex-PSO

Parameters	Values considered	PSO	Simplex-PSO
Slew rate (V/µs)	≥10	13.438	20
Phase margin (°)	>60	68.7	65.8
UGB (MHz)	≥10	23.14	12.64
Gain (dB)	>60	66.13	72
$V_{IC\,(min)}$ (V)	≥0.5	0.589	0.5093
$V_{IC\,(max)}$ (V)	≥1.5	1.7578	1.511
$V_{out\,(min)}$ (V)	≥0.1	0.515	0.1106
$V_{out\,(max)}$ (V)	≥1.75	1.775	1.996
Total MOS area (µm²)	<200	173	108.5

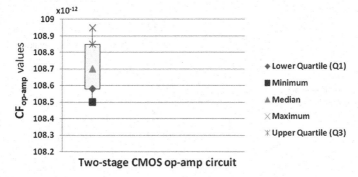

Fig. 11 Statistical plots for simplex-PSO based designed circuit

4.2 Profile of Convergence for the PSO and Simplex-PSO

Figure 12 shows the convergence profile plot of low-voltage CMOS op-amp circuit utilizing PSO. PSO computes total circuit area of 173 µm² in 3.746 s to execute 100 fitness evaluations. The convergence profile plot for the simplex-PSO based design of low-voltage CMOS op-amp circuit has been shown in Fig. 13. Simplex-PSO calculates total circuit area of 108.5 µm² in 3.148 s to execute 100 fitness evaluations. From Figs. 12 and 13, it is evident that both the PSO and simplex-PSO attain the optimal circuit area within the 100 number of evaluations.

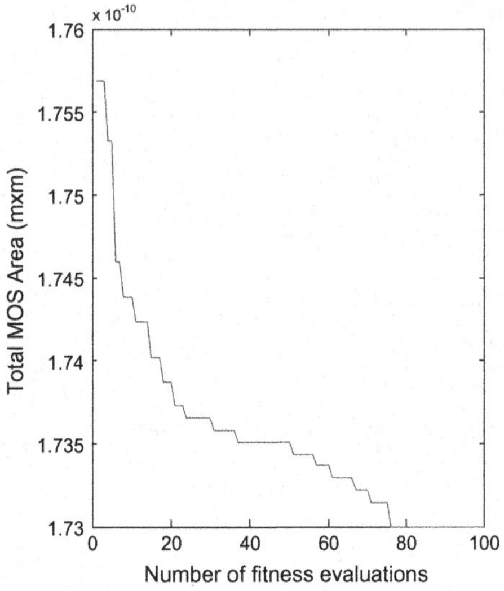

Fig. 12 Profile of
convergence of simplex-PSO
based CMOS op-amp circuit

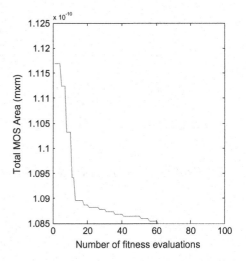

Fig. 13 Profile of
convergence of PSO-based
CMOS op-amp circuit

5 Conclusions

In this work, PSO and simplex-PSO methods are adopted to design CMOS operational amplifier circuit. Both the PSO and simplex-PSO are able to determine the parameters of optimal design for the circuit that is designed. SPICE simulation is done by using the parameters of optimal design obtained from PSO and simplex-PSO, individually. Simulation results establish that evolutionary techniques based circuit design the specified criteria of design and also optimize the area the circuit occupies. Simplex-PSO shows much improved results than PSO with respect to SR, gain, total

MOS area, and execution time for the designed circuit. So, the simplex-PSO is the better optimizer than PSO for the designed circuit. Seeing the simplex-PSO based results, it is confirmed that simplex-PSO is able to design the analog VLSI circuit. Future scope would be to explore the optimization efficiency of other metaheuristic methods for analog VLSI circuit sizing.

References

1. Kennedy, Eberhart, R.: Particle swarm optimization. In: Proceedings of IEEE International Conference On Neural Network, vol. 4, pp. 1942–1948 (1995)
2. Eberhart, R., Shi, Y.: Comparison between genetic algorithm and particle swarm optimization. Evolutionary Programming-VII, pp. 611–616. Springer (1998)
3. Vural, R.A., Yildirim, T.: Analog circuit sizing via swarm intelligence. AEU Int. J. Electron. Commun. 66(9), 732–740 (2012)
4. Vural, R.A., Yildirim, T.: Swarm intelligence based sizing methodology for CMOS operational amplifier. In: Proceedings of 12th IEEE Symposium on Computational Intelligence and Informatics, pp. 525–528 (2011)
5. Ceperic, V., Butkovic, Z., Baric, A.: Design and optimization of self-biased complementary folded cascode. In: Proceedings of IEEE Mediterranean Electrotechnical Conference (MELECON), pp. 145–148 (2006)
6. Liu, B., Wang, Y., Yu, Z., Liu, L., Li, M., Wang, Z., Lu, J., Fernandez, F.V.: Analog circuit optimization system based on hybrid evolutionary algorithms. Integr. VLSI J. 42, 137–148 (2009)
7. Hershenson, M., Boyd, S.P., Lee, T.H.: Optimal design of a CMOS op-amp via geometric programming. IEEE Trans. Comput. Aided Des. Integr. Circ. Syst. 20, 1–21 (2001)
8. Liu, B.D., Lee, J.Y., Wang, H.H.: Parameter extraction and optimization for MOSFET models. Int. J. Electron. 63, 873–884 (1987)
9. Chen, Y.-L., Wu, W.-R., Liu, C.-N.J., Li, J.C.-M.: Simultaneous optimization of analog circuits with reliability and variability for applications on flexible electronics. IEEE Trans. Comput. Aided Des. Integr. Circ. Syst. 33, 24-35 (2014)
10. Ling, S.H., Iu, H.H.C., Leung, F.H.F., Chan, K.Y.: Improved hybrid particle swarm optimized wavelet neural network for modelling the development of fluid dispensing for electronic packaging. IEEE Trans. Ind. Electron. 55(9), 3447–3460 (2008)
11. Biswal, B., Dash, P.K., Panigrahi, B.K.: Power quality disturbance classification using fuzzy c-means algorithm and adaptive particle swarm optimization. IEEE Trans. Ind. Electron. 56(1), 212–220 (2009)
12. Hong-feng, X., Guan-Zheng, T.: A novel particle swarm optimizer without velocity: Simplex-PSO. J. Cent. South Univ. 17(2), 349–356 (2010)
13. Allen, P., Holberg, D.: CMOS Analog Circuit Design, 2nd edn. Oxford University Press, New York (2002)

RGA-Based Wide Null Control for Compact Linear Antenna Array

Durbadal Mandal, Rajib Kar and Shrestha Bandyopadhyay

Abstract As the microwave spectrum becomes more and more crowded with users, interference rejection techniques become increasingly necessary. One way to reduce the interference is to introduce a point of minimum signal reception in the interference direction of antenna radiation pattern. This article presents the synthesis of non-uniformly spaced linear array with reduced side lobe levels while controlling beam width and points of minimum signal reception by applying Real-Coded Genetic Algorithm (RGA) technique. The algorithm ascertains an optimal separation between the antenna elements that contributes deeper nulls to the radiation pattern in a specified range of direction. The effectiveness of algorithm is compared and presented in the form of graphs and tables. Spacing between two successive elements, d is taken to be between $\lambda/4$ and $\lambda/8$ thus enabling the design of compact multiple antenna terminals.

Keywords Linear antenna array · Wide nulls · Non-uniform spacing
Side lobe level (SLL) · Half power beam width · Real-Coded genetic algorithm
(RGA)

1 Introduction

Linear antenna array synthesis has been broadly examined in the most recent decade. Common optimization objectives in synthesis of antenna array are the side lobe level (SLL) reduction (while maintaining the main beam gain) and controlling the minimum signal reception points to minimize interference effects. An antenna array can be used in diverse way to enhance the performance of a communications system [1–4]. An array working relies upon the case where the desired signal and unwanted co-channel interference should arrive from different directions. Different applica-

D. Mandal (✉) · R. Kar · S. Bandyopadhyay
Department of Electronics and Communication Engineering, National Institute of Technology
Durgapur, Durgapur 713209, India
e-mail: durbadal.bittu@gmail.com

© Springer Nature Singapore Pte Ltd. 2019   269
S. K. Bhatia et al. (eds.), *Advances in Computer Communication and Computational
Sciences*, Advances in Intelligent Systems and Computing 759,
https://doi.org/10.1007/978-981-13-0341-8_25

tions require antennas neither to emit nor to receive radiation from some specific directions because power radiated (or received) in (or from) these directions would be insignificant. It is notable that multiple-input multiple-output (MIMO) systems with multiple antenna elements both at the transmitter (Tx) and receiver (Rx) end offers very high capacity, whereas fading is independent of the links between various Tx and Rx antenna pairs. If antennas are closely spaced, the fading might be correlated because of limited space at the mobile terminal. Indoor analysis shows that capacity remains high, even for small value of d such as $\lambda/5$ [5, 6] or even less [7].

Synthesis techniques dealing with null control of antenna pattern are of significance because of two different but related reasons. First, because of increased pollution in the electromagnetic environment [8, 9], synthesis methods, are becoming important as they allow placing of more than one nulls at specified range of directions, in the radiation pattern. Second, they add vision into adaptive antenna systems. There are different methods to deal with the null control [10] for minimizing the effects of interference and jamming. Considering the linear array geometry, this can be obtained by designing spacing between elements and keeping excitation over the array aperture uniform. The goal of the paper is to optimize the inter-element spacing in order to introduce broad nulls, thus working in the direction of improvement of antenna array pattern. The need of broad nulls gets importance where the angle of arrival of undesired interference slightly time-varying or not exactly known, or when continuous steering is required for comparatively sharp null to obtain a feasible value of signal-to-noise-ratio. Optimization techniques such as Real-Coded Genetic Algorithm (RGA), Particle Swarm Optimization (PSO) algorithms are very relevant tools for searching the best antenna models as a great variety of parameters gets involved. However, since its development, the RGA algorithm and its variants have proved themselves as a better approach and this is reflected in the results obtained below.

The rest of the paper is organized as follows: Sect. 2 states the equation for the design of linear antenna array. Section 3 provides a description of Real-Coded Genetic Algorithm (RGA). Section 4 represents the computational result and finally the work is concluded along with a brief summary in Sect. 5.

2 Design Equation

An odd number $(2M+1)$ of isotropic components are symmetrically set along the z-axis as an array, as demonstrated in Fig. 1, (where, M is an integer). M elements are located on both sides of the origin. One element is situated at the origin itself.

Presuming symmetrical amplitude excitation about the origin, the array factor (AF) for non-uniformly spaced broadside array given is expressed as

Fig. 1 Schematic
representation of a
symmetric linear array
antenna structure for (*2
M + 1*) elements along z-axis

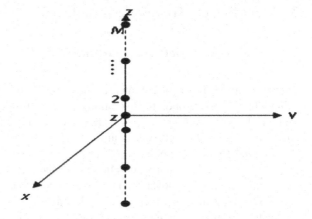

$$(AF)_{2M+1} = In + 2 \sum_{n=1}^{M} In \, \cos\left[\left(\frac{2n-1}{2}\right)k \, d_n \cos\theta + \beta\right], \qquad (1)$$

where,

β	Phase weight at nth element
In	Excitation of Amplitude for nth array element
$(2M + 1)$	Total number of elements in the antenna array
k	Propagation constant
d_n	Spacing between nth and $(n - 1)$th element
θ	Angle of radiation of electromagnetic plane wave

Antenna elements are considered to be isotropic. For null, cost function (CF) is expressed as

$$CF = \frac{\left|\prod_{i=1}^{m} AF(null_i)\right|}{|AF_{\text{max}}|}, \qquad (2)$$

where, "m" denotes the maximum possible number of locations for null being imposed. In this paper, the value of "m" is taken as one and two. $AF(null_i)$ represents the value of array factor at the ith null. AF_{max} gives the maximum value of array factor. Absolute values are taken in the cost function expression for the numerator as well as the denominator. If cost function attains a smaller value, then array factor at predefined positions becomes small. Therefore, RGA algorithm regulates the inter-element spacing for minimization of the cost function.

3   Evolutionary Techniques Employed

3.1   Real-Coded Genetic Algorithm

Fundamentally genetic algorithm is an exhaustive probabilistic random search algorithm which depends upon the evolutionary ideas of natural selection that imitates biological evolution. Here, each generation holds a population of individuals and each individual holds a possible solution to the problem in a coded form known as chromosome. Now, the problem is optimized to evaluate each chromosome with the help of fitness function commonly called as the objective function or the cost function. After that, selection crossover and mutation operations are performed for generating a new set of population which is more fit than its previous population. Selection mechanism aims at selecting more fit individuals (called parents) for crossover and mutation. Crossover is the reason for the exchange of genetic materials between the more apted individuals (parents) to produce offspring, while mutation is responsible for introducing modern advanced and unique Genetic characteristics in the offspring.

For the Genetic Algorithms, the terms used shown in Fig. 2 are given below

- *Population*: Uniform random variables are used for creating initial population. The different sets of inter-element spacing $d_1, d_2, \ldots, d_i, \ldots, d_M$ constitute the chromosomes of each generation. With i varying from 1 to M d_i represents the gene of each chromosome.
- *Evaluation of Fitness Function*: With the help of (1) and (2) Array Factor (AF) and the Cost Function (CF) are evaluated for each set of chromosomes in each generation. Based on the cost function value the sets of inter-element spacing are sorted and finally the set of chromosome which gives the best improved solution for the radiation pattern is obtained.
- *Crossover*: Here, single point crossover is used.
- *Mutation*: Mutation rate was set to 0.01.

Stopping Criteria: When the number of iterations/cycles exceeds the maximum limit, the operation will stop.

4   Computational Results

4.1   Analysis of Radiation Patterns

The simulated results are given in this section which shows the performance of the imposed nulls in the radiation pattern at various positions. Here, 41 elements are assumed for linear antenna array structures and every element being excited equally. The nulling performances are improved for the predefined antenna pattern. Similarly, the nulls are imposed at predefined peak positions. Results are obtained by using RGA

Fig. 2 Flowchart depicting genetic algorithm

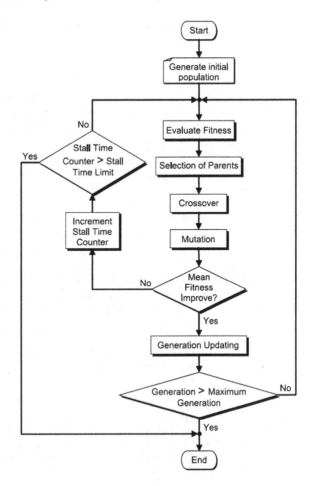

Table 1 (Initial parameters)

Window ranges considered	Worst SLL relative to the main beam (in dB)
1st Null-2nd Peak	−14.52
1st Null-3rd Peak	−14.52
1st Null-4th Null	−14.52
1st Peak-2nd Peak	−14.52

algorithm. Here RGA parameters should be chosen precisely as this technique is quite sensitive to certain parameters sometimes. So, best chosen maximum population pool size, $n_p = 120$, maximum iteration cycles for optimization, $N_m = 100$. The mutation rate set to 0.01 and we have considered uniform crossover. Random variables were used for initializing GA algorithm.

For the reference radiation pattern the initial parameters are enlisted in Table 1.

Table 2 Final parameters with wide null imposed in 1st null-2nd peak window range

Algorithm used	Inter-element spacing $(d_1, d_2, \ldots, d_M)$ in λ	Max. null level (dB)	Peak SLL (dB)	HPBW ($\theta°$)
RGA	0.8893 3.3453 5.4205 5.0786 4.9568 5.1610 3.7189 3.1542 3.2622 4.2613 5.4253 4.7880 4.0048 3.1687 3.0631 0.9450 4.6267 5.0078 3.4379 0.4879	−32.68	−15.27	8.9

Table 3 Final parameters with wide null imposed in the 1st null-3rd peak window range

Algorithm used	Inter-element spacing $(d_1, d_2, \ldots, d_M)$ in λ	Max. null level (dB)	Peak SLL (dB)	HPBW ($\theta°$)
RGA	4.2862 3.2656 5.9800 4.6265 4.8506 5.5451 3.5444 5.0594 3.5103 3.6742 4.7189 0.9790 4.4554 5.2605 3.7819 3.2554 5.8696 4.2931 4.4481 5.6416	−28.05	−13.33	7.0

Introduction of wide nulls is considered for four different cases. The different ranges considered are: 1st null to 2nd peak, 1st null to 3rd peak, 1st null to 4th null and 1st peak to 2nd peak. 1st null to 2nd peak is positioned over the range $\theta = 60.66°–78.12°$. Due to the symmetricity of the radiation pattern; wide null is obtained between $\theta = 101.9°–119.3°$. The next range considered for implementing wide nulls in between 1st null and 4th null, i.e., $\theta = 46.08°–78.12°$. When nulls are imposed over this window range, due to symmetricity of pattern deeper null is obtained at $\theta = 101.9°–133.9°$ also. 1st null to 4th null window is yet another range considered to introduce broader nulls. This imposes wide nulls between $\theta = 38.7°–78.12°$ and between $\theta = 101.9°–141.3°$. Similarly, imposing null between 1st peak and 2nd peak generates wide null between $\theta = 60.66°–78.12°$ and $\theta = 106.7°$ and $119.3°$.

Table 4 Final parameters with wide null imposed in the 1st null–4th null window range

Algorithm used	Inter-element spacing $(d_1, d_2, \ldots, d_M)$ in λ	Max. null level (dB)	Peak SLL (dB)	HPBW $(\theta°)$
RGA	4.2862 3.2656 5.9800 4.6265 4.8506 5.5451 3.5444 5.0594 3.5103 3.6742 4.7189 0.9790 4.4554 5.2605 3.7819 3.2554 5.8696 4.2931 4.4481 5.6416	−28.05	−13.33	7.0

Table 5 Final parameters with wide null imposed in the 1st peak–2nd peak window range

Algorithm used	Inter-element spacing $(d_1, d_2, \ldots, d_M)$ in λ	Max. null level (dB)	Peak SLL (dB)	HPBW $(\theta°)$
RGA	4.4310 5.4186 5.3628 3.5346 4.3865 5.6780 4.2547 3.1552 4.5122 5.8575 3.2814 3.6122 5.5710 4.1676 5.6260 3.0518 5.3652 0.8037 3.3448 5.5689	−31.48	−13.31	7.2

Tables 2, 3, 4, and 5 gives the optimized inter-element spacing values, the final maximum window null level, peak SLL value and Half Power Beamwidth (HPBW) values for the radiation pattern corresponding to Figs. 3, 4, 5, 6, 7, and 8 respectively.

Figure 3 depicts radiation pattern in case of imposing of broad nulls between 1st null and 2nd peak using RGA. The initial maximum Null Level for the reference radiation pattern is −14.52 dB. Using RGA the Final Maximum Null Level improves to −32.68 dB. The peak SLL obtained is −15.27 dB and the HPBW in resultant case is 8.9°. The mean SLL has improved to −34.14 dB implementing RGA.

Figure 4 shows the improvement in radiation pattern for the case of imposing broad nulls between 1st null and 3rd peak using RGA. The maximum null has been brought down to −28.05 dB using RGA. The HPBW obtained using RGA is 7.0°. The peak SLL has been reduced to −13.33 dB and the mean SLL is reduced to −27.71 dB for the case of RGA.

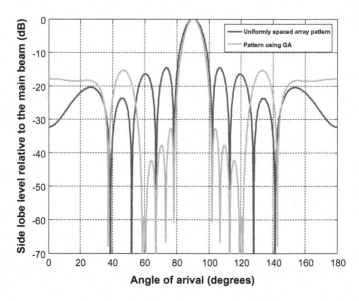

Fig. 3 Best array pattern found by RGA for the 41-element array from 1st null to 2nd peak i.e. θ = 60.66°–78.12° and θ = 101.9°–119.3°

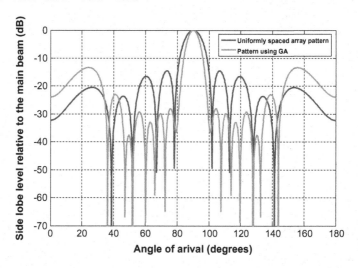

Fig. 4 Best array pattern found by RGA for the 41-element array from 1st null to 3rd peak i.e. θ = 46.08°–78.12° and θ = 101.9°–133.9°

Figure 5 depicts the improved radiation pattern when wide null is imposed between the 1st null and 4th null using RGA. The maximum value of null obtained with RGA is −28.05 dB. With RGA the resultant HPBW obtained in this case is 7.0°. The mean SLL has improved to −32.9 dB while the peak SLL goes down to −13.3 dB.

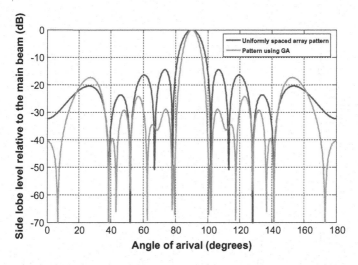

Fig. 5 Best array pattern found by RGA for the 41-element from 1st null to 4th null i.e.θ = 38.7°–78.12° and 101.9°–141.3°

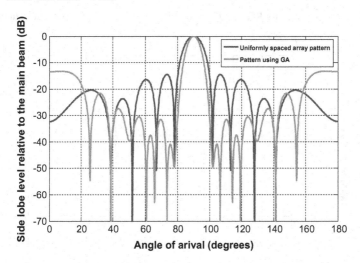

Fig. 6 Best array pattern found by RGA for the 41-element from 1st peak to 2nd peak i.e. θ = 60.66°–73.44° and θ = 106.6°–119.3°

The resultant radiation pattern obtained for the case of window nulling between 1st peak and 2nd peak using RGA algorithm is shown in Fig. 6. Using RGA the final maximum value of window null is −31.48 dB. The HPBW obtained is 7.2°. The peak SLL in the final radiation pattern is −13.31 dB while the mean SLL is–28.97 dB.

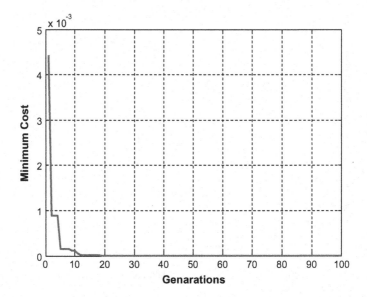

Fig. 7 Convergence curve for RGA for the case of wide nulling between 1st null and 2nd peak

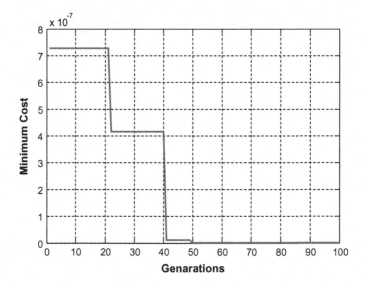

Fig. 8 Convergence curve for RGA for the case of wide nulling between 1st null and 4th null

4.2 Convergence Profiles

For the optimization technique, we need to obtain the convergence profile. Therefore, minimum values of CF are framed against count of iteration cycles. MATLAB

R2016a version on core (TM) 2 duo processor, 3.00 GHz with 2 GB RAM is used to write the optimization programs. Figures 7 and 8 show the convergence curve of RGA for the cases of wide nulling between 1st null and 2nd peak and between 1st null and 4th null respectively.

5   Conclusion

Transmission of data needs to be less effected by the losses and noises, so antenna array systems must be designed to reduce these negative effects. Experimental results show that the synthesis of non-uniformly spaced Linear Antenna Array offers an appreciable broad null, better peak SLL along with the reduction of 3-dB BW as compared to the corresponding uniformly spaced Linear Array. The major contribution of the paper is threefold (i) reduction of the inter-element spacing to less than $\lambda/4$ resulting in a compact design, (ii) obtaining deeper null over a specified window range, (iii) improvement of HPBW. There has been a considerable reduction of peak SLL also for most of the cases. The well-known method of Particle Swarm Optimization Algorithm is taken up as solution for this problem design. This algorithm generates a radiation pattern with wide null in the specified window range. However, the comparative analysis of the maximum null level of the window considered shows RGA to be a better approach for wide null control.

Acknowledgements This work is funded by The Science and Engineering Research Board (SERB), New Delhi, Government of India, project no. SB/EMEQ-319/2013, dated July 31, 2013.

References

1. Ballanis, C.A.: Antenna Theory Analysis and Sesign, 2nd edn. John Willey and Son's Inc., New York (1997)
2. Godara, L.C.: Applications of antenna arrays to mobile communications, Part I: performance improvement, feasibility, & system considerations. In: Proceedings of the IEEE, vol. 85(87) (1997)
3. Winters, J.H., Salz, J., Gitlin, R.D.: The impact of antenna diversity on the capacity of wireless communication systems. IEEE Trans. Commun. **42**, 1740–1751 (1994)
4. Barrett, M., Arnott, R.: Adaptive antennas for mobile communications. Electron. Commun. Eng. J. **6**, 203–214 (1994)
5. Jungnickel, V., Pohl, V., von Helmolt, C.: Capacity of MIMO systems with closely spaced antennas. IEEE Commun. Lett. **7**(8) (2003)
6. Pohl, V., Jungnickel, V., Haustein, T., von Helmolt, C.: Antenna spacing in MIMO indoor channels. In: Proceedings of the IEEE Vehicular Technology Conference Birmingham, AL (2002)
7. Vaughan, R.G., Scott, N.L., Closely spaced monopoles for mobile communications. Radio Sci. **28**, 1259–1266 (1993)
8. Roje, V.: Wire antenna theory applied to the assessment of the radiation hazard in the vicinity of the GSM base stations. Serbian J. Electrical Eng. **1**(1), 15–26 (2003)

9. Danker, B.: Dipoles, unintentional antennas and EMC. Serbian J. Electrical Eng. **5**(1), 31–38 (2008)
10. Styeskal, H., Shore, R.A., Haupt, R.L.: Methods for null control and their effects on the radiation pattern. IEEE Trans. Antennas Propag. **AP-34**(3) (1986)

A Design of Highly Stable and Low-Power SRAM Cell

P. Upadhyay, R. Kar, D. Mandal, S. P. Ghoshal and Navyavani Yalla

Abstract A novel architecture of low power and highly stable SRAM cell has been proposed in this paper. The proposed design contains a voltage source (VS). This voltage source minimizes the swing voltage required for switching activity. Swing voltage reduction controls the dynamic power consumption at high speed. One NMOS transistor is used to isolate the direct path between bit line and data storage points and increases the external noise tolerance level. Power consumptions at various frequencies and static noise margins of the proposed model are noted down and the comparison is made with the other reported SRAM cell designs. 45 nm CMOS technology and Microwind 3.1 software tool are used for circuit simulation.

Keywords CMOS technology · Active power consumption · Passive power consumption · SRAM · SNM · Swing voltage

1 Introduction

As the CMOS technology enhances, miniaturization of devices is required. But to avoid high electric fields device effect supply voltage is also goes down. This reduction in supply voltage also minimizes the power utilization in circuit because both power consumption and power supply are correlated with each other [1]. Dynamic power consumption is related with frequency, supply voltage, and swing voltage. Fast devices needed faster volatile memory. So static RAM (random access memory) is used because it stores the 1-bit information as well as its complement on its bit and

P. Upadhyay · R. Kar (✉) · D. Mandal · N. Yalla
Department of Electronics and Communication Engineering, National Institute of Technology Durgapur, Durgapur 713209, India
e-mail: rajibkarece@gmail.com

S. P. Ghoshal
Department of Electrical Engineering, National Institute of Technology Durgapur, Durgapur 713209, India
e-mail: aspghoshalnitdgp@gmail.com

© Springer Nature Singapore Pte Ltd. 2019
S. K. Bhatia et al. (eds.), *Advances in Computer Communication and Computational Sciences*, Advances in Intelligent Systems and Computing 759,
https://doi.org/10.1007/978-981-13-0341-8_26

bit bar lines. At high speed dynamic power utilization of SRAM circuit increases. And this reduces the battery life. So low-power SRAM cell design is required. Lot of literature has been present related with power consumption and stability of SRAM cell.

In [2], both passive and active power saving techniques has been described. A power saving techniques based on bit line is used during active mode and in passive mode a digitally controllable retention approach is used. A 7T SRAM cell with single bit line structure for low power has been described in [3].

In [4], the spin-transfer torque (STT)-SRAM has been discussed. This technique would help to minimize the power consumption in the communication bus and provides a longer time of storage with very low power consumption. This SRAM model is related with dynamic power consumption. A high speed, high density and low power 5T structure is discussed in [5]. The cell model is working on the single-ended storage principle. A capacitive charge-sharing write assist FinFET based SRAM has been reported in [6].

In [7], a leakage current compensation based SRAM design is described. SRAM cell is based on the sensor which measures the leakage and if the leakage is more than the predefined threshold value, the speed of the device decreases automatically to reduce the power consumption. Different SRAM cell architectures are also discussed in [8–13] to achieve low power solution.

In the present paper 8T SRAM architecture has been suggested which utilizes lesser power and contains high stability. Voltage source in the proposed design provides the less swing voltage at higher frequency. NMOS transistor (M7) helps to minimize the sub-threshold current and improves stability.

The organization of paper is described as: Sect. 2 gives the brief discussion on conventional static RAM architecture; Sect. 3 presents architecture and operation of the suggested static RAM. Section 4 described the simulated power and SNM values, Sect. 5 discussed the final conclusion.

2 Conventional SRAM Cell Architecture

A standard six transistor static RAM cell design is presented in Fig. 1. In this architecture two CMOS cross coupled inverters are connected in a way to store 1-bit of information. Two access transistors (T2 and T5) are connected for accessing information from the data busses [8].

At the time of write operation word line (WL) makes high and allow access transistors (T2 and T5) to receive the data. If BL = "1" then BBL will be zero because both are complement of each other. Both are cannot be same potential at the same time while write action is performed.

Pre-charge action is performed on BBL and BL before starting the read operation. Then one bit line is automatically release its charge because SRAM circuit is either in position of "1" or in "0". This causes a potential difference between both the bit lines. Then sense amplifier computes this potential difference.

Fig. 1 6T SRAM design

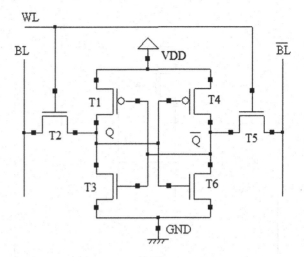

3 Proposed SRAM Architecture

The suggested circuit has a voltage source (VS) which connects with the bitbar line (BBL) output node. NMOS transistor M8 is connects or disconnects the voltage source according to the requirement of the cell operation. M7 MOS transistor is attached with both the CMOS inverters (M1–M3 and M4–M6). Write word line (WWL) signal controls the access transistor M2 at the time of write operation. Similarly, read word line (RWL) and control signal for read (CSR) activated at the time of read cycle. The suggested 8T SRAM cell design has been presented in Fig. 2.

Feedback connection between the inverters is disconnected during write operation and it is reconnected before started the read operation. This feedback connection and disconnection is performed by making CSR signal High or Low. For performing write operation only one bit line (BBL) is used. New data is transferred into the cell with the use of BBL and transistor M2. The BL and transistors M5 and M7 are used for reading operation. For write operation first make M7, M5 is in OFF condition and M2 is in ON condition. Point Q stores complement of the input information at BBL. A inverter is connected with the BBL input signal. This inverter provides a compliment of BBL input value. This complement value is stored on point Q and gives the driving voltage to CMOS inverter (M1–M3).

At the time of write "0", BBL goes Low and BBL BAR signal goes high and this makes the Q point High. High state at Q turns M3 ON and turns M1 OFF. So any charge stored on QB discharge through M3 to ground. While writing "1", BBL is goes High and BBL BAR signal goes Low and due to that Q point goes Low. Low state at Q turns OFF the transistor M3 and turn ON the transistor M1. So charge is stored on QB through M1 transistor. BBL high signal also turns ON the M8 transistor and voltage source VS reduces the swing voltage at node QB. Input and Output signal diagram for write operation has been shown Fig. 3.

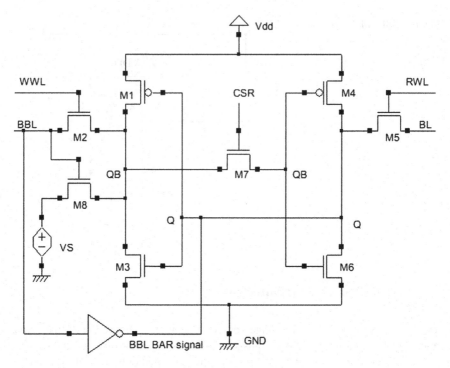

Fig. 2 The proposed model SRAM cell

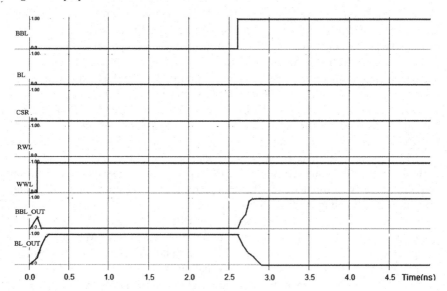

Fig. 3 Write operation input and output signal diagram of proposed cell

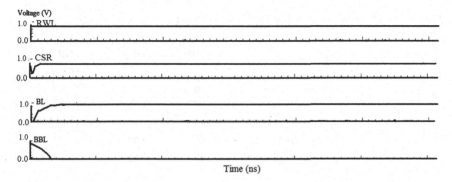

Fig. 4 Read operation input and output signal diagram of proposed cell

Table 1 Simulation parameters

Parameters	Values
Supply voltage	0.5 V
Temperature	25 °C
Bitline capacitance	15 pF
Process technology	45 nm

Read operation of proposed design is very much similar to conventional design (discussed in Sect. 2). M7 and M5 make ON with the help of CSR and RWL signals, respectively. Input and Output signal states for read action of suggested SRAM circuit is clearly presented in Fig. 4.

4 Results and Discussion

This section examines the simulated results of proposed architecture. Dynamic power consumption at various frequencies and read/write static noise margins has been measured. Dsch tool is used for circuit design and BSim4 model is used for simulation calculation. Table 1 shows the parameters which are used at the time of simulation.

4.1 Dynamic Power Dissipation Analysis

During change of state from "0" to "1" or "1" to "0" for SRAM cell, swing voltage plays a key role and this is directly related with dynamic power dissipation [14], which is shown in (1).

$$P_{dynamic} = \alpha C V_{dd} V_{Swing} f, \tag{1}$$

SRAM cell	Dynamic power dissipation at various frequencies (μW)		
	500 MHz	1 GHz	2 GHz
5T [5]	8.553	15.127	31.339
6T [8]	10.890	19.002	39.979
7T [3]	5.578	11.249	21.861
8T [12]	9.562	19.015	37.472
9T [13]	7.715	14.372	29.482
Proposed 8T	2.193	3.629	4.286

Table 2 Dynamic power consumption

where C = Load capacitance; α = Activity factor; f = Clock frequency; V_{Swing} = Swing Voltage at the output node; V_{dd} is the power supply voltage.

As frequency of the circuit increases, the activity of switching is enhanced and this increases the swing voltage value. High value of swing voltage means higher dynamic power consumption.

So for that purpose voltage source VS is used in the proposed model which reduces swing voltage at higher switching activity state. So for high speed devices the dynamic power dissipation is reduced.

Dynamic power consumption at various frequencies has been shown in Table 2.

The proposed model utilizes the least dynamic power 2.193 μW at 500 MHz, 3.629 μW at 1 GHz and 4.286 μW at 2 GHz as compared with the other reported SRAM cells. Also consumption of power for proposed model is not increased sharply than the other models when the frequency increases from 500 MHz to 2 GHz.

4.2 Static Noise Margin Analysis

Least value of dc noise voltage which is required to change the state is called static noise margin (SNM). Stability of any digital circuit is measured through SNM value [9]. SNM is depends on the threshold voltage of MOS transistors present in the circuit [10]. For increasing SNM value, the threshold voltages of the MOS transistors should be minimized. The reason behind is that MOS transistors having high threshold voltages are hard to flip the operation from ON to OFF or vice a versa.

Read SNM is the measurement of voltage which is required for changing the state during reading cycle. Figure 5 shows the waveform curve of read SNM of the proposed cell design. This curve is generated through the test schematic. As the supply voltage decreases the value of read static noise margin also reduced [11] and this increases instability in the transistors. This causes the failure of read operation. Figure 5 shows the RSNM curve of proposed model.

Fig. 5 Read SNM curve of proposed SRAM model

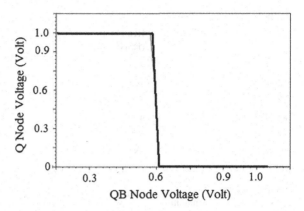

Fig. 6 Write SNM curve of proposed SRAM model

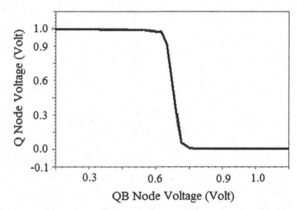

Write SNM value shows the write stability of circuit at the time of write operation. The ability of an SRAM cell to change its state is called write SNM. The write SNM curve has presented in Fig. 6.

Table 3 shows the RSNM and WSNM for the proposed SRAM cell is 581 mV and 602 mV, respectively. Voltage source provides some extra voltage during read/write operation and increases the RSNM and WSNM values. The M7 transistor operation also improves the stability value of proposed model.

For the proper operation of SRAM cell the width and length ratios of MOS transistor are chosen very carefully. The length of all the transistors is taken identical according to the CMOS technology used. The simulation results at different width ratios has been observed and this shows that for providing proper driving voltage the width ration of access transistor (M2) with pull up transistor (M1) should be approximately equivalent to 1.5. Similarly the ration of width of pull down transistor (M3) with access transistor (M2) should be approximately equivalent to 1.5. The same is true for other inverter circuit.

Table 3 Static noise margin

SRAM Cell	Read Static Noise Margin (RSNM) (mV)	Write Static Noise Margin (WSNM) (mV)
5T [5]	493	513
6T [8]	425	448
7T [3]	537	557
8T [12]	420	467
9T [13]	529	539
Proposed 8T	581	602

Table 4 Widths of transistors used for simulation

Transistors	Width (nm)
M1 & M4	75
M2 & M5	115
M3 & M6	175
M7	350
M8	350

$$\frac{W_2}{W_1} \approx \frac{W_3}{W_2} \approx 1.5 \text{ and } \frac{W_5}{W_4} \approx \frac{W_6}{W_5} \approx 1.5 \tag{2}$$

Transistor Widths chosen for simulation has been shown in Table 4. Proper sizing of transistors full fill the driving voltage need to the SRAM circuit at the time of ON and OFF operation. Read and write access times of SRAM cell are the total times required for completing one cycle of read and write, respectively. For the proposed SRAM cell the value of write access time is 9.831 ns and the read access time is 11.171 ns which is poorer than the conventional SRAM. Leakage current 7.899 pA is measured during standby mode.

5 Conclusions

Miniaturization of the CMOS transistors is required as the technology advances. But power and stability are the key concerns as the channel length of MOS transistors is reduced. In the present paper a highly stable and less power eating SRAM model has been suggested. The suggested architecture has voltage source which is responsible for minimizing the swing voltage at the time of any switching activity. Minimization in swing voltage restricts dynamic power utilization at high speed switching activities. Proper width ratio improves the stability of the proposed SRAM. So the proposed 8T SRAM model provides a low power and highly stable architecture for power hungry high-speed devices.

References

1. De, V.: Energy-efficient computing in nanoscale CMOS. IEEE Des. Test **33**(2), 68–75 (2016)
2. Tachibana, F., Hirabayashi, O., Takeyama, Y., Shizuno, M., Kawasumi, A., Kushida, K., Suzuki, A., Niki, Y., Sasaki, S., Yabe, T., Unekawa, Y.: A 27% active and 85% standby power reduction in dual-power-supply SRAM using BL power calculator and digitally controllable retention circuit. IEEE J. Solid-State Circuits **49**(1) (2014)
3. Madiwalar, B., Kariyappa, D.: Single bit-line 7T SRAM cell for low and high SNM. In: International Multi Conference on Automation, Computing, Communication, Control and Compressed Sensing, pp. 223–228 (2013)
4. Halawani, Y., Mohammad, B., Homouz, D., Al-Qutayri, M., Saleh, H.: Modeling and optimization of memristor and STT-RAM-Based memory for low-power applications. IEEE Trans. Very Large Scale Integr. VLSI Syst. **24**(3), 1003–1014 (2016)
5. Gupta, R., Gill, S.S.: A novel low leakage and high density 5T CMOS SRAM cell in 45 nm technology. In: Proceedings Recent Advances in Engineering and Computational Sciences, pp. 1–6 (2014)
6. Karl, E., Guo, Z., Conary, J., Miller, J., Ng, Y.G., Nalam, S., Zhang, K.: A 0.6 V, 1.5 GHz 84 Mb SRAM in 14 nm FinFET CMOS technology with capacitive charge-sharing write assist circuitry. IEEE J. Solid-State Circuits **51**(1), 222–229 (2016)
7. Wang, C.C., Wang, D.S., Liao, C.H., Chen, S.Y.: A leakage compensation design for low supply voltage SRAM. IEEE Trans. Very Large Scale Integr. VLSI Syst. **24**(5), 1761–1769 (2016)
8. Akashe, S., Sharma, S.: Leakage current reduction techniques for 7T SRAM cell in 45 nm technology. J. Wirel. Personal Commun. **16**(4), 147–165 (2012)
9. Kang, K., Jeong, H., Yang, Y., Park, J., Kim, K., Jung, S.O.: Full-swing local bit line SRAM architecture based on the 22-nm FinFET technology for low-voltage operation. IEEE Trans. Very Large Scale Integr. VLSI Syst. **24**(4), 1342–1350 (2016)
10. Rajanna, V.K., Amrutur, B.: A variation-tolerant replica-based reference-generation technique for single-ended sensing in wide voltage-range SRAMs. IEEE Trans. Very Large Scale Integr. VLSI Syst. **24**(5), 1663–1674 (2016)
11. Pal, S., Islam, A.: Variation tolerant differential 8T SRAM cell for ultralow power applications. IEEE Trans. Comput. Aided Des. Integr. Circuits Syst. **35**(4), 549–555 (2016)
12. Pasandi, G., Fakhraie, S.M.: An 8T low-voltage and low-leakage half-selection disturb-free SRAM using bulk-CMOS and FinFETs. IEEE Trans. Electron Devices **61**(7), 2357–2363 (2014)
13. Tu, M.H., Lin, J.Y., et al.: A single ended disturb-free 9T subthreshold SRAM with cross-point data-aware write word-line structure, negative bit-line, and adaptive read operation timing tracing. IEEE J. Solid State Circuits **47**(6), 1469–1482 (2012)
14. Weste, N.H.E., Harris, D., Banerjee, A.: CMOS VLSI Design, 3rd edn., pp. 55–57. Pearson Education (2005)

Optimal Design of 2.4 GHz CMOS LNA Using PSO with Aging Leader and Challenger

S. Mallick, R. Kar, D. Mandal, Tanya Dasgupta and S. P. Ghoshal

Abstract This paper presents in front of us a novel approach for the optimal design of a Low Noise Amplifier (LNA) with inductive source degeneration circuit using a recently proposed evolutionary optimization technique called PSO with Aging Leader and Challenger (ALC-PSO). The proposed ALC-PSO based approach has succeeded in dealing with the disadvantages faced by PSO algorithm and is employed in this paper for the optimal design of LNA circuit. The MOSFET widths and component's values are optimized by using ALC-PSO algorithm in order to maximize the gain, minimize the Noise Figure (NF) and to optimize the overall performance of the LNA circuit. The simulation results obtained for the designed LNA circuit confirm the effectiveness of the ALC-PSO based approach over PSO in terms of solution quality, design specifications, and design objectives. The optimally implemented LNA circuit in 0.35 μm CMOS technology yields the gain of 18.64 dB, noise figure of 1.779 dB and power dissipation of 10.60 mW.

Keywords CMOS · Circuit sizing · LNA · Noise figure · PSO with aging leader and challenger · Evolutionary optimization techniques

S. Mallick · R. Kar (✉) · D. Mandal · T. Dasgupta
Department of Electronics & Communication Engineering, National Institute of Technology, Durgapur, Durgapur, India
e-mail: rajibkarece@gmail.com

S. Mallick
e-mail: soumenju@gmail.com

S. P. Ghoshal
Department of Electrical Engineering, National Institute of Technology, Durgapur, Durgapur, India

© Springer Nature Singapore Pte Ltd. 2019   291
S. K. Bhatia et al. (eds.), *Advances in Computer Communication and Computational Sciences*, Advances in Intelligent Systems and Computing 759,
https://doi.org/10.1007/978-981-13-0341-8_27

1   Introduction

To design wireless systems mostly in the radio frequency (RF) region, CMOS technology is used extensively due to the low cost and simple integration. RF components are the basic building blocks of transceivers operating in GHz frequency range. A lot of efforts are required for the designing of RF circuit components. Even after performing rigorous calculations and finding the parameter values it is not guaranteed that the circuit will perform as expected. In Radio Frequency Integrated Circuits (RFIC), LNA is considered as a black box because of its superior response at greater frequencies. Because of the mismatch of the output and the input impedances the maximum power transfer is not possible. For a wireless receiver, the first building block is LNA. The main function of LNA is to amplify the incoming signal which is wireless, without much involvement of distortion and noise. The performance of the LNA, with respect to noise, radically improves the overall system's performance in terms of noise. The inductive source degenerated LNA [1–4] is commonly used because of its better noise performance.

Generally, optimization problems are formulated with various design parameters, specifications, and objective functions for the analog circuits to be designed. For the composite problems, more time and space are required for the optimization technique. Recent bio-inspired optimization methods called swarm intelligence [5] overcomes the limitations of the above said optimizing methods. PSO is one of the popular optimization methods among various bio-inspired methods.

For wideband WLAN, 2.4 GHz low-IF receiver is reported in [6], dual band concurrent CMOS LNA is presented in [7], comparison of CMOS and SiGe LNA is given in [8], 2.4 GHz single-ended LNA in [9] and in [10] noise lowering and technique of improvement in linearity for differential cascode LNA is discussed.

In Sect. 2, design procedure of inductive source degeneration CMOS LNA and the objective function is put up. PSO and ALC-PSO algorithms are discussed briefly in Sect. 3. Performance results verified by CADENCE are given in Sect. 4. Conclusions are drawn in Sect. 5.

2   Design Aim and Objective Function

The best design of an inductive source degenerated CMOS LNA circuit (Fig. 1) is carried out in this paper. To do this, the design specifications/optimizing variables considered are source resistance (R_S), load resistance (R_L), load capacitor (C_L), blocking capacitor (C_S), drain inductance (L_d), source inductance (L_s), gate inductance (L_g), and centre frequency (f_0).

The cost/objective function is developed in terms of overall gain obtained from the proposed CMOS LNA with inductive source degeneration circuit, which would be optimized for the design specifications, also called optimizing variables, as mentioned above, with the help of ALC-PSO algorithm and validated by Cadence Spectre-RF simulator.

The goal of optimization is usually to minimize the noise figure and to maximize the gain.

2.1 Design Steps of CMOS LNA with Degeneration of Inductive Source

The design steps are given in [4] for the CMOS LNA with degeneration of inductive source.

$$Z_{in} \approx s\left(L_g + L_s\right) + \frac{1}{sC_{gs}} + \frac{g_m L_s}{C_{gs}}, \tag{1}$$

where L_g is the gate inductance and L_s is the transistor source inductance and both are assumed to be ideal inductors. The gate to source capacitance is denoted by C_{gs} and the transconductance of M1 by g_m.

To achieve the input match at the resonance the following condition is to be satisfied.

$$R_s = \frac{g_m L_s}{C_{gs}} = 50\,\Omega \tag{2}$$

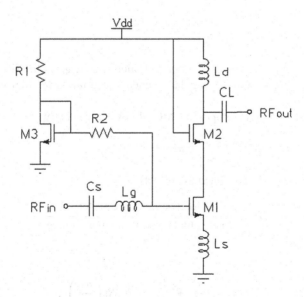

Fig. 1 CMOS LNA with inductive degeneration of source

In other words, at resonance frequency ω_0, to provide the input match once L_s is selected, L_g an also be chosen to make Z_{in} in real term and is equal to (2). For L_g, the following equation must hold good.

$$s\left(L_g + L_s\right) + \frac{1}{sC_{gs}} = 0 \tag{3}$$

$$\omega_0 = \frac{1}{\sqrt{(L_g + L_s)C_{gs}}} \tag{4}$$

The quality factor (Q) of the input circuit at the resonance is given as

$$Q_{in} = \frac{1}{\omega_0(R_s + \omega_T L_s)C_{gs}} = \frac{1}{2\omega_0 R_s C_{gs}} \tag{5}$$

where ω_0 is the resonant frequency. The voltage across the capacitor at resonance is expressed as

$$V_{gs} = Q_{in}V_{in} \tag{6}$$

Hence, the effective transconductance can be given as

$$G_m = g_m Q_{in} = \frac{\omega_T}{\omega_0\left(1 + \frac{\omega_T L_s}{R_s}\right)R_s} \tag{7}$$

where,

$$\omega_T \cong \frac{g_m}{C_{gs}} \tag{8}$$

and Q_{in} is the effective Q of the input amplifier circuit; g_m is transistor transconductance; ω_0 is the operating angular frequency and ω_T is the transition angular frequency.

The overall gain of the proposed CMOS LNA is given by (9).

$$Gain = G_m R_L \tag{9}$$

The noise factor (F) is defined as [1, 2, 4]

$$F = \frac{\text{Total output noise power}}{\text{Total output noise due to input source}}$$

Or,

$$F = 1 + \frac{R_l}{R_s} + \frac{R_g}{R_s} + \frac{\gamma}{\alpha} g_m R_s \left(\frac{\omega_0}{\omega_T}\right)^2, \tag{10}$$

where $\alpha = \frac{g_m}{g_{d0}}$; g_{d0} is the drain conductance in zero-bias condition; R_l is the series resistance of inductors; Rs is the resistance of source; Rg is the resistance of gate and γ is the substrate-bias coefficient.

The noise figure (NF) is to be calculated by (11).

$$F_{min,P} \approx 1 + 2.4\frac{\gamma}{\alpha}\left(\frac{\omega_0}{\omega_T}\right) \qquad (11)$$

The absolute minimum noise figure is denoted by (12).

$$F_{min} = 1 + \frac{2}{\sqrt{5}}\frac{\omega_0}{\omega_T}\sqrt{\gamma\delta(1 - |c|^2)}, \qquad (12)$$

where δ is the coefficient of gate noise, c is the coefficient of correlation.

$$F_{min} \approx 1 + 2.3\left(\frac{\omega_0}{\omega_T}\right) \qquad (13)$$

Optimum width of the MOS is given by (14).

$$W_{opt} = \frac{1}{3\omega_0 L C_{ox} R_s} \qquad (14)$$

The population matrix is taken as 100×7. The number of particles is 100 and the dimension of particle vector is 7. The particle vector formation is shown in (15).

$$X_{LNA} = [R_s, R_L, C_S, C_L, L_s, L_g, L_d] \qquad (15)$$

The following parameters have been considered for ALC-PSO. The coefficients of acceleration, $C_1 = C_1 = 2$; the initial value for lifespan, $\Phi_0 = 3$; the probability is set to $pro = 1/d$. The Objective Function (OF) is defined as the sum total gain of the proposed optimized LNA circuit. The OF used in this work is given in (16).

$$OF = \frac{\omega_T}{\omega_0}\left(\frac{R_L}{2R_s}\right), \qquad (16)$$

where, ω_T is the transition angular frequency; ω_0 is the operating angular frequency; R_S and R_L are the source and load resistance, respectively. The target value of OF is needed to be greater than 15 dB for the optimally designed CMOS LNA circuit with degeneration of inductive source.

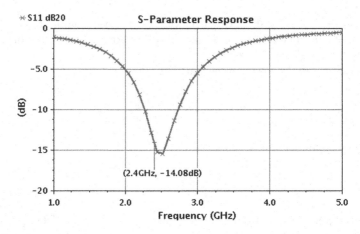

Fig. 2 S11 plot

3 ALC-PSO Algorithm

ALC-PSO algorithm [11] is well known and is not explained here in detail. The subsections of ALC-PSO are given below

1. Procedures of ALC-PSO
2. Controller of Lifespan
3. Challenger generation
4. Criterion for the acceptance of the new leader

Figure 2 shows the flowchart of ALC-PSO algorithm.

4 Performance Results

ALC-PSO algorithm is executed and simulated in MATLAB R2013a on Intel core™ i5-2430M CPU @3.00 GHz. The algorithm has been run for 100 epochs.

The target of this work is to improve the gain of the proposed LNA by keeping the noise figure (NF) as low as possible. After scompiling all the equations with the design conditions and the optimizing constraints specified in (15) the optimum design problem is formulated as the objective function maximization as specified in (16).

Table 1 Technology parameters of the proposed CMOS LNA design

Technology parameters	Values considered
Channel length (L)	0.35 μm
Oxide capacitance (C_{ox})	4.6 nF/μm^2
Supply voltage (V_{dd})	2.5 V
Transconductance parameter (K_n)	170 μA/V^2
Substrate-bias coefficient (γ)	2
Transconductance ratio (α)	0.85
Coefficient of gate noise (δ)	4
Correlation coefficient (lcl)	0.395

Table 2 Optimized components values of the proposed CMOS LNA

Component	Value
MOSFET (M1) width	245 μm
MOSFET (M2) width	245 μm
MOSFET (M3) width	65 μm
Reference resistance (R_1)	2 KΩ
Bias resistance (R_2)	1 KΩ
Gate inductance (L_g)	8.1 nH
Source inductance (L_s)	0.4 nH
Drain inductance (L_d)	4.5 nH
Blocking capacitance (C_s)	10 pF
Load capacitance (C_L)	1 pF

4.1 Simulated Results for CMOS LNA

For the optimal design of CMOS LNA, the technology parameters used are shown in Table 1. MOSFET lengths are selected as 0.35 μm. The simulation results indicate that by using ALC-PSO, the maximum gain of 18.64 dB with the faithful values of design specifications and components (M1, M2, M3, R_1, R_2, L_d, L_s and L_g) are obtained in 0.9319 s (Table 2).

Figure 3 depicts the plot of loss of input return (S_{11}), Fig. 4 depicts the reverse isolation plot (S_{12}), Fig. 5 depicts the plot of gain (S_{21}), Fig. 6 depicts the plot of the output return loss (S_{22}), Fig. 7 depicts the plots of stability factors (K_f and B_{if}), Fig. 8 depicts the plots of noise figure (NF) and the minimum noise figure (NF_{min}).

The noise figure (NF) of 1.779 dB and the least noise figure (NFmin) of 1.634 dB are achieved. The LNA circuit becomes unstable in the presence of feedback paths for specific combinations of source inductance (L_s) and load inductance (L_d). A normally

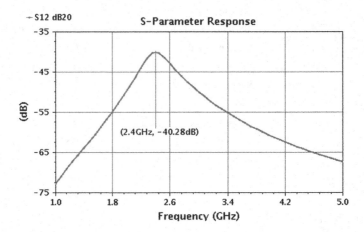

Fig. 3 S12 plot

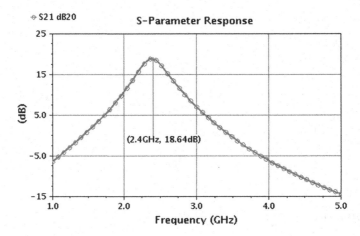

Fig. 4 S11 plot

stable LNA circuit may oscillate with the variation of voltages and frequencies. The
Stern stability factor is given by (17).

$$K_f = \frac{1 + |\Delta|^2 - |S_{11}|^2 - |S_{22}|^2}{2|S_{12}||S_{21}|}, \tag{17}$$

where, $B_{if}(\Delta) = S_{11}S_{22} - S_{12}S_{21}$.

The circuit becomes absolutely stable when K_f is greater than one and B_{if} is less
than one. For the S parameters (S_{11}, S_{12}, S_{21}, and S_{22}), the stability assessment has
been considered over a wide range of frequency to ensure that K_f remains greater
than one for the complete frequency range. With the increase in reverse isolation

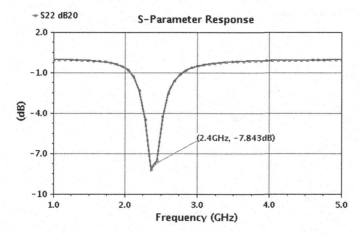

Fig. 5 S12 plot

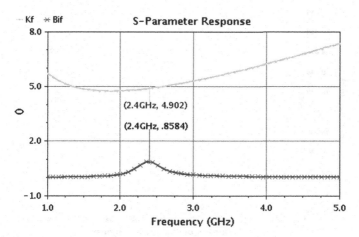

Fig. 6 Kf and B$_{if}$ plots

(S_{12}), the stability of LNA too increases. Stability analysis proves that $K_f = 4.902$ and B_{if} (Δ) = 0.8584 at 2.4 GHz, stating that designed LNA is completely stable.

The linearity confines the actual power drawn to the load by the LNA. The simulation of linearity is done using the analysis of periodic steady-state (PSS). The linearity analysis of the proposed LNA with 1 dB compression point (P1 dB) and input 3rd order intercept point (IIP3) are −10.2 dBm and −5.5 dBm, respectively.

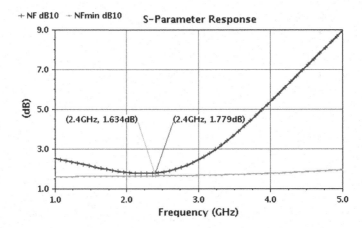

Fig. 7 NF and NF$_{min}$ plots

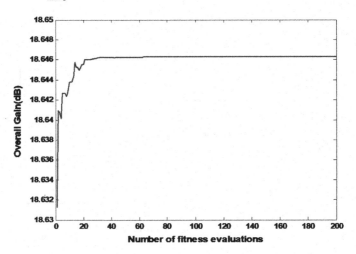

Fig. 8 Convergence curve of gain of CMOS LNA obtained by ALC-PSO

The DC power consumption of the proposed LNA is only 10.60 mW for 2.5 V power supply. The simulation results imply that the output and input stages of the proposed ALC-PSO based CMOS LNA with inductive source degeneration achieve a good matching, good stability factors. Meanwhile, DC power consumption (P_{DC}) and the noise figure (NF) are very small as well, which satisfy the RF circuit requirements. A widely employed figure of merit (FOM) for LNAs with low power is the ratio of power gain and DC power consumption (Gain/P_{DC}). The comparison of the overall performance of the LNA has been done by the authors using an additional FOM which includes the effect of amplifier gain, noise figure and DC power consumption (P_{DC}) as follows [12]:

Table 3 Performance comparison of the proposed CMOS LNA

Design parameters	Reference					
	[6]	[7]	[8]	[9]	[10]	This work
CMOS technology (μm)	0.6	0.35	0.25	0.35	0.35	0.35
Frequency (GHz)	2.4	2.4	2.45	2.4	2.2	2.4
Supply voltage V_{dd} (V)	3.3	2.5	3	3.3	1.8	2.5
S11 (dB)	−19	−25	<−14.2	<−33	<−13	−14.08
S12 (dB)	NR*	NR*	NR*	NR*	NR*	−40.28
S21 (dB)	17.5	14	15.1	6	8.6	18.64
S22 (dB)	NR*	NR*	NR*	NR*	NR*	−7.843
NF (dB)	2.3	2.3	2.88	4.8	1.92	1.779
P1 dB (dBm)	−9.0	−8.5	NR*	NR*	NR*	−10.2
IIP3 (dBm)	1.8	0	2.2	0.55	−2.6	−5.5
P_{DC} (mW)	26.4	10	24.3	NR*	16.2	10.60
FOM (mW^{-1})	0.51	1.08	0.33	N/A	0.58	2.26

$$FOM\,[\text{mW}]^{-1} = \frac{Gain\,[\text{abs}]}{(NF - 1)\,[\text{abs}] \cdot P_{DC}\,[\text{mW}]} \tag{18}$$

The results of simulation show that the proposed CMOS LNA based on ALC-PSO with inductive source degeneration has the best overall FOM in comparison with other literatures [6–10] (Table 3).

Simulation results as shown in Figs. 3, 4, 5, 6, 7 and 8 prove the proposed design based on ALC-PSO algorithm, satisfies each of the design goals and improves the overall gain, also reducing the noise figure of the proposed CMOS LNA. The performance comparison is given in Table 3 shows that the proposed design achieves the best performances in terms of different parameters in comparison with those of the reported literature.

4.2 Convergence Profile of the Gain of LNA Using ALC-PSO

Figure 9 shows the convergence profile of gain by employing the algorithm ALC-PSO for the design of CMOS LNA. The overall gain of 18.64 dB is obtained and the execution time is 0.9313 s for 100 number of iteration.

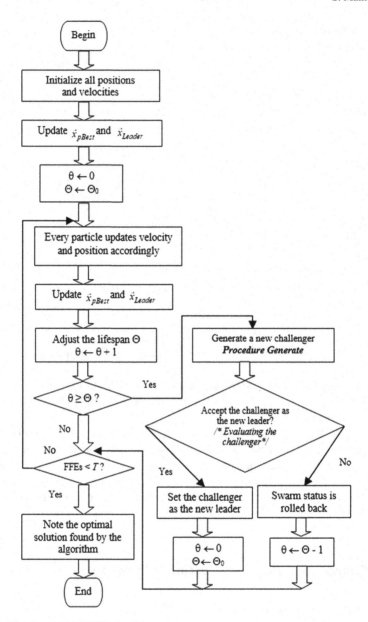

Fig. 9 Flowchart of ALC-PSO algorithm

5   Conclusions

The aspect ratios, inductor, and capacitor values are the main design variables optimized for the CMOS LNA design. Each and every design constraints are obtained with the exploitation of ALC-PSO algorithm and are verified by Cadence Spectre-RF simulator. The simulated results confirm that ALC-PSO based designed LNA yields the minimum noise figure and the maximum gain by satisfying all of the design criteria. In this paper, the CMOS LNA based on ALC-PSO with inductive source degeneration circuit design achieves the best gain (S_{21}) of 18.64 dB, the least noise figure (NF) of 1.779 dB and the best figure of merit (FOM) of 2.26 $[mW]^{-1}$.

References

1. Sheffer, D.K., Lee, T.H.: A 1.5 V, 1.5 GHz CMOS low-noise amplifier. IEEE J. Solid-State Circuits **32**(5), 745–759 (1997)
2. Sheffer, D.K., Lee, T.H.: Corrections to "A 1.5 V, 1.5 GHz CMOS low-noise amplifier". IEEE J. Solid-State Circuits **40**(6), 1397–1398 (2005)
3. Nguyen, T.K., Kim, C.H., Ihm, G.J., Yang, M.S., Lee, S.G.: CMOS low-noise amplifier design optimization tehcniques. IEEE Trans. Microw. Theory Tech. **52**(5), 1433–1442 (2004)
4. Lee, T.H.: The Design of CMOS Radio-Frequency Integrated Circuits. Cambridge University Press, Cambridge, U.K. (2006)
5. Bonabeau, E., Dorigo, M., Theraulaz, G.: Swarm Intelligence: From Natural to Artificial System. Oxford University Press, New York (1999)
6. Behbahani, F., Leete, J.C., Kishigami, Y., Roithmeier, A., Hoshino, K., Abidi, A.A.: A 2.4-GHz low-IF receiver for wideband WLAN in 0.6 μm CMOS-architecture and front-end. IEEE J. Solid-State Circuits **35**, 1908–1916 (2000)
7. Hashemi, H., Hajimiri, A.: Concurrent dual-band CMOS low noise amplifiers and receiver architectures. In: VLSI Circuits Symposium of Digest of Technical Papers, pp. 247–250, June 2001
8. Li, X., Brogan, T., Esposito, M., Myers, B., Kenneth, K.O.: A comparison of CMOS and SiGe LNA's and mixers for wireless LAN application. In: IEEE Custom Integrated Circuits Conference (CICC), pp. 531–534 (2001)
9. Su, J., Meng, C., Li, Y., Tseng, S., Huang, G.: 2.4 GHz 0.35 μm CMOS single-ended LNA and mixer with gain enhancement techniques. In: Asia-Pacific Microwave Conference (APMC), pp. 1550–1553, Dec 2005
10. Fan, X., Zhang, H., Sanchez-Sinencio, E.: A noise reduction and linearity improvement technique for a differential cascode LNA. IEEE J. Solid-State Circuits **43**(3), 588–599 (2008)
11. De Prasad, Bishnu, Kar, R., Mandal, D., Ghoshal, S.P.: Particle swarm optimization with aging leader and challengers for optimal design of analog active filters. Circuits Syst Signal Process. **34**(3), 707–737 (2015)
12. Linten, D., Aspemyr, L., Jeamsaksiri, W., Ramos, J., Mercha, A., Jenei, S., Thijs, S., Garcia, R., Jacobsson, H., Wambacq, P., Donnay, S., Decoutere, S.: Low-power 5 GHz LNA and VCO in 90 nm RF CMOS. In: VLSI Symposium Technology Digest, pp. 372–375 (2004)

Input-Output Fault Diagnosis in Robot Manipulator Using Fuzzy LMI-Tuned PI Feedback Linearization Observer Based on Nonlinear Intelligent ARX Model

Farzin Piltan, Manjurul Islam and Jong-Myon Kim

Abstract This paper proposes a model-based fault detection and diagnosis (FDD) technique for six degrees of freedom PUMA robot manipulator in presence of noise in actuator and sensor faults. The inverse modeling based on an adaptive method, which combines the fuzzy C-means clustering with the modified autoregressive eXternal (ARX) model, is presented for the system identification. The proposed adaptive nonlinear ARX fuzzy C-means (NARXNF) clustering technique obtains an improved convergence and error reduction than that of the traditional fuzzy C-means clustering algorithm. In addition, proportional integral (PI) feedback linearization observation is used for diagnosing the fault, where the convergence, robustness, and stability are validated by fuzzy linear matrix inequality (FLMI). Experimental results, in presence of 40% noise, show that the rate of root mean square (RMS) error for end-effector position is less than 0.00624. The proposed method also improves the rate of sensors and actuators FDD without additional hardware.

Keywords Inverse modeling · Fault diagnosis · Fuzzy C-means clustering
Autoregressive external model · PI feedback linearization observation

F. Piltan · M. Islam · J.-M. Kim (✉)
Department of Electrical, Electronics and Computer Engineering,
University of Ulsan, Ulsan, South Korea
e-mail: jongmyon.kim@gmail.com

F. Piltan
e-mail: piltan_f@iranssp.org

M. Islam
e-mail: m.m.manjurul@gmail.com

© Springer Nature Singapore Pte Ltd. 2019
S. K. Bhatia et al. (eds.), *Advances in Computer Communication and Computational Sciences*, Advances in Intelligent Systems and Computing 759,
https://doi.org/10.1007/978-981-13-0341-8_28

1 Introduction

Robot manipulators are widely used in industries for distinctive tasks. These manip-ulators are nonlinear, time-varying, and dynamically coupled due to having a com-plicated design, thus, they have nonlinear behaviors for control, identification, and fault diagnosis. In recent years, fault detection, fault estimation, and fault diagnosis techniques for industrial robots have been sharply increased. Abnormal industrial robot manipulators including fault detection and fault diagnosis are main challenges for avoiding the system breakdown and increasing the system's safety based on the early fault detection and identification. Thus, designing a stable and reliable fault detection and diagnosis (FDD) method for robot manipulator is an utmost priority in real-world applications [1, 2].

Fault diagnosis is one of the fundamental issues in various electrical and mechani-cal devices. Several techniques have been introduced to detect, identify, and estimate faults and these are divided into two main categories: functional-based and hardware-based fault diagnosis techniques. Hardware-based fault diagnosis is reliable and sta-ble for fault diagnosis, but it increases the price and the enlargement of the device. Because of the problem inherent in the hardware-based fault diagnosis, the use of function-based fault diagnosis in the industry has been increased. Functional-based fault diagnosis can be of three different categories namely, data-driven fault diagno-sis that works on massively available data, a knowledge-based method that works on systems knowledge, and a model reference-based method that works on known dynamics of the target system. In this paper, a model-based system identification observation technique is proposed for fault detection and diagnosis for six degrees of freedom (DOF) PUMA robot manipulator.

System identification-based observer techniques can be linear system identifi-cation and nonlinear system identification [3]. Linear system identification is used to identify the linear and basic nonlinear systems. In the earlier, the application of system identification for fault detection of robot manipulator was limited to linear techniques [4, 5] because of their simplicity in design structure and well-understood dynamic response, but it has a challenge to modeling the noisy and nonlinear sys-tems [6]. But, it is to be able to identify or estimate a nonlinear system, such as robot manipulator, we need to design a nonlinear system model. Artificial intelligent (AI)-based theory is widely applied for nonlinear system identification [3, 7]. Fuzzy-based identification method [8] and neural network-based method [6], are used to improve the performance of linear identification in robot manipulator. Although, fuzzy-based methods have been successfully applied in manifold applications of control engi-neering and intelligent decision-making, the linear fuzzy-based technique has some challenges such as guarantee stability and design accuracy rule base [9]. To improve the performance of linear fuzzy techniques, network-based adaptive methodology is suggested.

As this study is concerned with model-based fault diagnosis, the main challenges about fault diagnosis in the robot manipulator is to estimate the fault magnitude and the direction of fault to find the accurate fault location. For this propose, an appropriate thresholding technique is applied to estimate fault magnitudes on the residual signal, which is generated from the difference between actual and system's estimation signal, while system model performs its update operation [8, 10, 11]. In practices, these residuals are highly sensitive to the possible faults in the system that is caused to evaluate the fault diagnosis [12]. These signals are certainly independent of inputs and outputs processing in normal (or healthy) condition, because these signals should be zero or close to zero for normal condition. The magnitude of residual signals increases in faulty conditions and these residual signals are considerably different from zero [13]. Therefore, our proposed observer model for robot sensor and actuator faults diagnosis is based on an improved PI feedback linearization observer with fuzzy LIM, to ensure improve the convergence, stability, and robustness.

In overall, the key contributions of this paper are summarized as follows: (1) detecting faults and inverse modeling for six DOF robot manipulator based on the sensitive ARX fuzzy C-means clustering network, (2) to estimate and identify the fault based on proposed PI feedback linearization observation theory, (3) fuzzy LMI optimization technique to update the convergence, stability coefficients of observation model, and (4) further to generate an improved residual signal to hence the fault detection, and diagnosis rate.

This paper has following sections: Sect. 2 defines the problem statements. The proposed methodology including results of different case studies is in Sect. 3. In the final section, the conclusion has been presented.

2 Problem Statements

As the robot manipulator is a highly nonlinear, multi-input multi-output (MIMO), coupling effect, and multi-degrees of freedom system, if $x_1 = q$ and $x_2 = \dot{q}$, the Lagrange formulation of robot manipulator can be written in state space form as follows [1, 13]:

$$\begin{cases} X(k) = \ddot{q} = [AX(k-1) + b_y y(k-1) + b_u u(k-1)] + \beta_a f_a(k-1) \\ \ddot{q} = I^{-1}(q) \times \{[F(q) + F_d(q)] - [cor_{(q)}[\dot{q}\dot{q}] + cen_{(q)}[\dot{q}]^2 + F_{f(q,\dot{q})} + G_{(q)} + T_{ext}]\}, \\ Y(k) = (K)^T X(k) + \beta_s f_s(k-1) \end{cases} \quad (1)$$

where $F_{(q)}, F_{d(q)}, I_{(q)}, cor_{(q)}, cen_{(q)}, F_{f(q,\dot{q})}, G_{(q)}, T_{ext}, f_a, f_c$ are actuator torque vector, disturbance and uncertainty vector, inertia symmetric matrix, Coriolis matrix, centrifugal matrix, Viscose, and coulomb friction vector, gravity vector, actuator reaction torque vector, actual fault and sensor fault, respectively. According to (1), this paper defines its two main objectives: (1) inverse modeling and detecting the fault based on the adaptive nonlinear network Tsk fuzzy ARX method, (2) fault estimation and identification for robot manipulator based on model reference fuzzy Linear Matrix Inequality (FLMI) feedback linearization. Figure 1 illustrates the problem statements and methods to solve it.

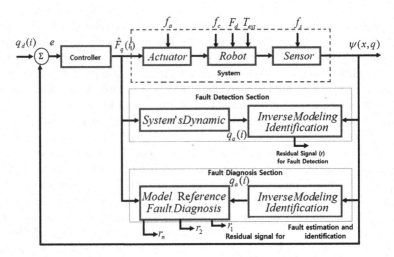

Fig. 1 The block diagram of the proposed fault detection and diagnosis scheme

3 Proposed Methodology

Robot manipulator is a nonlinear, multi-input multi-output (MIMO) and coupling effect system, thus to identify the system's behavior for fault detection and system's output performance estimation for fault diagnosis are the main consideration in this research.

3.1 Fault Detection

Inverse modeling (or kinematics) is a transformation function to find possible joints variable based on actual position and orientation [2, 4, 5]. The adaptive nonlinear autoregressive neuro-fuzzy (NARXNF) used for inverse kinematics modeling. This method estimates the output parameters of the robot manipulator. In the first step, a discrete ARX theory is applied to identify the robot manipulator since this method is efficient for system prediction. To detect the actual sensor faults in robot manipulator, the input-output filter TSK fuzzy ARX function is defined as below

The state space function for TSK fuzzy ARX is,

$$\begin{cases} \hat{X}(k) = A\hat{X}(k-1) + b_y \hat{y}(k-1) + b_u u(k-1) + b_{fuzzy} \xi(k-1) + \hat{f}_a(k-1) \\ \hat{Y}(k) = (K_{n,i})^T \hat{X}(k) + \hat{f}_c(k-1) \end{cases}$$

$$(2)$$

If the ARX-TSK fuzzy sampling rule defined by the following definition:

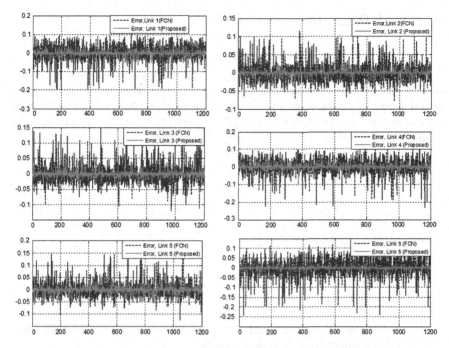

Fig. 2 RMS error for inverse model identification for proposed and FCN [6] in 6 DOF robot manipulator

If P_x is a_{11} and P_y is b_{11} then $\hat{q}_a(1) = P_x x_1 + P_y y_1 + r_1$,

If P_x is a_{21} and P_z is c_{21} then $\hat{q}_a(2) = P_x x_2 + P_z z_2 + r_2$,

to update the $\hat{q}_a(i)$, we have

$$\hat{q}_a(1, 2) = \frac{\omega_1 \hat{q}_a(1) + \omega_2 \hat{q}_a(2)}{\omega_1 + \omega_2} \tag{3}$$

The faults can be detected as below:

$$1 : if(f_a \neq \hat{f}_a) \rightarrow [y(k) - \hat{y}(k) \neq 0) \,\&\, (X(k) - \hat{X}(k)] \neq 0 \rightarrow r \neq 0$$
$$2 : if(f_s \neq \hat{f}_s) \rightarrow [(y(k) - \hat{y}(k) \neq 0) \,\&\, (X(k) - \hat{X}(k)) \neq 0] \rightarrow r \neq 0 \tag{4}$$

In order to verify the performance of proposed methodology for inverse model identification, this method is tested and compare with the state-of-the-art fuzzy C-means clustering network (FCN) [6]. The comparison error performance between the proposed method and FCN is depicted in Fig. 2 for a 6-DOF robot manipulator. According to the results in the figure, the error performance of the proposed method is better than FCN. The rate of RMS error for a proposed method for all six joints is $0.04°, 0.02°, 0.04°, 0.03°, 0.02°, 0.01°$ than that of FCN method are $0.2°, 0.15°, 0.15°, 0.1°, 0.2°, 0.13°$ respectively.

3.2 Fault Diagnosis

As the fault detection phase is defined in the previous section, there are two types of faults such as an actuator, and sensor in robot manipulator are analyzed in the section. The actuator and sensor faults are created by different factors, for instance, power supply or wiring problems for creating actuator faults, and bias and drift in measurements for creating the sensor fault. To improve the diagnosis accuracy, the mathematical equations for sensor based fault and actuator based fault are presented. According to the Fig. 1, the state space PI feedback observation technique is

$$
\begin{cases}
S(\hat{X}, F) = \hat{I}^{-1}[(\hat{F}) - (co\hat{r}[\dot{q}\dot{q}] + ce\hat{n}[\dot{q}]^2 + \hat{G})] \\
\hat{X}(k) = A\hat{X}(k-1) + b_y \hat{y}(k-1) + b_u u(k-1) + S(\hat{X}, F) + \beta_a \hat{f}_a(k-1) + K_p(e(k-1)) \\
e_a(k) = (q_a(k) - \hat{q}_a(k)) \\
e_s(k) = (\tau_s(k) - \hat{\tau}_s(k)) \\
\hat{f}_a(k) = \hat{f}_a(k-1) + K_{ia}(e_a(k-1)) \\
\hat{f}_s(k) = \hat{f}_s(k-1) + K_{i_s}(e_s(k-1)) \\
\hat{y}(k) = (K)^T \hat{X}(k) + \beta_s \hat{f}_s(k-1)
\end{cases}
\tag{5}
$$

To optimize the coefficient based on fuzzy LMI, the state error, actuator fault error, and sensor fault error are defined as follows:

$$
\begin{cases}
e_{state}(k) = X(k) - \hat{X}(k) \\
e_a(k) = (q_a(k) - \hat{q}_a(k)) \\
e_s(k) = (\tau_s(k) - \hat{\tau}_s(k))
\end{cases}
\tag{6}
$$

Based on LMI and fuzzy theories, to have the best convergence, the coefficients defined by the following formulation:

$$
K_{i_a} = \frac{K_p}{T_{i_a}}, \, K_{i_s} = \frac{K_p}{T_{i_s}} \rightarrow K_p' = \frac{K_p - K_{p(\min)}}{K_{p(\max)} - K_{p(\min)}} \in [0, 1]
$$

$$
K_{i_s}' = \frac{K_{i_s} - K_{i_s(\min)}}{K_{i_s(\max)} - K_{i_s(\min)}} \in [0, 1], 2 \le \alpha \le 5
\tag{7}
$$

(K_p, K_{i_a}, K_{i_s}), T_{i_a} and T_{i_s} are main coefficients for fault diagnosis, rise time for actuator fault, and rise time for sensor fault, respectively. To find above coefficients PD like fuzzy logic have been designed based on the following equations:

$$
K_p' = \left(\sum \theta^T \xi_p(x) \right) \in [0, 1], \, K_{i_s}' = \left(\sum \theta^T \xi_{i_s}(x) \right) \in [0, 1], \alpha = \left(\sum \theta^T \xi_\alpha(x) \right),
\tag{8}
$$

To isolate and estimated the actuator fault in robot manipulator, when $\hat{f}_a(k) = f_a(k)$

$$(y(k) - \hat{y}(k) = 0) \, \& \, (X(k) - \hat{X}(k)) \neq 0 \Rightarrow \left[X_1^T(k) \; X_2^T(k) \right]^T -$$

$$\left[\hat{X}_1^T(k) \; \hat{X}_{2,f_a}^T(k) \right]^T \neq 0 \Rightarrow \begin{cases} x_{N(i),y}(k) - \hat{x}_{N,\hat{y}}(k) = 0 \\ x_{N(i),u}(k) - \hat{x}_{N(i),u+f_a}(k) \neq 0 \end{cases} \tag{9}$$

To analyze the sensor fault in robot manipulator, if $\hat{f}_s(k) = f_s(k)$

$$(y(k) - \hat{y}(k) \neq 0) \& (X(k) - \hat{X}(k)) \neq 0 \Rightarrow \left[X_1^T(k) \; X_2^T(k) \right]^T -$$

$$\left[\hat{X}_{1,f_s}^T(k) \; \hat{X}_2^T(k) \right]^T \neq 0 \Rightarrow \begin{cases} x_{N(i),y}(k) - \hat{x}_{N,\hat{y}+f_s}(k) \neq 0 \\ x_{N(i),u}(k) - \hat{x}_{N(i),u+f_a}(k) = 0 \end{cases} \tag{10}$$

3.3 Robot Manipulator Case Studies

To analyze the robustness of proposed fault detection and diagnosis algorithm in robot manipulator three types of case are presented. To test the power of actuator fault diagnosis (case 1), following actuator faults are applied to a robot manipulator.

$$\tau_{1_{fa}}, \tau_{4_{fa}}(N \cdot m) = \begin{cases} 4500, \, 8 \leq t \leq 15 \\ 0, \, otherwise \end{cases}, \tau_{2_{fa}}, \tau_{5_{fa}}(N \cdot m)$$

$$= \begin{cases} 6500, \, 4 \leq t \leq 8 \\ 0, \, otherwise \end{cases}, \tau_{3_{fa}}, \tau_{6_{fa}}(N \cdot m)$$

$$= \begin{cases} 2000, \, 9 \leq t \leq 13 \\ 0, \, otherwise \end{cases}$$

Figure 3 illustrates six actuators faulty signals, actuator faulty estimation and normal threshold (500 Nm). According to the figure, the thresholds are exceeded in less than 0.25 s for first, third, fourth, and sixth actuators, for the second actuator is less than 0.15 s, and for the fifth actuator is less than 0.2 s. To test the power of sensor fault diagnosis (case 2), following sensor faults are applied to a robot manipulator.

$$\theta_{1_{fa}}, \theta_{4_{fa}}(Rad) = \begin{cases} 0.8, \, 5 \leq t \leq 8 \\ 0, \, otherwise \end{cases}, \theta_{2_{fa}}, \theta_{5_{fa}}(Rad)$$

$$= \begin{cases} 0.4, \, 7 \leq t \leq 12 \\ 0, \, otherwise \end{cases}, \theta_{3_{fa}}, \theta_{6_{fa}}(Rad)$$

$$= \begin{cases} 1.5, \, 9 \leq t \leq 13 \\ 0, \, otherwise \end{cases}$$

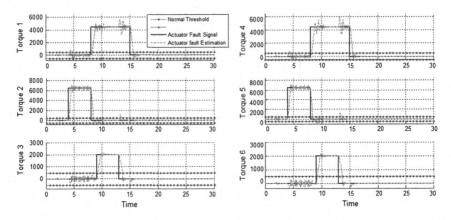

Fig. 3 Residual of actuator fault signals and normal selection threshold (Case 1)

Sensor faulty signal, the faulty signal estimation and the selection normal threshold ($\pm 0.017^{Rad} = 1°$) is shown in Fig. 4. According to Fig. 4, the rise time for signal estimation for all joints are less than 0.2 s. To test the power of fault diagnosis in robot manipulator based on model reference technique, following sensor and actuator faults (case 3) are applied to a robot manipulator.

$$\tau_{1_{fa}}, \tau_{4_{fa}}(N \cdot m) = \begin{cases} 4500, 8 \leq t \leq 15 \\ 0, otherwise \end{cases}, \tau_{2_{fa}}, \tau_{5_{fa}}(N \cdot m)$$

$$= \begin{cases} 6500, 4 \leq t \leq 8 \\ 0, otherwise \end{cases}, \tau_{3_{fa}}, \tau_{6_{fa}}(N \cdot m)$$

$$= \begin{cases} 2000, 9 \leq t \leq 13 \\ 0, otherwise \end{cases}$$

$$\theta_{1_{fa}}, \theta_{4_{fa}}(Rad) = \begin{cases} 0.8, 5 \leq t \leq 8 \\ 0, otherwise \end{cases}, \theta_{2_{fa}}, \theta_{5_{fa}}(Rad)$$

$$= \begin{cases} 0.4, 7 \leq t \leq 12 \\ 0, otherwise \end{cases}, \theta_{3_{fa}}, \theta_{6_{fa}}(Rad)$$

$$= \begin{cases} 1.5, 9 \leq t \leq 13 \\ 0, otherwise \end{cases}$$

Figures 5 and 6 have been focused on the power of actuator plus sensor fault diagnosis respectively in robot manipulator.

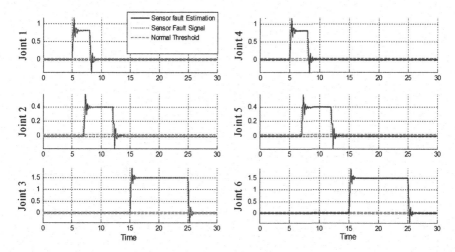

Fig. 4 Residual signal for sensor fault and normal selection threshold (Case 2)

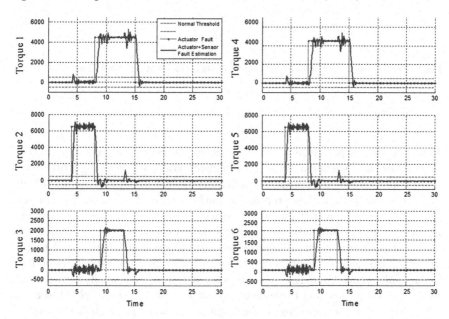

Fig. 5 Residual signals for sensor + actuator fault and normal state threshold (Case 3)

Based on Fig. 5, we can detect and isolate the actuator fault through analyzing the torque performance. To detect and isolate the sensor fault in case 3, the joint analysis is recommended. Figure 6 shows the power of joint analysis to detect and isolate the sensor fault in a robot with multiple faults.

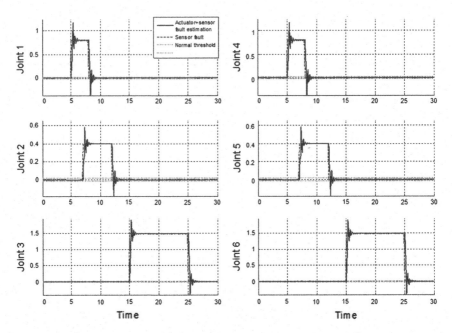

Fig. 6 Residual signals for sensor + actuator fault and normal state threshold (Case 3)

4 Conclusions

In this paper, we presented a new fault detection, estimation, and identification scheme with an application of robot manipulator based on developing a fuzzy LMI optimization for PI feedback linearization observation theory and a nonlinear intelligent ARX inverse model. The proposed method was verified with six degrees of freedom PUMA robot manipulator, proving qualified efficiency in detecting, isolating, and identifying concurrent faults on sensors and actuators. The proposed intelligent nonlinear ARX inverse modeling was proposed to solve the challenge of inverse modeling and fault detection in multi-DOF robot manipulator as well. According to the results, proposed inverse modeling system identification significantly reduced the rate of error modeling from about 0.3° in state-of-the-art—FCN [6] to about 0.03° to fault detection. To improve the diagnostic accuracy for multi-faults systems, fuzzy LMI feedback linearization observation is employed instead of adding extra hardware (e.g, sensors).

Acknowledgements This work was supported by the Korea Institute of Energy Technology Evaluation and Planning (KETEP) and the Ministry of Trade, Industry & Energy (MOTIE) of the Republic of Korea (Nos. 20162220100050, 20161120100350, 20172510102130). It was also funded in part by The Leading Human Resource Training Program of Regional Neo industry through the National Research Foundation of Korea (NRF) funded by the Ministry of Science, ICT and future Planning (NRF-2016H1D5A1910564), and in part by the Basic Science Research Program through the National Research Foundation of Korea (NRF) funded by the Ministry of Education (2016R1D1A3B03931927).

References

1. Siciliano, B., Khatib, O. (eds): Springer Handbook of Robotics. Springer (2016)
2. Ngoc Son, N., Anh, H.P.H., Thanh Nam, N.: Robot manipulator identification based on adaptive multiple-input and multiple-output neural model optimized by advanced differential evolution algorithm. Int. J. Adv. Robot. Syst. **14**(1), 1729881416677695 (2016)
3. Anh, H.P.H., Nam, N.T.: Novel adaptive forward neural MIMO NARX model for the identification of industrial 3-DOF robot arm kinematics. Int. J. Adv. Robot. Syst. **9**(4), 104 (2012)
4. Weyer, Erik: Finite sample properties of system identification of ARX models under mixing conditions. Automatica **36**(9), 1291–1299 (2000)
5. Zhao, Wen-Xiao, Chen, Han-Fu: Recursive identification for Hammerstein system with ARX subsystem. IEEE Trans. Autom. Control **51**(12), 1966–1974 (2006)
6. Al-Dabbagh, R.D., Kinsheel, A., Mekhilef, S., Baba, M.S., Shamshirband, S.: System identification and control of robot manipulator based on fuzzy adaptive differential evolution algorithm. Adv. Eng. Softw. **78**, 60–66 (2014)
7. Alavandar, Srinivasan, Nigam, M.J.: Neuro-fuzzy based approach for inverse kinematics solution of industrial robot manipulators. Int. J. Comput. Commun. Control **3**(3), 224–234 (2008)
8. Wu, L., et al.: Fault detection for underactuated manipulators modeled by Markovian jump systems. IEEE Trans. Ind. Electron. **63**(7), 4387–4399 (2016)
9. Aleksovski, D., et al.: A comparison of fuzzy identification methods on benchmark datasets. IFAC-PapersOnLine **49**(5), 31–36 (2016)
10. Jami'in, M.A., et al.: Quasi-ARX neural network based adaptive predictive control for nonlinear systems. IEEE Trans. Electr. Electron. Eng. **11**(1), 83–90 (2016)
11. Van, M., Franciosa, P., Ceglarek, D.: Fault diagnosis and fault-tolerant control of uncertain robot manipulators using high-order sliding mode. Math. Probl. Eng. 2016 (2016)
12. Sarıoğlu, A., Kural, A.: Modeling and ARX identification of a quadrotor MiniUAV. In: 2015 9th International Conference on Electrical and Electronics Engineering (ELECO. IEEE (2015)
13. Hartmann, A., et al.: Identification of switched ARX models via convex optimization and expectation maximization. J. Process Control **28**, 9–16 (2015)

Robot Path Planning by LSTM Network Under Changing Environment

Masaya Inoue, Takahiro Yamashita and Takeshi Nishida

Abstract Path planning is an important function for executing autonomous moving robots, and many path planning methods that satisfy various constraints, such as avoiding obstacles and energy efficiency, have been proposed. However, these conventional methods have several difficulties for apply to the actual applications, such as the instability, low reproducibility, huge training data set required. Therefore, we propose a novel robot path planning method that combines the rapidly exploring random tree (RRT) and long short-term memory (LSTM) network. In this method, numerous and good paths are generated in the robot configuration space by the RRT method, a convolutional autoencoder and LSTM combination network is trained by them. The proposed method overcomes the difficulty of general methods with neural networks, i.e., "the acquisition of a large amount of training data." Moreover, the difficulty of general random based methods, i.e., "the reproducible path generation" is resolved with high-speed.

Keywords Path planning · LSTM · Convolutional autoencoder

1 Introduction

Path planning is an important function for executing autonomous moving robots, and many path planning methods that satisfy various constraints, such as avoiding obstacles and energy efficiency, have been proposed. There are many random sampling algorithms widely used for robot path planning, such as the probabilistic roadmap

M. Inoue (✉) · T. Yamashita · T. Nishida
Kyushu Institute of Technology, 1-1 Sensui, Tobata, Kitakyushu,
Fukuoka 804-8550, Japan
e-mail: q344203m@mail.kyutech.jp

T. Yamashita
e-mail: yamashita.takahiro610@mail.kyutech.jp

T. Nishida
e-mail: nishida@cntl.kyutech.ac.jp

© Springer Nature Singapore Pte Ltd. 2019   317
S. K. Bhatia et al. (eds.), *Advances in Computer Communication and Computational Sciences*, Advances in Intelligent Systems and Computing 759,
https://doi.org/10.1007/978-981-13-0341-8_29

[1], rapidly exploring random tree (RRT) [2], and their improved algorithms. These methods are used for connecting nodes to create a path and create a strong feature so that the path can always be found under the condition that the starting point and goal point can be connected. Moreover, in situations where obstacles or other factors exist in the target area, or when the environment dynamically changes, it is difficult to set nodes to be prescribed in space. Therefore, it is known that these methods are more advantageous in such situations compared to Dijkstra's algorithm and A* method [3] that use fixed nodes. However, since these algorithms use random numbers, some problems occur, such as fluctuations of the found path for each search time and finding a redundant path when the search time is not sufficient.

On the other hand, path planning methods by machine learning such as methods using a neural network (NN) [4] and deep Q-network combining reinforcement learning [5] have been proposed. Although these methods require large amounts of learning beforehand, they can generate paths at high speed after learning. In addition, unlike the abovementioned methods, the same path is generated for the same starting point and goal point. Owing to the generalization ability of NN, path generation is possible even for conditions not used during learning. However, these methods cannot positively take into account the robot's physicality as prior knowledge, and trial and error is required to acquire a large amount of training data for learning. It is often impossible to perform many trial and error processes on site.

Therefore, we propose a new path planning method that combines the random sampling method and NN method to overcome "the fluctuation and redundancy of found path" with the random sampling algorithm and "the acquisition of a large amount of training datasets" for the NN method. In the proposed method, a large amount of high-quality paths are generated by RRT in the configuration space by considering the robot's physicality and environment, and learning by the NN is executed using these datasets. After learning, the NN can generate a high-quality path quickly and stably. This network also develops generalization ability.

2 Problem Setting

First, let C be the configuration space of a robot. In this study, to simplify the analysis, we set $C \subset \mathbb{R}^2$ and the region is set as

$$C = \left\{ \boldsymbol{p} \triangleq [p_1\ p_2]^T \,\middle|\, p_{1l} \leq p_1 \leq p_{1u},\ p_{2l} \leq p_2 \leq p_{2u} \right\}, \tag{1}$$

where p_{1l}, p_{1u}, p_{2l}, and p_{2u} are constant. The starting points $\boldsymbol{p}_s$ and the goal points $\boldsymbol{p}_g$ of the paths are generated using random numbers in the specific regions. The regions of $C_s \subset C$ and $C_g \subset C$ are set as follows:

$$C_s = \left\{ \boldsymbol{p}_s \triangleq [p_{1s} \ p_{2s}]^T \Big| p_{1sl} \leq p_{1s} \leq p_{1su}, \ p_{2sl} \leq p_{2s} \leq p_{2su} \right\},$$

$$C_g = \left\{ \boldsymbol{p}_g \triangleq [p_{1g} \ p_{2g}]^T \Big| p_{1gl} \leq p_{1g} \leq p_{1gu}, \ p_{2gl} \leq p_{2g} \leq p_{2gu} \right\}.$$

We consider the problem of generating a path that connects arbitrarily set starting points $\boldsymbol{p}_s \in C_s$ and goal points $\boldsymbol{p}_g \in C_g$ without colliding with obstacles.

Next, the path generated using the starting point and goal point pair is expressed as follows:

$$P^{(n)} \triangleq \left\{ \boldsymbol{p}_s^{(n)}, \boldsymbol{p}_1^{(n)}, \ldots, \boldsymbol{p}_e^{(n)}, \ldots, \boldsymbol{p}_E^{(n)}, \boldsymbol{p}_g^{(n)} \right\}, \tag{2}$$

where $n = 1, \ldots, N$ represents the path number; $m = 1, \ldots, E$ is the node number; and $\boldsymbol{p}_e^{(n)} \in C$ is a node of the path. The sum of the Euclidean distances between the nodes is called the total length of the path.

In this research, we consider the problem based on $C \subset \mathbb{R}^2$; hence, we describe the configuration space as an image with obstacles. The image is represented as matrix $\boldsymbol{I}^{(r)} \in \mathbb{R}^{A \times B}$ composed of pixels, where $r = 1, \ldots, R$ is the index of the image, and A and B are the number of pixels of the height and width of the image, respectively. The obstacle area is represented by setting the value of matrix elements as 1, and there are no obstacles in C_s and C_g. Also, the value of matrix elements representing the region where the obstacle does not exist is 0. The path $P^{(n)}$ used for training must be set so that there are no obstacles on the node and the line segment connecting them.

3 Proposed Path Planning Method

3.1 Architecture of Proposed Network

The proposed network for path generation consists of the convolutional autoencoder (CAE) [6] and long short-term memory (LSTM) network [7]. The overview is shown in Fig. 1. The CAE section is integrated into the LSTM for feeding the environmental information, and the LSTM section executes the path generation.

The proposed method has four phases as follows:

1. Training of the CAE section (CAE training phase).
2. Generating training data sets using trained CAE and RRT (Training data generating phase).
3. Path training by LSTM network (LSTM training phase).
4. Generating the new path (Recalling phase).

This process is illustrated in Fig. 2.

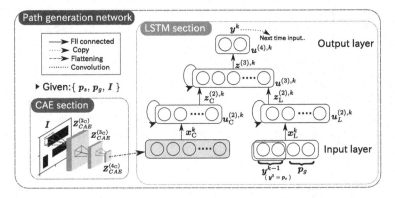

Fig. 1 Overview of the proposed network. This shows signal flow in the recalling phase

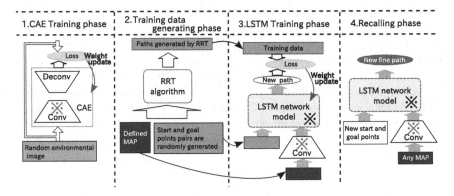

Fig. 2 Flow of the phases of the proposed method

3.2 CAE Section

Structure of CAE Section In this section, the information of obstacles are extracted from $I^{(r)}$ and the size is compressed by the CAE. Then, the output from the CAE is integrated with the path information in the input layer of the LSTM section. By compressing the image $I^{(r)}$, the size of the LSTM section is reduced. Furthermore, by inputting information on the environment to the LSTM, it becomes possible to generate a path suitable for various environments. If there is no CAE section, the LSTM only learns the statistics of the paths without information on obstacles and avoids it. In this a case, it is impossible to acquire latent knowledge that obstacles must be avoided.

The CAE section is an L_C layered convolutional NN. Convolution processing is known to be very effective in image recognition. The convolution in the l_C-th layer is executed against the element z_{abd} of the output tensor $Z_{CAE}^{(l_C-1)} \in \mathbb{R}^{C \times V \times Q}$ using an element h_{vqcd} of the filter set $H \in \mathbb{R}^{V \times Q \times C \times D}$

$$u_{abd}^{(l_C)} = \sum_{c=1}^{C} \sum_{v=1}^{V} \sum_{q=1}^{Q} z_{sa+v,sb+q,c}^{(l_C-1)} h_{vqcd}^{(l_C)} + b_{abd}^{(l_C)} \tag{3}$$

$$z_{abd}^{(l_C)} = \eta^{(l_C)} \left(u_{abd}^{(l_C)} \right) \tag{4}$$

where V and Q are the sizes in the horizontal and vertical axis direction of the filter, respectively; $c = 1, \ldots, C$ represents the channel number of input tensor; $d = 1, \ldots, D$ is the number of filters; s is the pixel width scanned by the filter; b_{abd} is a bias; and $\eta^{(l_C)}(\cdot)$ is called the *activation function*. The output tensor of the first layer corresponds to $\boldsymbol{I}$. In addition, each element z corresponds to each pixel.

During training, restoration is performed using an L_C layered deconvolution layer, and the error backpropagation method is applied to the error between the restored image and original image. The deconvolution layer is the layer that performs convolution processing after padding target matrix components with arbitrary values [9]. It also has the function of restoring the compressed tensor by convolution layer.

CAE Training Phase The CAE section is trained using environmental datasets with the following procedure:

(1) Using randomly positioned obstacles, a large amount of environmental images $\boldsymbol{I}^{(r)}$ are generated.
(2) The images $\boldsymbol{I}^{(r)}$ are divided into mini-batch sets, and CAE training is executed using these.
(3) Learning is aborted after sufficiently repeating (2).

Training Data Generating Phase The training dataset for LSTM section is generated with the following procedure:

(1) Give $\boldsymbol{I}^{(r)}$ ($r = 1, \ldots, R$) and N sets of start $\boldsymbol{p}_s^{(rn)} \in C_s$ and goal $\boldsymbol{p}_g^{(rn)} \in C_g$ ($n = 1, \ldots, N$) for each image.
(2) Generate paths $P^{(rn)}$ using $\boldsymbol{p}_s^{(rn)}$ and $\boldsymbol{p}_g^{(rn)}$ in $\boldsymbol{I}^{(r)}$ by applying RRT.
(3) Modify the generated paths by the trajectory refinement method.
(4) Extract high-quality paths based on the number of nodes E.

It should be noted that in (4), the mode value of the number of nodes of generated paths is found; paths with number of nodes that deviate from this value are deleted. With these procedures, various paths for environmental images containing various obstacles are generated. Then, the image information $\boldsymbol{Z}_{CAE}^{(r)}$ compressed by the CAE and path $P^{(rn)}$ are combined to become the training datasets of the LSTM.

3.3 LSTM Section

LSTM network is a recurrent neural network architecture. It can learn the dynamic feature of time series signals. Moreover, the effectiveness and high ability of RNN for path generation of moving robots are suggested in [8].

Let us assume an LSTM network consisting of L_L layers. There are two input layers and hidden layers that receive outputs from them. Moreover, their output signals are integrated at the one upper hidden layer, and it connected to the output layer. Specifically, in the simulation described later, the input of the input layer that accepts the CAE output is called $x_C^k \in \mathbb{R}^{108}$, while the input of the input layer that accepts information on the route is called $x_L^k \in \mathbb{R}^4$

$$x_C^k = z_{CAE}^{(r)}, \ x_L^k = \left[y^{(k-1)T} \ p_g^T \right]^T, \ y^k = p_k, \tag{5}$$

where $z_{CAE}^{(r)}$ is a flattened vector of $Z_{CAE}^{(r)}$ output of the CAE section, $y^0 = p_s$, and k represents the recalling step corresponding to node number. When x^k is input into the LSTM network, the network recalls the next node of path y^k. The number of cells in the input layers are 108 and 4, and they are fully connected to the corresponding hidden layers. The number of cells in each hidden layer is larger than the number of cells in the input layers (e.g., 120). The output of the hidden layers are integrated into the next hidden layer.

LSTM Network The input and output vectors of the l_L-th layer at time k are represented by

$$u^{(l_L),k} \triangleq \left[u_1^{(l_L),k} \cdots u_j^{(l_L),k} \cdots u_J^{(l_L),k} \right]^T, \tag{6}$$

$$z^{(l_L),k} \triangleq \left[z_1^{(l_L),k} \cdots z_j^{(l_L),k} \cdots z_J^{(l_L),k} \right]^T \tag{7}$$

where j is the unit number, J is the total number of units, and J is not common to all layers. The input vector is $x^k = u^{(1),k}$ in the input layer, and the output vector is $y^k = z^{(L_L),k}$ in the output layer. The unit number and total number of units of the $(l_L - 1)$-th layer are represented by j^- and J^-, respectively. The input propagation

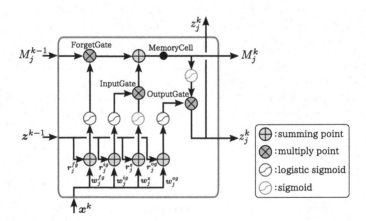

Fig. 3 Structure of a LSTM unit

weight from the $(l_L - 1)$-th layer to the l-th layer is represented by $\boldsymbol{W}^{(l_L)} \in \mathbb{R}^{J \times J^-}$. The recurrent weight in the l_L-th layer ($l_L = 2, \ldots, L_L - 1$) is represented by $\boldsymbol{R}^{(l_L)} \in \mathbb{R}^{J \times J}$. The bias is represented by $\boldsymbol{b}^{(l)}$, and each component is represented by $w, r,$ and b. Here, j' is an arbitrary unit number corresponding to the output signal of the time $k - 1$ of the l_L-th layer The components of $\boldsymbol{u}^{(l_L),k}$ are given by

$$u_j^{(l_L),k} = \sum_{j^-}^{J^-} w_{jj^-}^{(l_L)} z_{j^-}^{(l_L-1),k} + \sum_{j'}^{J} r_{jj'}^{(l_L)} z_{j'+}^{(l),k-1} + b_j^{(l_L)}. \tag{8}$$

The elements of the output vector of the l-th layer are represented by

$$z_j^{(l_L),k} = \eta^{(l_L)} \left(u_j^{(l_L),k} \right). \tag{9}$$

To summarize the discussion above, the output of the l-th layer is represented as

$$\boldsymbol{z}^{(l_L),k} = \eta^{(l_L)} \left(\boldsymbol{W}^{(l_L)} \boldsymbol{z}^{(l_L-1),k} + \boldsymbol{R}^{(l_L)} \boldsymbol{z}^{(l_L),k-1} + \boldsymbol{b}^{(l_L)} \right), \tag{10}$$

and the output of the output layer is represented as

$$\boldsymbol{y}^k = \eta^{(L_L)} \left(\boldsymbol{W}^{(L_L)} \boldsymbol{z}^{(L_L-1),k} \right). \tag{11}$$

LSTM Unit The LSTM network is constructed by LSTM units (Fig. 3) [10], and the following calculation is executed for each unit. First, the forget gate opening ratio of the d-th unit is calculated as

$$f_d^k = \sigma \left(\boldsymbol{w}_d^{fg} \boldsymbol{x}^k + \boldsymbol{r}_d^{fg} \boldsymbol{z}^{(k-1)} + b^{fg} \right), \tag{12}$$

where $\sigma(\cdot)$ is the logistic sigmoid function, $\boldsymbol{w}_d^{fg}$ is a weight vector corresponding to the input vector $\boldsymbol{x}^k$ from the previous layer, $\boldsymbol{r}_d^{fg}$ is a weight vector corresponding to the input vector $\boldsymbol{z}^{k-1}$ from previous time, and b^{fg} is a bias.

Next, the opening ratio of the input gate is calculated as

$$i_j^k = \sigma \left(\boldsymbol{w}_j^{ig} \boldsymbol{x}^k + \boldsymbol{r}_j^{ig} \boldsymbol{z}^{(k-1)} + b^{ig} \right). \tag{13}$$

The signal through the input gate is

$$M_j^k = f_j^k \cdot M_j^{k-1} + i_j^k \cdot \tanh \left(\boldsymbol{w}_j^z \boldsymbol{x}^k + \boldsymbol{r}_j^z \boldsymbol{z}^{(k-1)} + b^z \right). \tag{14}$$

This value is transmitted as the internal state of the unit at the next time. Then, the opening ratio of the output gate is calculated as

$$o_j^k = \sigma \left(\boldsymbol{w}_j^{og} \boldsymbol{x}^k + \boldsymbol{r}_j^{og} \boldsymbol{z}^{(k-1)} + b^{og} \right), \tag{15}$$

and the output value z_j^k is derived as

$$z_j^k = o_j^k \cdot \tanh(M_j^k). \tag{16}$$

LSTM Training Phase The LSTM network is trained using the path dataset as follows:

(1) Shuffle the datasets $\{I^{(r)}, P^{(rn)}\}$ and divide them into mini-batch sets.
(2) Execute the following processes on the first mini-batch.

    1. x^k is constructed from the dataset and input into the LSTM at time k.
    2. y^k is recalled by forward propagation of x^k.
    3. Distance between y^k and $P^{(rn)}$ is accumulated as recalling error.
    4. Construct x^{k+1} using y^k, time step is replaced $k := k + 1$, and return to 2.
    5. Loop until the final time.

(3) Modify connection weights in the network using backpropagation through time method [7].
(4) Execute the processing of (2) and (3) for the next mini-batch, and execute all mini-batches as well (This is called 1 epoch).
(5) Repeat processes (1)–(4) and execute a sufficient number of epochs.

3.4 Generating the New Path

When a combination of an environmental image and a desired starting point and goal point is input into the network where learning has been completed, a path is generated. The network continues to generate nodes y^k until it approaches the goal point. Stop the path generation when it approaches the goal point beyond a certain threshold and combine it with the goal point.

4 Simulation

4.1 Conditions

Examples of configuration space C or environmental image $I^{(r)}$ supposed in experiments are shown in Fig. 4. The vertical and horizontal sizes of this region were set as $p_1 \in [-1, 1]$ and $p_2 \in [-1, 1]$, respectively.

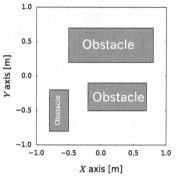
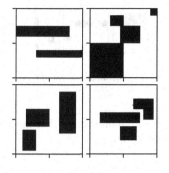

(a) Two-dimensional region including obstacles.

(b) Examples of environmental images.

Fig. 4 Configuration space C or environmental image $I^{(r)}$

4.2 Training

Training of CAE Table 1 shows the parameters of the constructed CAE section. With the CAE section, the environmental information $I^{(r)} \in \mathbb{R}^{100 \times 100}$ was compressed to $Z_{\text{CAE}}^{(r)} (A = 6, B = 6, D = 3)$ by three convolution processes.

The training of the CAE section was executed by restoring the compressed signal in the deconvolution layer and applying the error backpropagation method to the error from the original image. In order to confirm the generalization ability of the CAE section after the training phase, the environmental image which was not used for learning was restored after compression. The results of the generalization test,

Table 1 Parameters of the CAE section

Setting items	Detail
Size of environmental image $I^{(r)} \in \mathbb{R}^{A \times B}$	$A = B = 100$
Number of convolutional layers L_C	3
Number of filters in each layer	12, 6, 3
Number of stride sizes in each layer s	1, 3, 5
Filter size $(V \times Q)$	$V = Q = 5$
Activation function $\eta(\cdot)$	Rectifier function
Error function	Mean squared error
Mini-batch size	200
Initial weights	Random number in $[-0.1, 0.1)$
Initial bias	None
Learning rate adjustment	Adam [12]

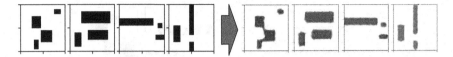

Fig. 5 Results of generalization test of CAE

shown in Fig. 5, demonstrate that the CAE section acquired the ability to extract environmental features.

4.2.1 Training Data Generation and Training of LSTM

Ten types (i.e., $r = 10$) of environmental images with different obstacle arrangements were prepared. We also set $p_{1sl} = -1.0$, $p_{1su} = -0.5$, $p_{2sl} = 0.5$, and $p_{2su} = 1.0$ to set C_s, and $p_{1gl} = 0.5$, $p_{1gu} = 1.0$, $p_{2gl} = -1.0$, and $p_{2gu} = -0.5$ to set C_g. Next, we sampled $n = 3500$ combinations of start and goal points from C_s and C_g, and generated paths $P^{(rn)}$ using RRT implemented in the Open Motion Planning Library [11]. We carried out processing to refine the generated paths using the path pruning method, shortcut method [13], and B-spline method [14]. Through these procedures, we constructed the training dataset $\{I^{(r)}, P^{(rn)} | r = 10, n = 3500\}$.

The main parameters of the LSTM section are listed in Table 2.

4.3 Recalling of Path

Path generation in trained environment The network was trained in 10 environments, and the path generation ability of the trained network was investigated against

Table 2 Parameters of the LSTM section

Setting items	Detail
Input dimension	108 (CAE), 4
Output dimension	2
Number of hidden layers	2
Number of units in one hidden layer	120
Activation function of output layer	Identity function
Error function	Mean squared error
Mini-batch size	200
Initial weights	Random $[-0.1, 0.1)$
Initial bias	None
Learning rate adjustment	Adam [12]

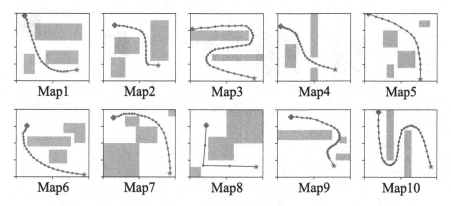

Fig. 6 Examples of the generated paths in trained environments. The symbol ◇ represents the starting point, and the symbol ⋆ represents the goal point

Table 3 Success rates of path generation in trained environment

Environment	Map1	Map2	Map3	Map4	Map5	Map6	Map7	Map8	Map9	Map10
Success rate (%)	98.23	99.99	99.61	70.13	99.97	98.31	99.32	100	94.38	99.46

arbitrary chosen starting and goal points in the trained environment. Examples of generated paths are shown in Fig. 6. A route was generated by preparing 10,000 random pairs of starting points and goal points for each environment. Table 3 shows the success rates of path generation without collision with obstacles. The success rates were over 98% in 8 out of 10 environments.

Path generation in unknown environment The path generation ability of the trained network was investigated against arbitrary starting and goal points in an unknown environment. Examples of generated paths are shown in Fig. 7. In the unknown environment, similar to the learned environment, the success rate was relatively high and path generation often failed when the environment was significantly different from the trained environment. Therefore, in order to verify the generalization ability of the trained network, the positions of all the obstacles in a specific trained environment were randomly fluctuated, and the change in the success rate of route generation was observed. The vertical and horizontal positions of all the obstacles in the environment shown in map 1 of Fig. 6 were fluctuated from ±0.01 to ±0.2 (m) randomly and individually. The results shown in Fig. 8 demonstrate that sufficiently high performance or generalization ability can be expected if fluctuations in the position of the obstacle are within 10%. It is found from these results that when the given environment is similar to the trained data set, the network realizes high path generation capability. Namely, since the network outputs the appropriate number of nodes and paths based on the training data set, it will be possible to improve the generalization ability by extending the training data set.

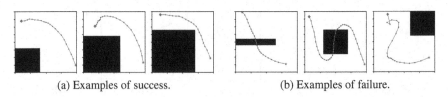

(a) Examples of success. (b) Examples of failure.

Fig. 7 Examples of the generated paths in unknown environments

Fig. 8 Relationship between magnitude of environmental change and success rate

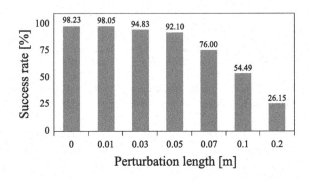

5 Conclusion

The proposed method has the following advantages: (1) In trained environments, paths can be generated without colliding with obstacles even if the starting and goal points change. (2) Even for unknown environments similar to trained environments, the method has the ability to generalize path generation. (3) In the composition of the dataset, path selection based on arbitrary evaluation index is possible. (4) It is possible to prevent the generation of poor quality or redundant trajectory caused by the trajectory generation method such as RRT. (5) The generated path does not fluctuate unlike with methods based on random numbers. (6) Path generation after training can be done quickly. (7) It is possible to configure the proposed method as hardware that operates at low power and high speed. We proposed a method to construct an LSTM network that recalls the path of a robot by training with a large number of paths generated by RRT. The simulation results confirm that the proposed network achieves high learning and generalization abilities.

References

1. Kavraki, L.E., Svestka, P., Latombe, J.-C., Overmars, M.H.: Probabilistic roadmaps for path planning in high-dimensional configuration spaces. IEEE Trans. Robot. Autom. **12**(4), 566–580 (1996)
2. LaValle, S.M., Kuffner, J.J.: Rapidly-exploring random trees: progress and prospects. Algorithmic and Computational Robotics: New Directions, pp. 293–308. Wellesley (2001)

3. Choset, H., Lynch, K., Hutchinson, S., Kantor, G., Burgard, W., Kavraki, L., Thrun, S.: Principles of Robot Motion: Theory, Algorithms, and Implementations. MIT Press, Cambridge (2005)
4. Yang, S.X., Meng, M.: An efficient neural network approach to dynamic robot motion planning. Neural Netw. **13**(2), 143–148 (2000)
5. Gu, S., et al.: Deep reinforcement learning for robotic manipulation with asynchronous off-policy updates (2016). arXiv:1610.00633
6. Masci, J., Meier, U., Cireşan, D., Schmidhuber, J.: Stacked convolutional auto-encoders for hierarchical feature extraction. In: Proceedings of Artificial Neural Networks and Machine Learning ICANN 2011, pp. 52–59 (2011)
7. Ilya, S.: Training recurrent neural networks, University of Toronto (2013)
8. Bin, N., Xiong, C., Liming, Z., Wendong, X.: Recurrent neural network for robot path planning. Parallel and Distributed Computing: Applications and Technologies, pp. 188–191. Springer Berlin Heidelberg (2004)
9. Zeiler, M.D., Krishnan, D., Taylor, G.W., Fergus, R.: Deconvolutional networks. In: 2010 IEEE Conference on Computer Vision and Pattern Recognition, pp. 2528–2535 (2010)
10. Gers, F.A., Schmidhuber, J., Cummins, F.: Learning to forget: continual prediction with LSTM. Neural Comput. **12**(10), 2451–2471 (2000)
11. Sucan, I.A., Moll, M., Kavraki, L.E.: The open motion planning library. IEEE Robot. Autom. Mag. **19**(4), 72–82 (2012)
12. Diederik, D., Ba, J.: Adam: a method for stochastic optimization (2014). arXiv:1412.6980
13. Harada, K.: Optimization in robot motion planning. JRSJ **32**(6), 508–511 (2014)
14. Harada, K., Hattori, S., Hirukawa, H., Morisawa, M., Kajita, S., Yoshida, E.: Two-stage time-parametrized gait planning for humanoid robots. IEEE/ASME Trans. Mech. **15**(5), 694–703 (2010)

Optimizing Database Storage Architecture with Unstructured Dataset on GPGPU for Real-Time System

Toan Nguyen Mau and Yasushi Inoguchi

Abstract With the rise of today's GPU supercomputers, the database manager becomes more important for many real-time systems. The problem is that the memory size is limited by the each single GPU device on system. This leads to the requirement of such database that can access for each device and the uniformity of all device's database on system (Toan and Yasushi, Audio fingerprint hierarchy searching on massively parallel with multi-GPGPUs using k-modes and lsh, 49–54, 2016) [1]. In this research, we focus the management of unstructured dataset on single GPU device and the communication between sub-database that are stored separate GPU device. Specifically, we considered the database manager for audio fingerprint storage for a real-time retrieval system, where each audio fingerprint is a binary sequence that holds the audio information for a source audio track. The system uses the memory on GPGPU devices for storing the sub-databases, that can support fast selecting, inserting, removing. With the significant result, our method can be applied to different datasets with high performance in the field of parallel processing on GPGPU.

Keywords GPU · CUDA · Database management · Parallel processing
Database · Audio processing · Audio fingerprint

1 Introduction

Using GPGPU for high performance computing becomes more and more popular nowadays because of the significant capacity of numerous of GPU cores. Data parallel is the formal method for increasing the throughput processing on GPGPU system

T. Nguyen Mau (✉)
Inoguchi Laboratory, School of Information Science, JAIST, Nomi, Japan
e-mail: nmtoan91@jaist.ac.jp

Y. Inoguchi
Research Center for Advanced Computing Infrastructure, JAIST, Nomi, Japan
e-mail: inoguchi@jaist.ac.jp

© Springer Nature Singapore Pte Ltd. 2019
S. K. Bhatia et al. (eds.), *Advances in Computer Communication and Computational Sciences*, Advances in Intelligent Systems and Computing 759,
https://doi.org/10.1007/978-981-13-0341-8_30

331

which uses the single kernel for all threads. In this research, we take advantage of the simplicity of NVIDIA's graphics processor, which uses CUDA programming language for deploying the data transferring and kernel processing. In general, CUDA uses the texture memory of GPU for the global memory and the graphic processors unit for streaming multiprocessors on general purpose use. NVIDA also uses different memory types for different uses of parallel processing, using correctly the memory for the suitable variables which can increase significantly the performance of system. GPGPU is the cheap device with high performance but its limitation is its small memory size, To build the system using GPGPU devices, we need to consider the memory storage of every single device.

Distributed database architecture is appropriate for huge database which has well-divided clusters. And with the use of multiple device memory for storing the sub-database, which can easily support for the instant access to the certain sub-database. In case of audio searching problem, the audio fingerprint can be clustered by the audio/music type, which can group the similar audio on same sub-database and can self-process on single device. In this paper, we handle the process of single sub-database on GPGPU with support for selecting, inserting, and deleting the data and used the SQL (Structured Query Language) for demonstrating these features on single GPGPU with the audio fingerprint dataset.

A typical real-time information retrieval system requires the huge dataset for matching with all the queries. The system needs to handle numerous of queries from different users at same time and the processing time needs to immediate response to all users. In another hand, the database of system must be updated consecutive for satisfying the data's accuracy. Specific to audio fingerprint searching system, when a user sends an audio fingerprint to the system to get the audio information, the time for searching must be within seconds and on same query session. The users for audio searching are from the Internet, so the system must tolerate multiple requests at once. The database of audio fingerprint searching system must be updated continuously for getting information for new release songs.

In this paper, we proposed the architecture that can store the unstructured database on NVIDA's GPGPU architecture. The database on GPGPU can be used on real-time system, which can solve most of the problems of real-time information retrieval system on single device and multiple devices. In addiction, our system has feature for clustering and has the ability for scaling the machine with multiple GPGPU devices.

2 Research Background

2.1 GPGPU Architecture and CUDA (Compute Unified Device Architecture)

Compute Unified Device Architecture (Cuda) is NVIDIA's architecture for GPGPU cards. It supports in managing the GPGPU organizational structure by itself.

Fig. 1 Memory hierarchy in CUDA device [2]

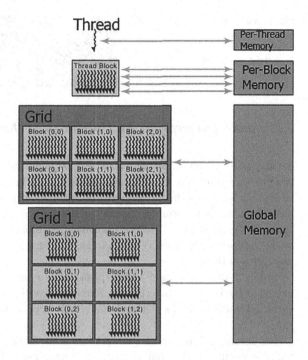

Programmer can access and handle the NVIDIA's GPGPU by CUDA C/C++ or CUDA Fortran, those are the extensions of C/C++ and Fortran programming language [2]. In Fig. 1, every kernel of CUDA is handled by a grid with blocks with 2D addressing $(0, 0)$ to $(block_{max}^x, block_{max}^y)$. And every block also has a 3D structure for threads indexing from $(0, 0, 0)$ to $(thread_{max}^x, thread_{max}^y, thread_{max}^z)$. Every threads in grid are deployed by one kernel sent from host. Threads are only distinguished by thread index tid including $(tid.x, tid.y, tid.z)$. The indexes of thread are the keys for variability purpose of CUDA [2].

Another important feature of CUDA is memory architecture. Figure 1 proves that CUDA device has different structure with main memory. Each thread has a small memory (per-thread memory) for its processing flow such as indexing numbers or temporary variables. Global memory is very similar to main memory, global memory can be accessed by every thread and block in grid. Basically, programmer can access this by transferring input data from main memory for storing output data to copy to main memory. Per-block shared memory is a special feature in CUDA architecture, it helps to gain performance by using the same resource of parallel threads in the same block. Besides that, CUDA also has texture memory (read only) for cache optimized 2D spatial access pattern and a high-speed constant memory for storing data accessed with high frequency [2].

Fig. 2 A example of
4096-bits fingerprint
extracted by HiFP2.0
fingerprint extraction
algorithm

2.2 Audio Fingerprint: An Unstructured Dataset

An audio fingerprint is a digital vector that extracted from the audio/song waveform
and able to standardize the content of audio/song source. The audio fingerprint can
easily help for comparing the similarities and differences of songs. In addition, using
audio fingerprint for storing can reduce the size of original audio/song with the
standard structure. In our system's database, we storage the audio fingerprints and
its meta information for every songs/track, we considered instead of storage the real
waveform of the song.

In Fig. 2, there is an audio fingerprint that is represented by a binary sequence
extracted by HiFP2.0 fingerprint extraction algorithms. This audio fingerprint is
extracted from first 2.97 s of a song and that can reduce the size of song by 512 times
[3].

For description the whole content of an audio input, with the different of audio
length, we will have different sizes of audio fingerprint [4]. With the different length
of audio fingerprint, there is some problem about storage for fast searching [1].

Using audio waveform for comparing have many difficulties because that is un-
normalize, so we favor to using audio fingerprint for storing and processing in our
system.

3 Related Works

3.1 Accelerating SQL Database Operations on GPU
with CUDA [5]

In [5], authors use SQLITE open source for the optimization the database management of SQL on GPGPU using CUDA [6]. With the dataset with five million row data
sets and each row is the numeric data type. The database that authors used is fixed-
column width (structured) for all dataset, that make the management and transform
is simpler than dealing with unstructured that each row has different width.

Global memory is used for storing the dataset, which have enough space for five
million data rows with the constant memory but have low latency. Authors choose
to use it to store the dataset information because it has high frequency accessed
by all threads. For every SQL query, it needs to convert into multiple opcodes that
will transfer to GPU for the SQLite virtual machine on GPU. Like all other GPU

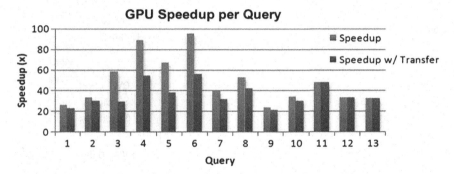

Fig. 3 The speedup of query execution on the GPU for each of the 13 queries considered, both including and excluding the results transfer time [5]

application, the after getting the results from GPU's virtual machine, it is necessary for transferring the results into host machine [6].

In Fig. 3, the speed up for the processing time for each query is faster 20–40 times than processing on CPU. The authors also showed that their method has similar result when increasing the throughput queries.

4 Proposed Method

In this paper, we use HIFP3.0 audio fingerprint for implementing the datasets that have different lengths. The first approach uses the tag-variable for description the HIFP3.0 audio fingerprint. With the helpful of these tags, we can easily access the information of this audio fingerprint.

4.1 Storing Strategy for Single Data Row

The problem is that the unstructured datasets have different sizes for every data row. There is need for such strategy that can store the single data row but supports the fast indexing so that it can easily stored in the same database.

In Fig. 4, a single file for storing an audio fingerprint includes the header with the meta information, the file length (the total length of current audio fingerprint), and header length: the size of the header, version: the version of audio fingerprint format. Each value of header is value with 32 bits variable. We use the version variable to control the version of file structure for the audio fingerprint reader. The sampling frequency is the information of source audio, where the DWT level is the decomposition level of current audio fingerprint which are the important detail for comparing the similarity of audio fingerprints. The information of 'next index' is for

Fig. 4 The strategy for
storing single audio
fingerprint with its meta
information

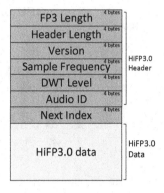

the storing the large audio fingerprint on database and is not necessary for single file storing. The rest of the file is the audio fingerprint sequence. In case of HiFP2.0, the audio fingerprint sequence is the binary sequence. This file structure can be used for another kind of data than the binary sequence.

4.2 Database Storing Strategy for GPGPU

Continuing with the data storing structure for single data, we use the same structure for storing multiple data rows on GPGPU device. Since data row has different file lengths, the problem is to determine the audio fingerprint when it is stored on GPGPU's global memory.

We proposed to use the indexing table for index all the data row on GPGPU. In general, the function of indexing table is similar to the memory pointer in C/C++ programming language. In Fig. 5, each value of pointer in indexing table points to the first memory address of an audio fingerprint. Follow the indexing value of every audio fingerprint is the variable for storing the status of the indicated audio fingerprint.

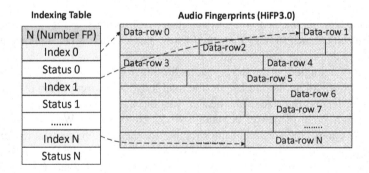

Fig. 5 The strategy for storing audio fingerprint database on GPGPU

Using the structure for every audio fingerprint, when reaching the first memory address of the fingerprint, we can read all the header information and the audio fingerprint content of current data row.

4.3 Selective the Data Rows on GPGPU

Selection is the most used query for every database, especially for information retrieval system. In this section, we proposed the serialization method for transfer the selective from host to GPGPU.

In Fig. 6, for the query serialization, we need to put the information of the properties that need to return after kernel running. The second important is the list of condition that the returned data row must be met. For the simple model, each condition has three variable which are the concerned property, the condition and the parameter value, so it needs 12 bytes for serializing a single condition. We implemented a simple SQL-parser for parsing the SQL statement to our serialization structure, this procedure has to be done on CPU and does not affect much to the querying process. For example in Fig. 6a, the query much have the Query Size for the length of current query. The next value is the type of query, it could be SELECT, UPDATE, and DELETE. The properties and conditions need to be represented as its sizes and indexes. Using the same approach, the Fig. 6b showed the result serialization to transfer the output from GPGPU to CPU. The number of data rows much be stored at the first index. The continue data are the property values of all return data row.

Transferring the result with multiple data rows from GPPGU to host also needs to be considered. In Fig. 6, a sequence of result must include the number of data row, and the values of each data row are the values of corresponding property from the query. Note that, when transferring the result from GPGPU to host with this structure, we need to copy the array twice because the length of result array is unknown for host.

Fig. 6 The serialization for data transfer between host ad GPGPU

Query Size	Number Row
Query Type	Row 1: Pro.1
Properties Size	......
Property 1	Row 1: Pro.n
.....	Row 2: Pro.1
Property n	......
Condition Size	Row 2: Pro.n
Condition 1 12 bytes	Row k: Pro.1
.....	......
Condition m 12 bytes	Row k: Pro.n

(a) Query Serialization   (b) Result Serialization

4.4 Deleting the Data Rows on GPGPU

Using the status variable of each audio fingerprint on GPGPU, we can easily delete the data row by changing the status into deleted label. In this case, not only the data row of GPGPU memory but the corresponding audio fingerprint has also been deleted. This method is similar to the FAT32/NTFS file system and limits the change from database to minimal. To do this, we can slow down the system by removing all the memory of the audio fingerprint and it also has the ability to recover the deleted data row in case of no new data row is taken over its memory location. When selecting the data row on GPGPU, the kernel needs to read the status variable and skip the deleted data row.

4.5 Inserting the Data Rows on GPGPU

After deleting the data rows, the database will have separate empty segments. In this section, we proposed the method for reusing the memory of deleted data row, that can reduce the memory used of GPPGU.

Algorithm 1 Inserting the data row for GPGPU Database

1: **procedure** INSERTROW($DataRow$)
2:    **if DB** is full **then**
3:       Extend the **DB**, then add $DataRow$
4:       Update Indexing Table and **DB**
5:    **else**
6:       **if** Exist Size($bucket$) $\geq$ Size($DataRow$) **then**
7:          Add $DataRow$ to bucket $bucket$
8:          Update Indexing Table
9:       **else**
10:          $bucket$ = FirstEmptyBucket()
11:          int size = Size(**BN**)
12:          (Head, Tail) = SPLITROW($DataRow$, size)
13:          INSERTROW(Head)
14:          INSERTROW(Tail)
15:       **end if**
16:    **end if**
17: **end procedure**

In Algorithm 1, the device function 'INSERTROW' (line 1) tries three ways to put the $DataRow$ into GPPGU database. Firstly (line 2), when the database **DB** is full, the database must append the space and store the new $DataRow$ at the newly allocated memory. Secondly (line 6), the database manager tries to find the enough space for the $DataRow$ and hold the $DataRow$ on it. In another way (line 9), the **DB** is not full and there is empty bucket(s) but not enough space for $DataRow$. in this case, the system will find the first empty $bucket$ and store the first part of $DataRow$

(line 12: Head) by calling the procedure 'INSERTROW' (line 13). For the rest of *DataRow*, we need to copy the meta information from the *Head* for it and it can be inserted into **DB** as another *DataRow*. The procedure for insert new *DataRow* is exactly same with its original *DataRow*. The storing of a sub-data row is exactly same with the single data rows on the dataset memory. To distinguish it, we can reuse the status variable for storing the information of the corresponding sub-data rows. The selective query must be changed for adaptive with new data storing strategy.

5 Results and Comparison

5.1 Evaluation Condition

We examined the performance by calculating the speed up ratio for our system with the original method. The specifications of hardware are shown in Table 1. The size of dataset is also being considered for turning from small to larger.

In Table 2, we showed the exact database size, loading time and transferring time from host to GPPGU device. Note that, this database management system focuses on real-time system, so the loading and transferring stages are processed once at the initialization steps. However, the data transferring from host and device during the testing process are used for comparing the performance between CPU and GPGPU. Otherwise, Table 3 demonstrates the tested queries that were used for testing our system. We tested for use with the normal query that finds the existing data row on

Table 1 Specification of testing machine

CPU	Intel core i7 6700K
CPU clocks	4.0 GHz
GPGPU	Tesla P100
GPGPU clocks	1.33 GHz
GPGPU memory	16 GB
CUDA	8.0
CUDA cores	3840

Table 2 Testing database information

Data rows	Size (KB)	Load time (ms)	Transfer time (ms)
300	12,250	6.06	2.40
1,000	36,749	5.90	4.95
3,000	122,497	22.70	14.00
10,000	404,239	68.92	43.91
30,000	1,224,965	176.43	132.40

Table 3 Testing queries used for examining the performance

Query type	SQL statement
Normal result	select AUDIO_ID from FP where AUDIO_ID = 30
Multiple results	select AUDIO_ID from FP where AUDIO_ID <30
Not found result	select AUDIO_ID from FP where AUDIO_ID = 1200
Inserting query	insert into FP filename S0030.fp3
Deleting query	delete from FP where AUDIO_ID = 30

database. And also the worst case when the query tries to find the data row does not exist in the database. We also use the query for updating the database on GPPGU in our experiments.

5.2 Evaluation Result

In Fig. 7, we carried out experiment between CPU and GPGPU. On CPU we use one main processing thread for sequentially handing 2048 selective queries with database stored on main memory. Comparing with system with database stored on GPGPU's memory, we have parallel 2048 threads with single kernel on GPGPU. We can see that the parallelism of using GPPGU have better nearly 100 times than using CPU. In this result, the CPCPU have higher performance in tern of throughput queries but the memory of GPPGU is limited by the hardware device.

In case of testing of inserting and deleting query. Our system has more worse performance than used CPU. This problem can not be eliminated by parallel processing because multiple threads are trying to modify the database at same time. Each thread needs to wait for its turn to change the indexing table or the dataset. However, in the typical case, the frequency of using the inserting or deleting is less more than of frequency of selecting queries. In another hand, the procedure of deleting data rows based on our proposed method does not effect the dataset and it can access to different location of indexing table, so that make the deleting queries are suitable for parallel with few conflict data accessing.

5.3 Comparing Results

Comparing with Accelerating SQL Database Operations on a GPU with CUDA [5], our proposed method can adaptive with unstructured dataset which we storing a lot today such as texts, audios, images. With ability to store more kind of dataset, our

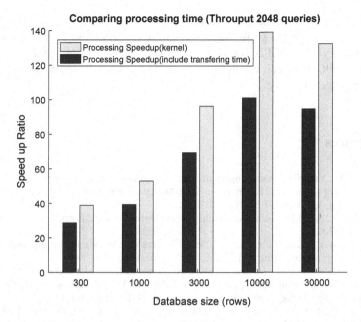

Fig. 7 The speed up ratio of testing 2048 parallel queries

method can be improved for similarity searching system or information retrieval system. In another hand, our method can achieve higher performance by not using the GPU's virtual machine. There was no result of the performance of updating and deleting query on [5], but due the complexity of fragment storing method, our method has not good performance than GPU in this case. However, with the structured database, it will more simple to modify the database.

6 Conclusion and Future Work

Our methods can handle multiple kinds of unstructured dataset with the labeling of header. With the optimization of indexing table, we can archive significant performance when processing with parallel on GPGPU. With storing separating data rows, we can use effectively the small memory amount of GPGPU device. In our experiments, we can achieve 100 times increasing the speed by using GPPGU comparing with CPU which have specifications on Table 1.

However in our method, the meta information of database is important for querying. That make the command for update the database slow down due to the data dependency. Otherwise, the frequency of deleting and inserting data rows will make database fragment over the time.

In future works, we need to propose the defragment method for database on GPPGU to increase the performance of selective query. And for the problem of small memory size GPGPU, we can improve our method for storing the database on different GPPGU devices, which make our system can handle big database and also increase the performance.

References

1. Toan, N.M., Yasushi, I.: Audio fingerprint hierarchy searching on massively parallel with multi-gpgpus using k-modes and lsh. In: Eighth International Conference on Knowledge and Systems Engineering (KSE), pp. 49–54. IEEE (2016)
2. Hill, F.S., Kelley, S.M.: Computer Graphics: using OpenGL, vol. 2. Prentice hall Upper Saddle River, NJ (2001)
3. Yang, F., Sato, Y., Tan, Y., Inoguchi, Y.: Searching acceleration for audio fingerprinting system. In: Joint Conference of Hokuriku Chapters of Electrical Societies (2012)
4. Toan, N.M., Yasushi, I.: Robust optimization for audio fingerprint hierarchy searching on massively parallel with multi-gpgpus using k-modes and lsh. In: International Conference on Advanced Engineering Theory and Applications, pp. 74–84. Springer (2016)
5. Bakkum, P., Skadron, K.: Accelerating sql database operations on a gpu with cuda. In: Proceedings of the 3rd Workshop on General-Purpose Computation on Graphics Processing Units, pp. 94–103. ACM (2010)
6. SQLite.About sqlite. https://www.sqlite.org/about.html. Accessed 19 May 2017

A Structure–Behavior Coalescence Approach for Model Singularity

Guo-Peng Qiu, Shuh-Ping Sun, Zijun Lee and William S. Chao

Abstract Generally, a system is very intricate that it includes many views, such as data, structure, function, behavior, and so on. There are two different ways to model such many views. The multi-model approach separately chooses a distinct model for every view. On the contrary, the single model approach, instead of choosing many unassociated models, will use only one integrated model. In this paper, we proposed the channel-based infinite-queue structure–behavior coalescence (SBC) process algebra which is based on model singularity. The multiple models are unassociated and therefore inconsistent with each other, which become the primary reason for the model multiplicity problems of the multi-model approach. Being able to think about a system in one single integrated model, the structure–behavior coalescence (SBC) approach truly avoids the model multiplicity problems.

Keywords Structure–Behavior coalescence · Process algebra
Model singularity · Model multiplicity

G.-P. Qiu · Z. Lee
Strait Animation College of Sanming University, Sanming, China
e-mail: 78421339@qq.com

Z. Lee
e-mail: 10427609@qq.com

S.-P. Sun (✉)
Department of Digital Media Design, I-Shou University, Kaohsiung, Taiwan
e-mail: spsun@isu.edu.tw

W. S. Chao
SBC Architecture International@, Los Angeles, USA
e-mail: architectchao@gmail.com

© Springer Nature Singapore Pte Ltd. 2019
S. K. Bhatia et al. (eds.), *Advances in Computer Communication and Computational Sciences*, Advances in Intelligent Systems and Computing 759,
https://doi.org/10.1007/978-981-13-0341-8_31

1   Introduction

Generally, a system is extremely intricate that it includes many views, such as data, structure, function, behavior, and so on. There are two different ways to model such many views. The multi-model [1] approach, such as Unified Modeling Language (UML), separately chooses a distinct model for every view as shown in Fig. 1.

On the contrary, the single model [1] approach such as object-process methodology (OPM), instead of choosing many unassociated models, will use only one integrated model as shown in Fig. 2.

The multiple models are unassociated and therefore inconsistent with each other, which becomes the primary reason for the model multiplicity problems of the multi-model approach [2]. Being able to think about a system in one single integrated model, the single model approach truly avoids the model multiplicity problems.

This paper proposes a structure–behavior coalescence (SBC) approach which is based on model singularity. The rest of this article is arranged as follows. Section 2 introduces the channel-based value-passing interaction which forms the foundation of the SBC approach. Section 3 illustrates the channel-based infinite-queue SBC process algebra. Section 4 details results and discussions. Section 5 is a conclusion.

Fig. 1 The multi-model approach

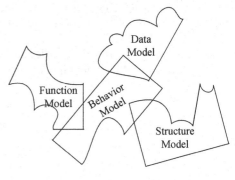

Fig. 2 The single model approach

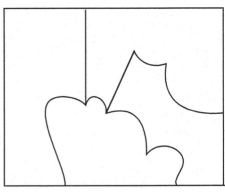

2 Channel-Based Value-Passing Interaction

An interaction represents an indivisible and instantaneous handshake or communication between two agents [3]. In the channel-based value-passing interaction approach as shown in Fig. 3, the caller agent (either external environment's actor or component) interacts with the callee agent (component) through the channel interaction.

The external environment uses a "Type-1 interaction" to interact with a component. We formally describe a channel-based value-passing Type_1 interaction as a 3-tuple TYPE_1_INTERACTION = < actor, channel_formula, callee_component >, where "actor" stands for the name of an external environment's actor, "channel_formula" stands for a channel formula and "callee_component" stands for the name of a callee component as shown in Fig. 4.

Two components use a "Type-2 interaction" to interact with each other. We formally describe a channel-based value-passing Type_2 interaction as a 3-tuple TYPE_2_INTERACTION = < caller_component, channel_formula, callee_component >, where "caller_component" stands for the name of a caller component, "channel_formula" stands for a channel formula and "callee_component" stands for the name of a callee component as shown in Fig. 5.

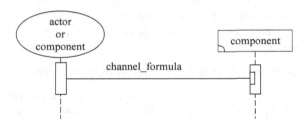

Fig. 3 Channel-based value-passing interaction

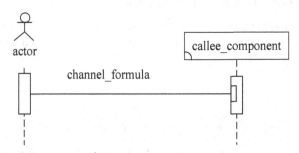

Fig. 4 Formal description of a channel-based Type_1 interaction

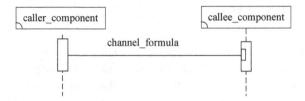

Fig. 5 Formal description of a channel-based Type_2 interaction

3 Channel-Based Infinite-Queue SBC Process Algebra

Language Constructs of Channel-Based Infinite-Queue SBC Process Algebra

The set of channel-based infinite-queue SBC processes [4] is defined by the following Backus-Naur Form (BNF) grammar, as shown in Fig. 6.

Rule 1 describes that a system is defined by the parallel composition of one or more replication (i.e., !) of the interaction flow diagrams (i.e., IFD) which is sequentially followed (i.e., ●) the null process (i.e., *STOP*).

Rule 2 describes that an interaction flow diagram (i.e. IFD) is defined by a Type_1 interaction (i.e., Type_1_Interaction) which is sequentially followed (i.e., ●) by zero or more Type_1_or_Type_2 interactions (i.e., Type_1_Or_2_Interaction).

Rule 3 describes that the Type_1_or_2 interaction (i.e. Type_1_Or_2_Interaction) is either a Type_1 interaction (i.e. Type_1_Interaction) or a Type_2 interaction (i.e. Type_2_Interaction).

Parallel Composition of One or More IFDs Defining a System

In the channel-based infinite-queue SBC process algebra [5], the parallel composition of one or more interaction flow diagrams (IFDs) defines a system as shown in Fig. 7.

A Type_1 Interaction Followed by Zero or More Type_1_or_Type_2 Interactions Defining an Interaction Flow Diagram

In the channel-based infinite-queue SBC process algebra, a Type_1 interaction followed by zero or more Type_1_or_Type_2 interactions defines an interaction flow diagram (IFD) as shown in Fig. 8.

(1) <System> ::= "! ("<IFD> " ● " *STOP* ")" {"|| ! (" <IFD> "● " *STOP* ")"}

(2) <IFD> ::= <Type_1_Interaction> {"●" <Type_1_Or_2_Interaction>}

(3) <Type_1_Or_2_Interaction> ::= <Type_1_Interaction>

                                                    | <Type_2_Interaction>

Fig. 6 BNF of channel-based infinite-queue SBC processes

Fig. 7 Parallel composition of one or more defines a system

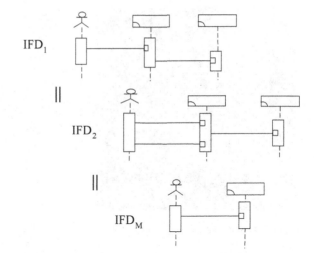

IFD_1

$\parallel$

IFD_2

$\parallel$

IFD_M

Fig. 8 A Type_1 interaction followed by zero or more Type_1_or_Type_2 interactions defines an IFD

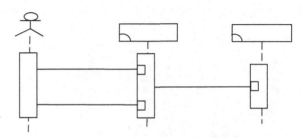

Summary of Channel-Based Infinite-Queue SBC Process Algebra

In the C-I-SBC-PA, the process of a system is defined as $\parallel_{i \in I} ! (IFD_i \bullet STOP)$ and the process of IFD_i is defined as $\bullet_{j \in J}$ interaction$_{ij}$. To combine them together, we summarize that in the C-I-SBC-PA a system is then formally described as $\parallel_{i \in I} ! (\bullet_{j \in J}$ interaction$_{ij} \bullet STOP)$.

Examining the C-I-SBC-PA model, we found that it has two parts. The first part describes the process of a system. The second part describes the interactions that occur in the process of a system.

4 Results and Discussions

Example of Smart Agriculture Innovative Service System

To illustrate the SBC approach of model singularity, let us see an example below that defines the Smart Agriculture Innovative Service System (SAISS). SAISS operates autonomously with the farm owners, farmers, or consumers. Not only farm owners and farmers benefit from the use of SAISS, consumers also.

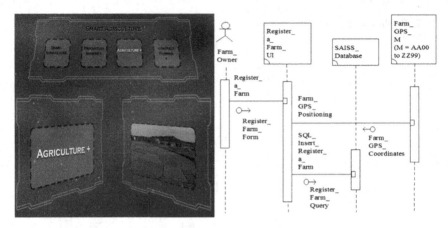

Fig. 9 IFD of the *Registering_a_Farm* behavior

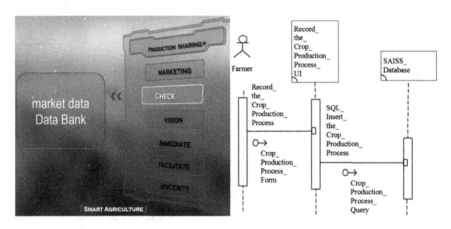

Fig. 10 IFD of the *Recording_the_Crop_Production_Process* behavior

There are four interaction flow diagrams (IFD) in the Smart Agriculture Innovative Service System. The first IFD models the *Registering_a_Farm* behavior as shown in Fig. 9.

The second IFD models the *Recording_the_Crop_Production_Process* behavior as shown in Fig. 10.

The third IFD models the *Viewing_the_Harvested_Products* Behavior as shown in Fig. 11.

The fourth IFD models the *Viewing_the_Product_Traceability* Behavior as shown in Fig. 12.

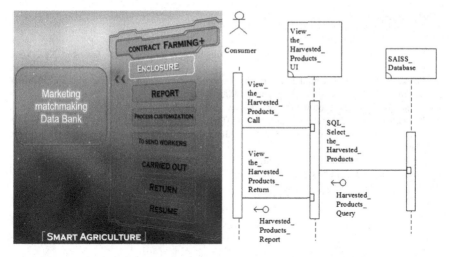

Fig. 11 IFD of the *Viewing_the_Harvested_Products* behavior

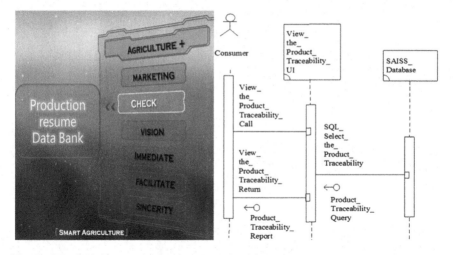

Fig. 12 IFD of the *Viewing_the_Product_Traceability* behavior

5 Conclusion

The multiple models are unassociated and therefore inconsistent with each other, which become the primary reason for the model multiplicity problems of the multi-model approach. Being able to think about a system in one single integrated model, the model singularity approach sincerely avoids the model multiplicity problems.

In this paper, the proposed channel-based infinite-queue SBC process algebra is a genuine model singularity approach. Besides detailing the channel-based infinite-queue SBC process algebra, we also demonstrated how to use the SBC approach to model the Smart Agriculture Innovative Service System.

References

1. Dori, D.: Object-Process Methodology: A Holistic Systems Paradigm. Springer, New York (2002)
2. Peleg, M., Dori, D.: The model multiplicity problem: experimenting with real-time specification methods. IEEE Trans. Softw. Eng. **26**(8) (2000)
3. Milner, R.: Communication and Concurrency. Prentice Hall (1989)
4. William S. Chao. Generalized SBC Process Algebra for Communication and Concurrency: The Structure-Behavior Coalescence Approach, 3rd Edition. CreateSpace Independent Publishing Platform, 2017
5. Chao, W.: Channel-based Infinite-queue SBC Process Algebra for Systems Definition. Create Space Independent Publishing Platform (2017)

Employment of Minimal Generating Sets and Structure of Sylow 2-Subgroups Alternating Groups in Block Ciphers

Ruslan Viacheslavovich Skuratovskii

Abstract In this article, the research of Sylow p-subgroups of A_n and S_n, which was started in Dmitruk and Suschansky (Structure of 2-Sylow subgroup of symmetric and alternating group, UMJ, N. 3, pp. 304–312, 1981, [1]), Skuratovskii (Cybern Syst Anal (1):27–41, 2009, [2]), Pawlik (Algebr Discret Math 21(2):264–281, 2016, [3]) is continued. Let $syl_2 A_{2^k}$ and $syl_2 A_n$ be Sylow 2-subgroups of the corresponding alternating groups A_{2^k} and A_n. We find a minimal generating set and the structure for such subgroups $syl_2 A_{2^k}$ and $syl_2 A_n$. The purpose of this paper is to research the structure of Sylow 2-subgroups and to construct a minimal generating set for such subgroups. The main result is to prove minimality of this generating set for the above indicated subgroups and also to describe their structure.

Keywords Minimal set of generators · Wreath product of groups · Sylow subgroups · Commutator subgroup · Semidirect product · Number of minimal generating sets and number of diagonal bases

1 Introduction

The goal of this paper is to research the structure of Sylow 2-subgroups of A_{2^k}, A_n and to construct a minimal generating set for $Syl_2 A_{2^k}$, $Syl_2 A_n$. The case of a Sylow subgroup where $p = 2$ is very special because the group $C_2 \wr C_2 \wr \ldots \wr C_2$ admits odd permutations, so $C_2 \wr C_2 \wr \ldots \wr C_2$ is not a subgroup of A_{2^k}. This case was not fully investigated in [1, 2]. This question is up till now under consideration. The authors of [1, 3] did not prove the minimality found by the system of generators for such Sylow 2-subgroups of A_n and structure of Sylow 2-subgroups was found by them not fully; for case $n = 2^k$, structure of $Syl_2 A_n$ was not found. There was a mistake in a statement about irreducibility of a set of $k + 1$ elements for $Syl_2(A_{2^k})$ that was in abstract [4] on Four Ukrainian conferences of young scientists in 2015 year.

R. V. Skuratovskii (✉)
Institute of Cybernetic of NAS of Ukraine, Maup Ikit, Kiev 21218, Ukraine
e-mail: ruslcomp@mail.ru

© Springer Nature Singapore Pte Ltd. 2019
S. K. Bhatia et al. (eds.), *Advances in Computer Communication and Computational Sciences*, Advances in Intelligent Systems and Computing 759,
https://doi.org/10.1007/978-981-13-0341-8_32

Let X be a finite alphabet. Sylow p-subgroups of A_{2^k} appear in the automaton theory, because if all states of an automaton A have output function that can be presented as a cycle $(1, 2, \ldots, p)$ then the profinite group $G_A(X)$ of this automaton is a Sylows p-subgroup of the restriction of the group of all automaton transformations $GA(X)$ also $G_A(X) < GA(X)$ [5]. In this case for the profinite groups, the following inclusion holds $Syl_p(Aut X) > FGA(X)$ [5]. Thus, finding the minimum cardinality of a generating set is important. Let X^* be the free monoid freely generated by X. All undeclared terms are from [6, 7].

2 Preliminaries

The set X^* is naturally a vertex set of a regular rooted tree, i.e., a connected graph without cycles and a designated vertex v_0 called the root, in which two words are connected by an edge if and only if they are of form v and vx, where $v \in X^*$, $x \in X$. The set $X^n \subset X^*$ is called the nth level of the tree X^* and $X^0 = \{v_0\}$. We denote by $v_{j,i}$ the vertex of X^j, which has the number i. The subtree of X^* induced by the set of vertices $\cup_{i=0}^{k} X^i$ is denoted by $X^{[k]}$. Note that the unique vertex $v_{k,i}$ corresponds to the unique word v in alphabet X. For every automorphism $g \in Aut X^*$ and every word $v \in X^*$, define the section (state) $g_{(v)} \in Aut X^*$ of g at v by the rule: $g_{(v)}(x) = y$ for $x, y \in X^*$ if and only if $g(vx) = g(v)y$. The restriction of the action of an automorphism $g \in Aut X^*$ to the subtree $X^{[l]}$ is denoted by $g_{(v)}|_{X^{[l]}}$. A restriction $g_{(v)}|_{X^{[1]}}$ is called the vertex permutation (v.p.) of g in a vertex v. Let introduce conventional signs for a v.p. state value of α in v_{ki} as $s_{ki}(\alpha)$ also put that $s_{ki}(\alpha) = 1$ if $\alpha_{(v_{ki})}|_{X^{[1]}}(x) = y$, $x \neq y$ such state of v.p. is active, and $s_{ki}(\alpha) = 0$ if $\alpha_{(v_{ki})}|_{X^{[1]}}(x) = x$ such state of v.p. is trivial. Let us label every vertex of X^l, $0 \leq l < k$ by sign 0 or 1 in relation to state of v.p. in it. Obtained by such way, a vertex-labeled regular tree is an element of $Aut X^{[k]}$.

The subtree of $X^{[k]}$ with a root in $v_{j,i}$ is denoted as $v_{j,i}X^{[k-j]}$. An automorphism of $X^{[k]}$ with nontrivial states of v.p. in some of $v_{1,1}$, $v_{1,2}$, $v_{2,1}$, $\ldots$, $v_{2,4}$, $\ldots$, $v_{m,1}$, $\ldots$, $v_{m,j}$, $m < k$, $j \leq 2^m$ is denoted by $\beta_{1,(i_{11},i_{12});\ldots;l,(i_{l1},\ldots,i_{l2^l});\ldots;m,(i_{m1},\ldots,i_{m2^m})}$ where the index that stands straight before parentheses is number of a level in parentheses; it can be written as tuple of states of v.p. of this level. In other words, it is defined $i_{mj} = 0$ if v.p. in v_{mj} is trivial, $i_{mj} = 1$ in other case, i.e., $i_{mj} = s_{mj}(\beta)$, where $\beta \in Aut X^{[k]}$, $m < k$. If for some l all $i_{lj} = 0$, then 2^l-tuple $l, (i_{l1}, \ldots, i_{l2^l})$ does not figure in indexes of β. But if numbers of active vertices are certain, for example, $v_{j,1}$ and $v_{j,s}$, which can be used more easily by notation $\beta_{j,(1,s)}$, where in parentheses, numbers of vertices are with active state of v.p. from a level j. If in parentheses only one index presents then parentheses can be omitted for instance, $\beta_{j,(s)}; = \beta_{j,s}$; Denote by $\tau_{i,\ldots,j}$ the automorphism of $X^{[k]}$, which has a nontrivial v.p. only in vertices $v_{k-1,i}$, $\ldots$, $v_{k-1,j}$, $j \leq 2^{k-1}$ of the level X^{k-1}. Denote by τ the automorphism $\tau_{1,2^{k-1}}$. Let us consider special elements such that $\alpha_0 = \beta_0 = \beta_{0,(1)}, \alpha_1 = \beta_1 = \beta_{1,(1)}, \ldots, \alpha_l = \beta_l = \beta_{l,(1)}$.

3   Main Result

Recall that a wreath product of permutation groups is an associative construction. Consider C_2 as additive group with two elements 0, 1. For constructing a wreath product, it is defined an action of C_2 by shift on $X = \{0, 1\}$. As well known

$$Aut\, X^{[k-1]} \simeq \underset{i=1}{\overset{k-1}{\wr}} C_2 \, [5].$$

Lemma 1 *Every automorphism that has active v.p. only on X^l, $l < k - 1$ acts by even permutation on X^k.*

Proof Actually, every transposition in vertex from X^l, $l < k - 1$ acts on even number of pair of vertexes because of binary tree structure. More precisely, it realizes even permutation on the set X^k with cyclic structure [8] $(1^{2^{k-1}-2^{k-l-l}}, 2^{2^{k-l-l}})$ because it is formed by the structure of binary tree.   □

Corollary 1 *Due to Lemma 1, automorphisms from $Aut\, X^{[k-1]} = \langle \alpha_0, \ldots, \alpha_{k-2} \rangle$ form a group $B_{k-1} = \underset{i=1}{\overset{k-1}{\wr}} C_2$ acting on X^{k-1} by even permutations. Size of B_{k-1} equals to $2^{2^{k-1}-1}$.*

The parity of an action follows from Lemma 1.

Let us denote by W_{k-1} the subgroup of $Aut\, X^{[k]}$ such that has active states only on X^{k-1} and number of such states is even, i.e., $W_{k-1} \lhd St_{Aut\, X^{[k]}}(k - 1)$ [6].

Proposition 1 *The order of W_{k-1} is equal to $2^{2^{k-1}-1}$, $k > 1$ and $W_{k-1} = C_2^{2^{k-1}-1}$.*

Proof There are 2^{k-1} vertices on X^{k-1}, where it can be elements of a group $V_{k-1} \simeq C_2 \times C_2 \times \cdots \times C_2 \simeq (C_2)^{2^{k-1}}$, but as a result of the fact that X^{k-1} contains only even number of nontrivial v.p. from X^{k-1}, there are only half of all permutations from $V_{k-1} \simeq St_{G_k}(k - 1)$ on X^{k-1}. So it is the subgroup $W_{k-1} \simeq C_2^{2^{k-1}} / C_2$ of V_{k-1}. So we can state that $|W_{k-1}| = 2^{2^{k-1}-1}$, W_{k-1} has $k - 1$ generators and W_{k-1} can be considered as a vector space of dimension $k - 1$.   □

For example, let us consider the subgroup W_{4-1} of A_{2^4} its cardinality is $2^{2^{4-1}-1} = 2^7$ and $|A_{2^4}| = 2^{14}$. Let us denote by G_k the subgroup of $Aut\, X^{[k]}$ such that $G_k \simeq B_{k-1} \ltimes W_{k-1}$.

Lemma 2 *The elements τ and $\alpha_0, \ldots, \alpha_{k-1}$ generate arbitrary element τ_{ij}. The set $\{\tau, \alpha_0, \ldots, \alpha_{k-1}\}$ is enough to generate a basis of W_{k-1}.*

Proof First, we shall prove the possibility of generating arbitrary $\tau_{ij}, 1 \le i, j \le 2^{k-1}$. According to [2, 7], the set $\alpha_0, \ldots, \alpha_{k-2}$ is the minimal generating set for group $Aut\, X^{[k-1]}$.

Since $Aut\, v_{1,1} X^{[k-2]} \simeq \langle \alpha_1, \ldots, \alpha_{k-2} \rangle$ acts on X^{k-1} transitively [2, 8, 14], then there exists a transposition of $v_{k-1,1}$ and $v_{k-1,j}$, $j \le 2^{k-2}$. For this goal, we act by α_{k-j} on τ: $\alpha_{k-j}\tau\alpha_{k-j} = \tau_{j,2^{k-2}}$. Similarly, we act on τ by the corespondent

α_{k-i} to get $\tau_{i,2^{k-2}}$ from τ: $\alpha_{k-i}\tau\alpha_{k-i}^{-1} = \tau_{i,2^{k-2}}$. Note that the automorphisms α_{k-j} and α_{k-i}, $1 < i, j < k-1$ act nontrivial only on subtree $v_{1,1}X^{[k-1]}$. To get $\tau_{m,l}$ from $v_{1,2}X^{[k-1]}$, i.e., $2^{k-2} < m, l \le 2^{k-1}$, we use α_0 to map $\tau_{i,j}$ in $\tau_{i+2^{k-2},j+2^{k-2}} \in v_{1,2}AutX^{[k-1]}$. To express an arbitrary transposition $\tau_{j,m}$ from W_{k-1}, we have to multiply $\tau_{1,j}\tau\tau_{m,2^{k-1}} = \tau_{j,m}$. To construct a permutation of $v_{k-1,1}$ and $v_{k-1,j}$, we need to realize a natural number j, $1 < j < 2^{k-2}$, in 2-adic set of presentation (binary arithmetic). Then, $j = \delta_{j_1}2^{m_j} + \delta_{j_2}2^{m_j-1} + \cdots + \delta_{j_{m_j+1}}$, $\delta_{j_i} \in \{0, 1\}$ which is a correspondence between δ_{j_i} that from such presentation and expressing of automorphisms: $\tau_{j,2^{k-1}} = \prod_{i=1}^{m_j}\alpha_{k-2-(m_j-i)}^{\delta_{j_i}}\tau \prod_{i=1}^{m_j}\alpha_{k-2-(m_j-i)}^{\delta_{j_i}}$, $1 \le m_j \le k-2$. Generating the basis of W_{k-1} by all τ_{ij} is clear. $\square$

Lemma 3 *Orders of groups* $G_k = \langle \alpha_0, \alpha_1, \alpha_2, \ldots, \alpha_{k-2}, \tau \rangle$ *and* $Syl_2(A_{2^k})$ *are equal to* 2^{2^k-2}.

Proof In accordance with Legendre's formula, the power of 2 in $2^k!$ is $\left[\frac{2^k}{2}\right] + \left[\frac{2^k}{2^2}\right] + \left[\frac{2^k}{2^3}\right] + \cdots + \left[\frac{2^k}{2^k}\right] = \frac{2^k-1}{2-1}$. We need to subtract 1 from it because we have only $\frac{n!}{2}$ of all permutations as a result: $\frac{2^k-1}{2-1} - 1 = 2^k - 2$. So $|Syl(A_{2^k})| = 2^{2^k-2}$. The same size has group $G_k = B_{k-1} \ltimes W_{k-1}$ and $|G_k| = |B_{k-1}| \cdot |W_{k-1}| = |Syl_2A_{2^k}|$. Since size of groups G_k according to Proposition 1 and the fact that $|B_{k-1}| = 2^{2^{k-1}-1}$ is 2^{2^k-2}, for instance, the orders of $Syl_2(A_8)$, B_{3-1} and W_{3-1} are such $|W_{3-1}| = 2^{2^{3-1}-1} = 2^3 = 8$, $|B_{3-1}| = |C_2 \wr C_2| = 2 \cdot 2^2 = 2^3$ and according to Legendre's formula, the power of 2 in $2^k!$ is $\frac{2^3}{2} + \frac{2^3}{2^2} + \frac{2^3}{2^3} - 1 = 6$ so $Syl_2(A_8) = 2^6 = 2^{2^k-2}$, where $k = 3$. Next example for A_{16}: $Syl_2(A_{16}) = 2^{2^4-2} = 2^{14}$, $k = 4$, $|W_{4-1}| = 2^{2^{4-1}-1} = 2^7$, $|B_{4-1}| = |C_2 \wr C_2 \wr C_2| = 2 \cdot 2^2 \cdot 2^4 = 2^7$. So we have the $|A_{16}| = |W_3||B_3|$ equality which endorse the condition of this Lemma. $\square$

An automorphisms group of the subgroup $C_2^{2^{k-1}-1}$ is based on permutations of copies of C_2. Orders of $\overset{k-1}{\underset{i=1}{\wr}} C_2$ and $C_2^{2^{k-1}-1}$ are equal. A homomorphism from $\overset{k-1}{\underset{i=1}{\wr}} C_2$ into $Aut(C_2^{2^{k-1}-1})$ is injective because a kernel of action $\overset{k-1}{\underset{i=1}{\wr}} C_2$ on $C_2^{2^{k-1}-1}$ is trivial, action is effective. The group G_k is a proper subgroup of index 2 in the group $\overset{k}{\underset{i=1}{\wr}} C_2$ [2, 9].

Theorem 1 *A **maximal 2-subgroup** of* $AutX^{[k]}$ *consisting of even permutations has the structure of the semidirect product* $G_k \simeq B_{k-1} \ltimes W_{k-1}$ *and is isomorphic to* $Syl_2A_{2^k}$.

Proof A maximal 2-subgroup of $AutX^{[k-1]}$ is isomorphic to $B_{k-1} \simeq \underbrace{C_2 \wr C_2 \wr \ldots \wr C_2}_{k-1}$ acting by even permutation on X^k according to Lemma 1. A maximal 2-subgroup which has elements with active states only on X^{k-1} is isomorphic to subgroup W_{k-1}. The construction of W_{k-1} contributes an action of W_{k-1} by even permutations. From Lemma 1 it follows that every element of B_{k-1} acts by even permutation on X^k. Thus, G_k acts by even permutations on X^k.

Using Corollary 1 and Proposition 1 about sizes of B_{k-1} and W_{k-1}, we get size of $G_k \simeq B_{k-1} \ltimes W_{k-1}$ is $2^{2^{k-1}-1} \cdot 2^{2^{k-1}-1} = 2^{2^k-2}$. A group G_k is subgroup of $Aut X^{[k]}$ and it is well known that $Aut X^{[k]} \simeq Syl_2 S_{2^k}$, so G_k is isomorphic to some subgroup $\widetilde{G}_k$ of $Syl_2 S_{2^k}$. A group $Syl_2 A_{2^k}$ is subgroup of $Syl_2 S_{2^k}$. As supplementary according to Lemma 3 order of $\widetilde{G}_k$ equals to order of $Syl_2(A_{2^k})$, hence, according to Sylow's theorems 2-subgroup, $G_k \simeq Syl_2 A_{2^k}$.

Since subgroups B_{k-1} and W_{k-1} are embedded in $Aut X^{[k]}$, then define an action of B_{k-1} on elements of W_{k-1} as $\tau^\sigma = \sigma \tau \sigma^{-1}$, $\sigma \in B_{k-1}$, $\tau \in W_{k-1}$, i.e., action by inner automorphism (inner action) from $Aut X^{[k]}$. Note that W_{k-1} is subgroup of stabilizer of X^{k-1}, i.e., $W_{k-1} < St_{Aut X^{[k]}}(k-1) \lhd Aut X^{[k]}$ and is normal too $W_{k-1} \lhd Aut X^{[k]}$, because conjugation keeps a cyclic structure of permutation so even permutation maps in even. Therefore, such conjugation induces an automorphism of W_{k-1} and $G_k \simeq B_{k-1} \ltimes W_{k-1}$. $\square$

Theorem 2 *The set* $S_\alpha = \{\alpha_0, \alpha_1, \alpha_2, \ldots, \alpha_{k-2}, \tau\}$ *of elements from* $Aut X^{[k]}$ *generates the group* G_k.

Proof As we see from Corollary 1, Lemmas 3 and 2 group G_k are generated by S_α and order of G_k equals order of $Syl_2(A_{2^k})$ according to Lemma 3. Isomorphism of $Syl_2(A_{2^k})$ and G_k is proved in Theorem 1. $\square$

Consequently, we construct a generating set, which contains k elements, that is less than in [4]. We will not distinguish $Syl_2(A_{2^k})$ and its isomorphic copy G_k in $Aut X^{[k]}$.

The structure of Sylow 2-subgroup of A_{2^k} is the following: $\wr_{i=1}^{k-1} C_2 \ltimes \prod_{i=1}^{2^{k-1}-1} C_2$, where we consider C_2 as group of action on two elements and this action is faithful. It adjusts with construction of normalizer for $Syl_p(S_n)$ from [10], where it was said that $Syl_2(A_{2^l})$ is self-normalized in S_{2^l}.

Definition 1 Let us call the index of automorphism β on X^l a number of active v.p. of β on X^l.

Definition 2 Define an element of type T as an automorphism $\tau_{i_0,\ldots,i_{2^k-1};j_{2^k-1},\ldots,j_{2^k}}$ that has even index at X^{k-1} and has exactly m active states, $m \equiv 1 \bmod 2$, in vertexes of form $v_{k-1,j}$, $1 \leq j \leq 2^{k-2}$ and m active states in vertices of form $v_{k-1,j}$, $2^{k-2} < j \leq 2^{k-1}$, $m \equiv 1 \bmod 2$. Set of such elements is denoted by T.

Definition 3 A combined generator is such an automorphism $\beta_{l;\tilde{\tau}}$ that the restriction $\beta_{l;\tilde{\tau}}|_{X^{k-1}}$ coincides with α_l and $Rist_{<\beta_{l;\tilde{\tau}}>}(k-1) = \langle \tau' \rangle$, where $\tau' \in$ T.

Definition 4 A combined element is such an automorphism $\beta_{1,i_1;2,i_2;\ldots;k-1,i_{k-1};\tilde{\tau}}$ that its restriction $\beta_{1,i_1;2,i_2;\ldots;k-1,i_{k-1};\tilde{\tau}}|_{X_{k-1}}$ coincides with one of the elements that can be generated by S_α and $Rist_{<\beta_{1,i_1;2,i_2;\ldots;k-1,i_{k-1};\tilde{\tau}}>}(k-1) = \langle \tau' \rangle$ [6] where $\tau' \in$ T. The set of such elements is denoted by C.

In other words, elements $g \in C$ on level X^{k-1} have such structure as elements and generators of type T, as well $\tau_{i_0,...,i_{2k-1}; j_{2k-1},...,j_{2k}} \in St_{Aut X^k}(k-1)$.

The minimum size of a generating set S of G we denote by rk$G$ and call the rank of G [11]. By the distance between vertices, we shall understand the usual distance at graph between these vertexes. By the distance $\rho(g)$ of automorphism $g \in Aut X^{[k]}$ (element), we shall understand the maximal distance between two vertexes with active v.p. of g.

Lemma 4 *An automorphism having a distance d_0 that has v.p. only on X^{k-1} cannot be generated by automorphism with a distance d_1 such that $d_1 < d_0$.*

Proof An element g with a distance $\rho(g) = d_0$, $d_0 < d_1$ can be mapped by automorphic mapping only in automorphism with a distance d_0 because automorphic mapping keeps incidence relation. So it possess property of isometry. Also, multiplication of portraits (labeled graphs) of automorphisms that have distance d_1 gives us portrait of element with distance no greater than d_1, and it follows from properties of group operation. For instance, $\tau_{1i}\tau_{1j} = \tau_{ij}$, where $i, j > 2^{k-2}$, $\rho(\tau_{1i}) = \rho(\tau_{1j}) = 2k - 2$ but $\rho(\tau_{ij}) < 2k - 2$.   □

Lemma 5 *An arbitrary automorphism $\tau' \in T$ can be expressed only as product of odd number of automorphisms from C or T.*

Proof Let us assume that there is no such element τ_{ij}, which has distance $2k - 2$, then accord to Lemma 4 and it is imposable to generate a pair of transpositions τ' with distance $\rho(\tau_{ij}) = 2k - 2$. If we consider product P of even number elements from T, then automorphism P has even number of active states in vertexes $v_{k-1,i}$ with number $i \leq 2^{k-2}$ so P does not satisfy the definition of generator of type T. An combined element $\beta_{i_l;\tau}$ can be decomposed in product $\beta_{i_l;\tau} = \tau \dot{\beta}_{i_l}$ so we can express it by using τ or using a product where odd number elements from T or C.   □

We have to take into account that all elements from T have the same main property that consists of odd number of active v.p. in vertices of form $v_{k-1,j}$, $j \leq 2^{k-2}$ and odd number of active v.p. in vertices with index $j : 2^{k-2} < j \leq 2^{k-1}$.

Let $S'_\alpha = \langle \alpha_0, \alpha_1, \ldots, \alpha_{k-2} \rangle$ so as it is well known [7] $\langle S'_\alpha \rangle = Aut X^{[k-1]}$. The cardinality of a generating set S is denoted by $| S |$ so $| S'_\alpha | = k - 1$. Recall that $rk(G)$ is the rank of a group [11].

Let $S_\beta = S'_\alpha \cup \tau_{i...j}$, $\tau_{i...j} \in T$ and $S'_\beta = \langle \beta_0, \beta_{1,(1);\tau^x}, \ldots, \beta_{k-2,(1);\tau^x}, \tau \rangle$, $x \in \{0, 1\}$, note if $x = 0$ then $\beta_{1,(1);\tau^x} = \beta_l$. In S'_β, it can be used τ instead of $\tau_{i...j} \in T$ because it is not in principle for the proof. So let $S_\beta = S_{\alpha'} \cup \tau$, and hence $\langle S_\beta \rangle = G_k$.

Let us assume that S'_β does not contain τ and so has a cardinality $k - 1$. To express element of type T from S'_β, we can use a word $\beta_{i,\tau}\beta_i^{-1} = \tau$ but if $\beta_{i,\tau} \in S'_\beta$ then $\beta_i \notin S'_\beta$ in contrary case $| S'_\beta | = k$. So we cannot express word $\beta_{i,\tau}\beta_i^{-1} |_{X^{[k-1]}} = e$ to get $\beta_{i,\tau}\beta_i^{-1} = \tau$. For this goal, we have to find relation in restriction of group G_k on $X^{[k-1]}$. We have to take into consideration that $G_k |_{X^{[k-1]}} = B_{k-1}$. Rest of relation in B_{k-1}. Really, in wreath product $\wr_{j=1}^{k} C_2^{(j)} \simeq B_{k-1}$ holds a constitutive relations $\alpha_i^{2^m} =$

e and $\left[\alpha_m^i \alpha_{i_n} \alpha_m^{-i}, \alpha_m^j \alpha_{i_k} \alpha_m^{-j}\right] = e$, $i \neq j$, where $\alpha_m \in S_\alpha'$, $\alpha_{i_k} \in S_\alpha'$ are generators of factors of $\wr_{j=1}^k C_2^{(j)}$ ($m < n$, $m < k$) [2, 12].

Lemma 6 *A generating set of G_k contains S_α' and has at least $k - 1$ generators.*

Proof The subgroup $B_{k-1} < G_k$ is isomorphic to $Aut X^{k-1}$ that has a minimal set of generators of $k - 1$ elements [7]. Moreover, the subgroup $B_{k-1} \simeq {}^{G_k}/W_{k-1}$, because $G_k \simeq B_{k-1} \ltimes W_{k-1}$, where $W_{k-1} \rhd G_k$. As it is well known that if $H \lhd G$ then $rk(G) \geq rk(^G/_H)$, because all generators of G_k may belong to different quotient classes [13]. □

As a corollary of last lemma, we see that generating set of size $k - 1$ does not exist because $S_\beta' \setminus \{\tau\}$ generates only a proper subgroup B_{k-1} of G_k as it was shown above.

Note that Frattini subgroup of any finite 2-group is equal to $\phi(G_k) = G_k^2 \cdot [G_k, G_k] = G_k^2$ because $G_k^2 > [G_k, G_k]$. Hence, generating sets of a 2-group G_k correspond to generating sets of 2-abelianization and to generating sets of quotient group by G_k^2.

Let $X_1 = \{v_{k-1,1}, v_{k-1,2}, \ldots, v_{k-1,2^{k-2}}\}$ and $X_2 = \{v_{k-1,2^{k-2}+1}, \ldots, v_{k-1,2^{k-1}}\}$. Let group $Syl_2 A_{2^k}$ acts on $X^{[k]}$.

Lemma 7 *An element g belongs to $G_k' \simeq Syl_2 A_{2^k}$ iff g is arbitrary element from G_k which has all even indexes. The set of all commutators K of Sylow 2-subgroup $Syl_2 A_{2^k}$ of the alternating group A_{2^k} is the commutant of $Syl_2 A_{2^k}$.*

Proof Let us prove the ampleness by induction of a number of level l. Recall that any automorphism $\theta \in Syl_2 A_n$ has an even index on X^{k-1}, so number parities of active v. p. on X_1 and on X_2 are the same. Conjugation by automorphism α from $Aut v_{11} X^{[k-1]}$ of automorphism θ has some number $x : 1 \leq x \leq 2^{k-2}$ of active v. p. on X_1 does not change x. Also, automorphism θ^{-1} has the same number x of v. p. on X_{k-1} as θ has. If α from $Aut v_{11} X^{[k-1]}$ and $\alpha \notin Aut X^{[k]}$, then conjugation $(\alpha\theta\alpha^{-1})$ permutes vertices only inside X_1 (X_2).

Thus, $\alpha\theta\alpha^{-1}$ and θ have the same parities of number of active v.p. on X_1 (X_2). Hence, a product $\alpha\theta\alpha^{-1}\theta^{-1}$ has an even number of active v.p. on X_1 (X_2) in this case. Moreover, a coordinate-wise sum by mod2 of active v. p. from $(\alpha\theta\alpha^{-1})$ and θ^{-1} on X_1 (X_2) is even and equal to $y : 0 \leq y \leq 2x$.

If conjugation by α permutes sets X_1 and X_2, then there are coordinate-wise sums of no trivial v.p. from $\alpha\theta\alpha^{-1}\theta^{-1}$ on X_1 (analogously on X_2) and have the following form: $(s_{k-1,1}(\alpha\theta\alpha^{-1}), \ldots, s_{k-1,2^{k-2}}(\alpha\theta\alpha^{-1})) \oplus (s_{k-1,1}(\theta^{-1}), \ldots, s_{k-1,2^{k-2}}(\theta^{-1}))$. This sum has even number of v.p. on X_1 and X_2 because $(\alpha\theta\alpha^{-1})$ and θ^{-1} have a same parity of no trivial v.p. on X_1 (X_2). Hence, $(\alpha\theta\alpha^{-1})\theta^{-1}$ has even number of v.p. on X_1 as well as on X_2.

An authomorphism θ from G_k was arbitrary, so number of active v.p. x on X_1 is arbitrary. And α is arbitrary from $Aut X^{[k-1]}$, so vertices can be permuted in such way that the commutator $[\alpha, \theta]$ has arbitrary even number y of active v.p. on X_1, $0 \leq y \leq 2x$.

A conjugation of an automorphism θ having arbitrary index x, $1 \le x \le 2^l$ on X^l by different $\alpha \in Aut X^{[k]}$ gives us all conjugated permutations of active v.p. that θ has on X^l. To construct arbitrary $\chi \in G'_k$, we use an induction by an index $2x$ of χ on X^l. So multiplication $(\alpha\theta\alpha^{-1})\theta$ generates a commutator having index y equal to coordinate-wise sum by $mod 2$ of no trivial v.p. from vectors $(s_{l1}(\alpha\theta\alpha^{-1}), s_{l2}(\alpha\theta\alpha^{-1}), \ldots, s_{l2^l}(\alpha\theta\alpha^{-1})) \oplus (s_{l1}(\theta), s_{l2}(\theta), \ldots, s_{l2^l}(\theta))$ on X^l. A indexes parities of $\alpha\theta\alpha^{-1}$ and θ^{-1} are same, so their sum by $mod 2$ is even. Choosing θ, we can choose an arbitrary index x of θ and also we can choose arbitrary α to make a permutation of active v.p. on X^l. Thus, we obtain an element with arbitrary even index on X^l and arbitrary location of active v.p. on X^l.

Check that property of number parity of v.p. on X_1 and on X_2 is closed with respect to conjugation. We know that numbers of active v. p. on X_1 as well as on X_2 have the same parities. So action by conjugation only can permute it; hence, we again get the same structure of element. Conjugation by automorphism α from $Aut v_{11} X^{[k-1]}$ automorphism θ, which has odd number of active v. p. on X_1, does not change its parity. Choosing the θ, we can choose arbitrary index x of θ on X^{k-1} and number of active v.p. on X_1 and X_2 and also we can choose arbitrary α to make a permutation active v.p. on X_1 and X_2. Thus, we can generate all possible elements from a commutant. Also, this result follows from Lemma 8 and structure of B'_k, where element $g \in B'_k$ has all possible even indexes on X^l, $l < k$.

Let us check that the set of all commutators K from $Syl_2 A_{2^k}$ is closed with respect to multiplication of commutators. Let $\kappa_1, \kappa_2 \in K$, then $\kappa_1\kappa_2$ has an even index on X^l, $l < k - 1$ because coordinate-wise sum $(s_{l,1}(\kappa_1), \ldots, s_{k-1,2^l}(\kappa_1)) \oplus (s_{l,\kappa_1(1)}(\kappa_2), \ldots, s_{l,\kappa_1(2^l)}(\kappa_2))$ of two 2^l-tuples of v.p. with an even number of no trivial coordinate has even number of such coordinate. Note that conjugation of κ can permute sets X_1 and X_2, so parities of x_1 and X_2 coincide. It is obvious that index of $\alpha\kappa\alpha^{-1}$ is even as well as index of κ.

Check that a set K is a set closed with respect to conjugation.

Let $\kappa \in K$, then $\alpha\kappa\alpha^{-1}$ also belongs to K, it is so because conjugation does not change index of an automorphism on a level. Conjugation only permutes vertices on level because elements of $Aut X^{[l-1]}$ act on vertices of X^l. But as it was proved, above elements of K have all possible indexes on X^l, so as a result of conjugation $\alpha\kappa\alpha^{-1}$ we obtain an element from K. Check that the set of commutators is closed with respect to multiplication of commutators. Let κ_1, κ_2 be an arbitrary commutators of G_k. The parity of the number of vertex permutations on X^l in the product $\kappa_1\kappa_2$ is determined exceptionally by the parity of the numbers of active v.p. on X^l in κ_1 and κ_2 (independently from the action of v.p. from the higher levels). Thus, $\kappa_1\kappa_2$ has an even index on X^l.

Hence, normal closure of the set K coincides with K. $\square$

Lemma 8 *An element* $(g_1, g_2)\sigma^i \in G'_k$ *iff* $g_1, g_2 \in G_{k-1}$ *and* $g_1 g_2 \in B'_{k-1}$.

Proof Indeed, if $(g_1, g_2) \in G'_k$, then indexes of g_1 and g_2 on X^{k-1} are even according to Lemma 7, thus $g_1, g_2 \in G_{k-1}$. A sum of indexes of g_1 and g_2 on X^l, $l < k - 1$ are even according to Lemma 7 too, so index of product $g_1 g_2$ on X^l is even. Thus, $g_1 g_2 \in B'_{k-1}$.

Let us prove the sufficiency via Lemma 7 and vice versa; if $g_1, g_2 \in G_{k-1}$, then indexes of these automorphisms on X^{k-2} of subtrees $v_{11}X^{[k-1]}$ and $v_{12}X^{[k-1]}$ are even as elements from G'_k have. The product $g_1 g_2$ belongs to B'_{k-1} by condition of this lemma, so sum of indexes of g_1, g_2 on any level $X^l, 0 \le l < k-1$ is even. Thus, the characteristic properties of G'_k described in Lemma 7 hold. $\qquad \square$

Proposition 2 *Frattini subgroup* $\phi(G_k)$ *acts by all even permutations on* X^l, $0 \le l \le k-1$ *and any element of* $\phi(G_k)$ *has even indexes on* X^{k-2} *of subtrees* $v_{11}X^{[k-1]}$ *and* $v_{12}X^{[k-1]}$. *Also* $\phi(G_k) = (G_k)'$.

Proof Since a group G_k^2 contains the subgroup G', then a product $G^2 G'$ contains all elements from the commutant. We need to prove that $G_k^2 \simeq G'$.

An index of the automorphisms α^2, $(\alpha\beta)^2$, and $\alpha, \beta \in G_k$ on $X^l, l < k-1$, are always even. In more detail, the indexes of α^2, $(\alpha\beta)^2$, and $\beta\alpha\beta^{-1}\alpha^{-1}$ on X^l are determined exceptionally by the parity of indexes of α and β on X^l (independently of the action of v.p. from the higher levels) and this parity is even. Since an index of $\alpha\beta$ on X^l is an arbitrary $x : 0 \le x \le 2^l$, then an index of $(\alpha\beta)^2$ is arbitrary even number that is between 0 and 2^l. As it was shown in Lemma 7, any $\gamma \in G_k$ has same parities of numbers of active v.p. on X_1 as well as on X_2. Then, γ^2 has an even number of active v.p. on each set X_1 and X_2. There are no elements with odd number of active v.p. on each set X_1 and X_2 in G_k. Thus, we can generate all possible elements from the commutant which was studied in Lemma 7. $\qquad \square$

We denote as $G_k(l)$ such subgroup of $Aut X^{[k]}$ that contains all v.p. from X^l, $l < k-1$. In other words, it contains all v.p. from $Stab_{Aut X^{[k]}}(l)$ and does not contain v.p. from $Stab_{Aut X^{[k]}}(l+1), l < k-1$. We denote as $G_k(k-1)$ such subgroup of $Aut X^{[k]}$ that consists of v.p. which are located on X^{k-1} and isomorphic to W_{k-1}. Let us construct a homomorphism from $G(l)$ onto C_2 in the following way: $\varphi_l(\alpha) = \sum\limits_{i=1}^{2^l} s_{li}(\alpha) \bmod 2$. Note that $\varphi_l(\alpha \cdot \beta) = \varphi_l(\alpha) \circ \varphi_l(\beta) = (\sum\limits_{i=1}^{2^l} s_{li}(\alpha) + \sum\limits_{i=1}^{2^l} s_{li}(\beta)) \bmod 2$.

Structure of subgroup $G_k^2 G_k' \lhd \overset{k}{\underset{1}{\wr}} S_2 \simeq Aut X^{[k]}$ can be described in next way. This subgroup contains the commutant G_k'. So it has on each X^l, $0 \le l < k-1$ all even indexes that can exist there. There does not exist v.p. of type T on X^{k-1}, rest of even the indexes are present on X^{k-1}. It is so, because the sets of elements of types T and C are not closed with respect to operation of raising to the even power. Thus, the squares of the elements do not belong to T and C. This implies the following corollary.

Corollary 2 *A quotient group* $\,^{G_k}/_{G_k^2 G_k'}$ *is isomorphic to* $\underbrace{C_2 \times C_2 \times \cdots \times C_2}_{k}$.

Proof The proof is based on two facts $G_k^2 G_k' \simeq G_k^2 \lhd G_k$ and $\left| G : G_k^2 G_k' \right| = 2^k$. Construct a homomorphism from $G_k(l)$ onto C_2 in the following way: $\varphi_l(\alpha) =$

$\sum_{i=1}^{2^l} s_{li}(\alpha) \bmod 2$. Note that $\varphi_l(\alpha \cdot \beta) = \varphi_l(\alpha) \circ \varphi_l(\beta) = (\sum_{i=1}^{2^l} s_{li}(\alpha) + \sum_{i=1}^{2^l} s_{li}(\beta))$
$\bmod 2$, where $\alpha, \beta \in \mathrm{Aut} X^{[n]}$. Index of $\alpha \in G_k^2$ on X^l, $l < k-1$ is even but index of $\beta \in G_k$ on X^l can be both even and odd. Note that $G_k(l)$ is abelian subgroup of G_k and $G_k^2(l) \trianglelefteq G_k$.

By virtue of the fact that we can construct the homomorphism φ_i from every subgroup $G_k(i)$ of this product to $G_k(i)/G_k^2(i)$, we have homomorphism from G_k to G_k/G_k^2. The group G_k/G_k^2 is elementary abelian 2-group because $g^2 = e$, $g \in G$ and $G' \triangleleft G^2$. Let us find a 2-rank of this group.

We use the homomorphism φ_l which is described above, to map $G_k(l)$ onto $G_k(l)/G_k^2(l)$ the $\ker \varphi_l = G_k^2(l)$. If α from $G_k(l)$ has odd number of active states of v.p. on X^l, $l < k-1$, then $\varphi_l(\alpha) = 1$ in $G_k(l)/G_k^2(l)$; otherwise, if this number is even than α from $\ker \varphi_i = G_k^2(l)$, so $\varphi_l(\alpha) = 0$. Hence, we have $G_k(l)/G_k^2(l) \simeq C_2$. Let us check that mapping $\varphi = (\varphi_0, \varphi_1, \ldots, \varphi_{k-2}, \phi_{k-1})$ is the homomorphism from G_k onto $(C_2)^k$.

Parity of index of $\alpha \cdot \beta$ on X^l is equal to sum by $\bmod 2$ of indexes of α and β; hence, $\varphi_l(\alpha \cdot \beta) = (\varphi_l(\alpha) + \varphi_l(\beta))$ because multiplication $\alpha \cdot \beta$ in G_k does not change a parity of index of β, $\beta \in G_k$ on X^l. Real action of element of active group $A = \underbrace{C_2 \wr C_2 \wr \ldots \wr C_2}_{l-1}$ from wreath power $\underbrace{(C_2 \wr C_2 \wr \ldots \wr C_2)}_{l-1} \wr C_2$ on element from passive subgroup C_2 of second multiplier from product gf, $g, f \in A \wr C_2$ does not change a parity of index of β on X^l, if index of β was even; then, under action it stands to be even and the sum $\varphi_l(\alpha) \bmod 2 + \varphi_l(\beta) \bmod 2$ will be equal to $(\varphi_l(\alpha) + \varphi_l(\beta)) \bmod 2$, and hence it does not change a $\varphi(\beta)$.

Indexes of $\alpha_{(v_{11})}$ and $\alpha_{(v_{12})}$ for arbitrary $\alpha \in G_k$ on X^{k-1} can be as even as well as odd. But these indexes of $\alpha_{(v_{11})}$ and $\alpha_{(v_{12})}$ are equal by $\bmod 2$.

The subgroup $G_k^2 G_k'$ admits only automorphisms α such that $\alpha_{(v_{11})}$ and $\alpha_{(v_{12})}$ have even indexes on X^{k-1}. So this set is a kernel of mapping from $G_k(k-1)$ onto C_2. This homomorphism can be obtained from the formula $\phi(\alpha) = \sum_{i=1}^{2^{k-2}} s_{k-1,i}(\alpha)(\bmod 2) \cdot$

$\sum_{i=2^{k-2}+1}^{2^{k-1}} s_{k-1,i}(\alpha)(\bmod 2)$. It follows from structure of G_k that $\sum_{i=1}^{2^{k-2}} s_{k-1,i}(\alpha)(\bmod 2) =$

$\sum_{i=2^{k-2}+1}^{2^{k-1}} s_{k-1,i}(\alpha)(\bmod 2)$. Thus, the image $\phi(G_k(k-1))$ consist of two elements: 0 and 1; we map these elements in different elements of C_2. Hence, homomorphism ϕ is surjective.

Hence, for an abelian subgroup $G_k(k-1) \simeq W_{k-1}$ such that $G_k(k-1) \rhd G_k^2$ $(k-1)$, it was constructed a homomorphism $\phi : (G_k(k-1)) \to G_k(k-1)/G_k^2(k-1) \simeq C_2$. This homomorphism is injective because for every j two different elements of $G_k(j)/G_k^2 G_k'(j)$, $0 \le j < k$, map in two different images in C_2. Element α that is in accord with a condition $\sum_{i=1}^{2^l} s_{li}(\alpha) \equiv 0(\mod 2)$ has an image 0 and if $\sum_{i=1}^{2^l} s_{li}(\alpha) \equiv$ 1(mod2) its image 1.

Since words of generators with no equal logarithms to any bases by mod 2 [14] belong to distinct cosets of the commutator, the subgroup $G_k^2(l)$ is the kernel of this mapping. The number of such bases is k because there are k generators. Hence, the homomorphism from G_k/G_k^2 onto $(C_2)^k$ is injective.

Let us check that the homomorphism φ is surjective. For this goal, we shall indicate preimage of arbitrary generator $g_l = (0, \ldots, 0, 1, 0, \ldots, 0)$ of $(C_2)^k$, where 1 1s on l coordinate. This preimage is $\alpha_{l,1} \in G_k$. As the result, we have $G_k/G_k^2 \simeq \underbrace{C_2 \times C_2 \times \cdots \times C_2}_{k}$. $\qquad \square$

Corollary 3 *The group* $Syl_2 A_{2^k}$ *has a minimal generating set of k generators.*

Proof Since quotient group of G_k by subgroup of Frattini $G_k^2 G_k'$ has minimal set of generators from k elements because $G_k/G_k^2 G_k'$ is isomorphic to linear p-space ($p = 2$) of dimension k (or elementary abelian group). Then, according to theorems from [15] $rk(G_k) = k$. Since $G_k \simeq A_{2^k}$, it means that A_{2^k} is a group with fixed size of minimal generating set. $\qquad \square$

Main Theorem. The set S_α is a minimal generating set for a group G_k that is isomorphic to Sylow 2-subgroup of A_{2^k}, $rk(Syl_2 A_{2^k}) = k$.

The existing of isomorphism between G_k and $Syl_2(A_{2^k})$ follows from Theorem 2. The minimality of S_β follows from Lemma 6 which says that the rank of $Syl_2(A_{2^k})$ is not less than $k - 1$ and Theorem 2. The fact that set of k elements is enough to generate G_k follows from Corollary 2 and from Theorem 2. Hence, we prove that $rk(Syl_2 A_{2^k}) = k$. Another way to prove the minimality of S_β is given in Corollary 2 because generating set of such group corresponds to generating sets of 2-abelianization.

For example, a minimal generating set of $Syl_2(A_8)$ may be constructed by the following way; for convenience, let us consider the next set:

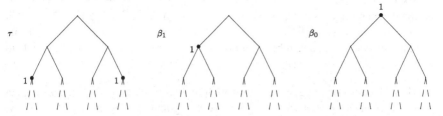

Consequently, in such a way, we construct the second k-element generating set for A_{2^k} that is less than in [4], and this set is minimal.

4 Applications in Braid-Based Cryptography

We will call *diagonal base* (S_d) for $syl_2S_{2^k} \simeq AutX^{[k]}$ that has the property $s_{jx}(\alpha_i) = 0$ iff $i \neq j$, (for $1 \leq x \leq 2^j$) and every α_i, $i < k$ has odd number of active v.p. This base is the similar to S_α that described in Theorem for $syl_2A_{2^k}$. A number of no trivial v.p. that can be on X^j is odd and the number of ways to chose tuple of no trivial v.p. on X^j for generator from S_d and equal to $2^{2^j} : 2^{2^j} - 1$. Thus, general cardinality of S_d for $syl_2S_{2^k}$ is 2^{2^k-k-1}. There is minimum one generator of type T in S_d for $syl_2A_{2^k}$. If m_1, m_2 mentioned above in Definition 2 are equal to 1, then this generator can be chosen in $C^1_{2^{k-2}}C^1_{2^{k-2}} = (2^{k-2})^2 = 2^{2k-4}$ ways. If $m_1 = m_2 = j$, then this generator can be chosen in $C^j_{2^{k-2}}C^j_{2^{k-2}}$ ways. In general case, $m_1 \neq m_2$ and there are $C^{m_1}_{2^{k-2}}C^{m_2}_{2^{k-2}}$ generators. Thus, general cardinality of S_d for $syl_2A_{2^k}$ is not less

then $2^{2^{k-1}-k-2}\sum\limits_{j=1}^{2^{k-2}-1}(C^j_{2^{k-2}})^2$.

Property 1 *The total number of minimal generating sets for* $syl_2A_{2^k}$ *is* $2^{k(2^k-k-2)} \cdot (2^k-1)(2^k-2)(2^k-2^2)\ldots(2^k-2^{k-1})$, *for* $syl_2S_{2^k}$ *it is* $2^{k(2^k-k-1)} \cdot (2^k-1) (2^k-2)\ldots(2^k-2^{k-1})$.

Proof Find total number of minimal generating sets for G_k and analogous sets for $AutX^{[k]}$. We take into account that the number of generating sets for $(C_2)^k \simeq G_k/G_k^2G_k'$ equals to $(2^k-1)(2^k-2)(2^k-2^2)\ldots(2^k-2^{k-1})$ and equals to the order of the group $GL(k, \mathbb{F}_2)$. Also, take into account that every element from $(C_2)^k$ has $|G_k|$: 2^k inverse images in G_k, because $(C_2)^k$ is a factor group. Hence, generating sets of a 2-group G_k correspond to generating sets of 2-abelization. Since $(C_2)^k$ is a quotient group of G_k by Frattini subgroup $\phi(G_k)$, any inverse image of quotient group generator is generator of G_k, so preimages number for each generator of $(C_2)^k$ is equal to size of normal subgroup $\phi(G_k)$.

There are k generators in a minimal generating set of $(C_2)^k$, therefore, to calculate number of preimages of the whole minimal generating set of $(C_2)^k$, the number $|\phi(G_k)| = |G_k| : 2^k$ should be raised to the power of k. So we can count a number of minimal generating sets of G_k. It equals to $(|G_k| : 2^k)^k = (2^{2^k-2} : 2^k)^k = 2^{k(2^k-k-2)}$. As a result, we have $2^{k(2^k-k-2)} \cdot (2^k-1)(2^k-2)(2^k-2^2)\ldots(2^k-2^{k-1})$. In the similar way, we obtain a number of minimal generating sets of $Syl_2S_{2^k}$. It equals to $2^{k(2^k-k-1)} \cdot (2^k-1)(2^k-2)(2^k-2^2)\ldots(2^k-2^{k-1})$. $\square$

Thus, if we associate generating set with alphabet and choice of generating set will be a privet key, then it can be applied in cryptography [16] what is important now [17, 18]. A group G_k can be used as a platform group G in the key establishment protocol [19] of generating common secret key (shared secret key). This group G_k satisfies all six properties from [19]. Moreover, our group G_k has exponential growth of different generating sets and diagonal bases that can be used for extension of key space. Diagonal bases are useful for easy constructing of normal form [19] of an element $g \in G_k$. As a private key, we choose one of generating sets. For every permutation π from $Syl_2 A_n$, we introduce a notion of canonical presentation in fixed generating set. We consider a classification of permutations in fixed generating set. For finding canonical presentations, we consider a set of words $\Lambda_n, k, k \geq n - 2$, elements of Λ_n, k are images of π after defined by us mapping ϕ and a tuple $V_m(\vec{v}, \vec{u})$, where $m > 1, v, u$—some vectors with elements from Z. On the basis of this notion, it proposed an algorithm of middle incline for constructing of canonical representation of any permutation. Conjugacy problem for this subgroup can be used as base in application for designing braid-based protocols. Also, researching of the structure of $Syl_2 A_n$ gives us possibility to solving of inclusion problem for set of elements of this subgroup. As well known, this problem is NP hard.

5 Conclusion

The proof of minimality of constructed generating sets was done, and also the description of the structure $Syl_2 A_{2^k}$, $Syl_2 A_n$ and its property was founded. In both cases, the underlying structures have a high combinatorial complexity. In this paper, it was proved that Sylow 2-subgroups $Syl_2 A_{2^k}$ of the alternating group have width by commutator subgroup $cw(Syl_2 A_{2^k}) = 1$ [20, 21]. Our results will be applied by us to key exchange protocol [19]. The complexity of brutforce attack on our system is $O(2^{k(2^{k-2})}(2^{k-1}))$. The cryptoresistability level is $k(2^k - 2)$.

References

1. Dmitruk, U., Suschansky, V.: Structure of 2-sylow Subgroup of Symmetric and Alternating Group, N. 3, pp. 304–312. UMJ (1981)
2. Skuratovskii, R.: Corepresentation of a Sylow p-subgroup of a group Sn. Cybern. Syst. Anal. (1), 27–41 (2009)
3. Pawlik, B.: The action of Sylow 2-subgroups of symmetric groups on the set of bases and the problem of isomorphism of their Cayley graphs. Algebr. Discret. Math. 21(2), 264–281 (2016)
4. Ivanchenko, V.: System of generators for 2-sylow subgroup alternating group. In: Four Ukraine Conference of Young Scientists, p. 60. KPI, Kiev (2015). http://matan.kpi.ua/uk/ysmp4conf.html
5. Grigorchuk, R., Nekrashevich, V., Sushchanskii, V.: Automata, dynamical systems, and groups. Trudy Mat. Inst. Imeny Steklova **231**, 134–214 (2000)
6. Nekrashevych, V.: Self-similar groups. International University Bremen, vol. 117, p. 230. American Mathematical Society. Monographs (2005)
7. Grigorchuk, R.I.: Solved and unsolved problems around one group. In: Infinite Groups: Geometric, Combinatorial and Dynamical Aspects, Basel, vol. 248. pp. 117–218. Progress in Mathematics (2005)
8. Sikora, V.S., Suschanskii, V.I.: Operations on Groups of Permutations, p. 256. Ruta, Cherniv (2003)
9. Skuratovskii, R.V.: Generators and relations of Sylow p-subgroups of symmetric groups S_n. Naukovi Visti KPI, pp. 93–101 (2014)
10. Weisner, L.: On the sylow subgroups of the symmetric and alternating groups. Am. J. Math. **47**(2), 121–124 (1925)

11. Bogopolski, O.: An Introduction to the Groups Theory European Mathematical Society, 189 pp. (2008)
12. Skuratovskii, R.V., Drozd, Y.A.: Generators and and relations for wreath products of groups. Ukr. Math. J. **60**(7), 1168–1171 (2008)
13. Magnus, V., Karras, A., Soliter, D.: Combinatorial Group Theory: Presentations of Groups in Terms of Generators and Relations, 453 pp. M.: Science (1974)
14. Kargapolov, M.I., Merzljakov, J.I.: Fundamentals of the Theory of Groups Springer. Softcover Reprint of the Original 1st ed., 1979 edn., 1st edn. 312 pp. Springer (1979)
15. Rotman, J.J.: An Introduction to the Theory of Groups, **XV**, 513 pp. Springer, New Yourk (1995)
16. Myasnikov, A.G., Shpilrain, V., Ushakov, A.: A practical attack on some braid group based cryptographic protocols. In: Crypto: Lecture Notes in Computer Science, vol. 3621, pp. 86–96. Springer (2005)
17. Holovatyi, M.: The state and society: the conceptual foundations and social interaction in the context of formation and functioning of states. Econ. Ann. **XXI**(9–10), 4–8 (2015)
18. Chornei, R., Hans Daduna, V.M., Knopov, P.: Controlled markov fields with finite state space on graphs. Stoch. Models **21**(4), 847–874 (2005). https://doi.org/10.1080/15326340500294520
19. Shpilrain, V., Ushakov, A.: A new key exchange protocol on the decomposition problem. Contemp. Math. **418**, 161–167 (2006)
20. Nikolov, N.: On the commutator width of perfect groups. Bull. Lond. Math. Soc. **36**, 30–36 (2004)
21. Skuratovskii, R.V.: Structure and minimal generating sets of Sylow 2-subgroups of alternating groups and their centralizers and commutators (in Ukrainian). In: 11th International Algebraic Conference in Ukraine, Kiev, p. 154 (2017). https://www.imath.kiev.ua/~algebra/iacu2017/abstracts
22. Dixon, J.D. Mortimer, B.: Permutation Groups, 410 pp. Springer, New York (1996)
23. Skuratovskii, R.V.: Minimal generating systems and properties of $Syl_2 A_{2^k}$ and $Syl_2 A_n$. X International Algebraic Conference in Odessa Dedicated to an Anniversary of Yu. A. Drozd, p. 104 (2015)
24. Skuratovskii, R.V.: Minimal generating systems and structure of $Syl_2 A_2^k$ and $Syl_2 A_n$. International Conference and Ph.D.-Master Summer School on Graphs and Groups, Spectra and Symmetries (2016). http://math.nsc.ru/conference/g2/g2s2/exptext/Skuratovskii-abstract-G2S2+.pdf

Self-repairing Functional Unit Design in an Embedded Out-of-Order Processor Core

Harini Sriraman and Venkatasubbu Pattabiraman

Abstract With increasing complexity of processor architectures and their vulnerability to hard faults, it is vital to have self-repairing processor architectures. This paper proposes the idea of autonomic repairing of permanent hard faults in the functional units of an out-of-order processor core using reconfigurable FPGA. The technique proposed utilizes the existing reconfigurable hardware in the new range of embedded systems like Intel Atom E6 × 5C series. The proposed technique includes an on-chip buffer, a fault status table (fully associative) and few control signals to the existing core. To perform self-repairing, decoder will identify reference to the faulty unit and initiate the reconfigurable hardware to be configured as the faulty unit referenced. Dispatch unit will help resolve the reservation station conflicts for the reconfigurable hardware. Execution of instruction that referenced the faulty unit gets executed in the reconfigurable unit. Dispatch unit and the buffers helps complete the out-of-order execution and in-order commit of the instructions that referenced a faulty unit. A hypothetical architecture that loosely resembles ALPHA 21264 is designed as a test bed for analyzing the proposed self-repairing mechanism. Area and time overhead analysis are done using Cadence NCVerilog simulator, Xilinx-Vivado ISE and FPGA Prototype board. Spatial and temporal costs of the proposed design are around 2% and 2.64% respectively. With recent increase in hybrid architectures that has FPGA tightly coupled with ASIC processor core, the proposed solution maximizes the reliability of the processor core without much area and time overhead.

Keywords Instruction interleaving · Self-repairing · Permanent hardware faults
Reconfigurable hardware · Fault tolerance

H. Sriraman (✉) · V. Pattabiraman
School of Computing Science and Engineering, VIT University
Chennai Campus, Chennai 600127, India
e-mail: harini.s@vit.ac.in

V. Pattabiraman
e-mail: pattabiraman.v@vit.ac.in

S. K. Bhatia et al. (eds.), *Advances in Computer Communication and Computational Sciences*, Advances in Intelligent Systems and Computing 759,
https://doi.org/10.1007/978-981-13-0341-8_33

365

1 Introduction

With increased complexity of processor chips, hard fault probability in a processor core increases. Hardware faults due to wearing out of the gate interconnection and leakage gets aggravated due to technology scaling. It is essential to handle these faults on the field to provide the expected reliability of the processor core. Root causes of permanent hardware faults is listed out in [1]. Effect of technology scaling on lifetime reliability of processors is discussed in [2]. The major causes of such faults are reduction in circuit size, on-chip power management techniques and increased power densities, as described in [3]. Ideal mean time for a processor to fail is 10^9 device hours. But due to the various conditions discussed, hard faults manifests during the useful lifetime of the processor. This will get aggravated in processor cores 22 nm and beyond. Especially due to fault mechanisms like electron migration, stress migration, time-dependent dielectric break down, thermal cycling and negative bias temperature instability, hard faults manifests themselves earlier than expected on the processor core. This affects the performance of the system and finally will make the system unusable. In an out-of-order processor core, the effects of these faults are more pronounced. So it is very essential to handle these faults for graceful degradation of the processor. In this paper, an effective intrinsic hard fault repair solution for functional units of out-of-order ASIC processor core using FPGA is proposed and analyzed.

2 Related Work

Hard faults handling includes fault detection and fault repair. Fault detection can be done using detection techniques or prediction techniques. Detection techniques identify faults after they affect the system state. Prediction will identify faults before it affects the system state. Hardware faults can be identified by tracing and analyzing software errors as given in [4]. Different set of instruction de-rating as given in [5] is done to identify the hard fault from software fault tracer. In [6], Instructions trace is used for diagnosing the hard faults and isolating them. Fault detection techniques proposed in [4, 5, 7] performs fault detection process after it has affected the system state. Circuit failure prediction specified in [4, 8], performs periodical tests of the circuits to find faults before it affects the system. In [9] self-testing for hierarchical processor architectures has been proposed.

Fault repair follows fault detection. Modular redundancy techniques given in [10], employ two or three duplications of the same circuit and the outputs are checked with a voter. The voter will output value that has repeated most. If no output has maximum count then fault is detected and the system will revert back to a safe check point. Modular redundancy technique will not perform repairing of the faulty unit. Instead they are fault tolerant techniques, best suited for transient errors. In case of permanent hardware faults, when a fault occurs in a unit, every time the unit is

Table 1 Existing fault handling categories

Technique	Fault handling	Fault coverage
Core cannibalization	Existing core components are cannibalized to work when fault occurs	Pipeline units
Redundancy	Redundant units used in the presence of faults	All units
Repair using array structures	Programmable arrays are used for repairing array structures like ROB, registers	ROB and registers

referenced, the fault will persist affecting the system state. So repairing of permanent faults is better rather than tolerating them every time they occur. Most of the existing literature like, [11–13] proposed self-repairing techniques for hardware faults in Field Programmable Gate Arrays (FPGA) as they are flexible. On the other hand, ASIC processor cores are rigid and configuring them on the field is impossible. As the recent trend sees an increase in hybrid architecture with ASIC and FPGA tightly coupled in the same processor chip [14–16], we in this paper tries to use the flexibility of FPGA to be reconfigured for handling permanent faults in out of order processor core. We utilize a processor similar to Intel Atom E6 × 5C series where the ASIC processor core is connected to the FPGA through high speed PCI express. Existing hardware fault handling techniques are categorized and listed in Table 1.

3 Proposed Work

The overview of the proposed hypothetical self-repairing architecture loosely based on ALPHA 21264 and OpenSARC T1 is given in Fig. 1. One Fault Repair Buffer (FRB), one Fault Repair Unit (FRU), and one Fault Status Table (FST) are included inside the out-of-order pipeline as shown in Fig. 1. FRU is connected to the processor core via PCIe bus. FRU is the FPGA unit and can be configured on the runtime to perform the function of the faulty unit. FRB is an on-chip buffer that stores the reconfiguration bit-stream to speed up the reconfiguring of FRU. The FST contains the fault status of units. Here we considered integer functional units. This can be extended for floating point unit also. The integer functional units include adder, multiplier, divider, and logical operations like compare, logical AND, OR, NOT. In this paper, we assume fault prediction by frequent testing of units using vector inputs. The design in Fig. 1 performs the fault repairing at sub-unit level granularity.

The control flow of the hypothetical processor core after FRU is configured as faulty unit is given in Fig. 2. The fetch performs the usual function of fetching. Decoder unit is modified to check for the referred unit in the fault status table, in addition to decode and register read.

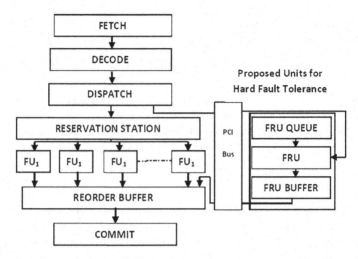

Fig. 1 Overall view of the test bed architecture

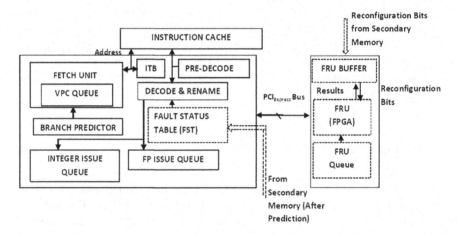

Fig. 2 Detailed control flow for self-repairing after FRU's reconfiguration

The dispatch unit is modified to accommodate the FRU initiator. It will also generate the FRU identifier whenever there are instructions referring to the faulty unit. The function of the FRU initiator is to check the Common Data Bus (CDB) for any updates on the operands it is waiting for. Only when both the operands are available, the inputs are transferred to FRU.

FRU includes queues and buffers to store input and output respectively. CDB will be used for transferring data from commit buffer to the corresponding registers as well as carry operand from dispatch unit to FRU queue. This transfer happens via the PCI express bus. CDB is also connected to FRU buffer via the PCIe in order to transfer output data from FRU to reservation station. In Fig. 2, dotted boxes are added to the existing design in the proposed architecture for tolerating hard faults.

3.1 Algorithm

```
MAIN ROUTINE
      WHILE Instructions fetched DO
        Call Fault-Prediction Routine periodically
        IF (Fault Predicted) THEN
          Call Fault-Repair Routine
        ELSE
          Continue Execution with existing Data Path
        ENDIF
      ENDWHILE
END MAIN

FAULT-PREDICTION ROUTINE
      WHILE TEST VECTORS Not Empty DO
          Apply test vectors to the functional unit &
          compare with Test Signatures
      IF (Signatures Mismatch)
          Update FST [FU] = 1
      ELSE
          Update FST [FU] = 0
      ENDIF
      ENDWHILE
END FAULT-PREDICTED

FAULT-REPAIR ROUTINE
      WHILE Instructions Fetched DO
          Decode (after decoding and register read per-
          form the following)
          Look-up FST for the presence of fault in the
          referred unit
          IF (FST [FU] = = 1)
```

BEGIN
I) Dispatch unit will generate control signals to update FRB with reconfiguration bits of referenced faulty unit.
II) Replace Reservation station name with the FRU Queue Number
III) Dispatch operands to FRU Queue and Initiate FRU to be configured as unit required
IV) Perform execution in the FRU
V) In the CDB, broadcast data with FRU reserved station for updating any waiting operands.
VI) Place output from FRU in the Reorder Buffer
VII) Complete out-of order execution with in-order commits.
ENDIF
ENDWHILE
END FAULT-REPAIR

4 Results and Discussion

The proposed design is analyzed with respect to spatial and temporal costs. FRU is designed as a coarse grained structure that is capable of reconfiguring any one of the unit like adder, multiplier, divider, or logical operation at a time.

4.1 Spatial Cost of Proposed Design

Spatial cost in the proposed design is due to the spatial overhead of FRB, FST and control signals added for the purpose of self-repairing. Equation 1 explains this.

$$SpatialCost = \sum SpatialOverhead(FRB, FST, FU) \tag{1}$$

To estimate the size of the FRB, the bit streams for configuring the FRU has to be finalized. We consider a coarse grained reconfiguration of functional unit in the FRU. The spatial cost of the proposed design is calculated by modifying the Verilog description of the existing architecture and synthesizing them using cadence NCVerilog.

4.1.1 FRB Spatial Overhead Estimation

The size of FRB is based on the maximum number of bits it can occupy. This in turn depends on the maximum number of reconfiguration bits required by any unit. The number of bits for reconfiguration is calculated using Eq. 2.

$$No.of.Configuration Bits = \sum \left(L * 2^{(No.of LUT - Input)}, M, X, F \right) \quad (2)$$

$$L : No.of.LUT \qquad M : No.of.Multipliers$$
$$X : No.of.XOR.Gates \quad F : No.of.Flip - Flops$$

After synthesis it is found that a maximum of 127830 bits are required for reconfiguring functional unit in the FRU. Equation 3 explains the estimation of FRB size.

$$FRB\_Size = Max(No.of.Configuration Bits(FU)) \quad (3)$$

4.1.2 FST Spatial Overhead Estimation

FST is designed to be a fully associative. The size of FST depends on the number of units that may become faulty. We have considered the integer functional units for self-repairing. Hence FST depends on the number of integer functional units. In the design proposed, faults are identified at decode stage by looking at the instructions. So, faulty units have to be identified through instructions. In the considered ISA (or any realistic ISA) there could be more than one instruction that could refer to the same unit. To identify the fault unit reference accurately, we have designed the address bits of the FST to include all the references to a particular unit. To perform this, the instruction format bits responsible for identifying the units, are analyzed. The instruction size of the proposed hypothetical architecture is 32 bits and is decoded based on the values of the bits 31 and 30. For all integer functional unit reference, the values of the instructions at bit position 30 and 31 has to be 1X. Apart from these two bits, five more bits from instruction bit position 19 through 24 in the instruction format are used to uniquely identify the functional unit. So totally six bits of the instruction (leaving bit 31) will identify the functional unit. The data bit of the FST will contain a "1" for a faulty unit and "0" for non-fault status. FST will have a compare port and one read write.

4.1.3 Total Spatial Cost Estimation

Total spatial cost estimation is done by modifying the existing architecture to represent the proposed test bed at HDL level description. Except for the FRU, other unit's spatial overhead is calculated by synthesizing the design in cadence NC Verilog. For

Table 2 Total spatial cost calculation

Units	Spatial overhead (μm^2)
Decoder unit: control logic to check FST	1054.557
Dispatch unit: FRU Initiator + control signals	2080.112
CDB accessing controls	1012.33
Area of existing architecture	16000000 μm^2
Spatial overhead of proposed architecture	**4573.6865 μm^2**
Spatial cost in percentage	**2%**

Table 3 Total temporal cost calculation

Assumed faulty unit	Number of faulty reference	Configuration bit loading time (ns)	Overhead when the corresponding unit alone fails (ns)
ADD/SUB	33.3	0.06	0.0012
CMP	11.1	1.95E-06	0.000043
EQL	1.6	9.60E-07	0.000001
MUL/DIV	4.8	319.57	0.6

FRU spatial estimation, Xilinx-Kintex is used. The total spatial cost calculation is shown in Table 2.

4.1.4 Temporal Cost of Proposed Design

Reconfiguration time should be maintained optimal otherwise the proposed design will become impractical. Temporal cost of the proposed design is affected by the parameters given in Eq. 4. Equation 5 gives details about reconfiguration time.

$$OverallTemporalCost = \sum (FSTST, RBLT, RcT, FRU\_Exe) \quad (4)$$

$$FSTST : FST - SearchTime \quad RBLT : ReconfigurationBitLoadingTime$$
$$RcT : Re - configurationTime \quad FRT\_Exe : ExecutionTimeInFRT$$

$$RcT = \sum (SetUpTime, SynchronizationTime, ConfigurationLoadingTime) \quad (5)$$

Table 3 gives the detailed estimation of temporal overhead of the proposed design. Temporal overhead is obtained by executing the Dhrystone benchmark with FPGA prototype of the proposed design. PCIe bus considered here can transfer 500 Mb/sec.

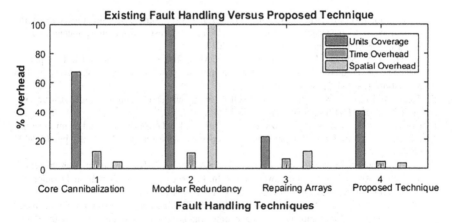

Fig. 3 Performance comparisons between proposed and existing techniques

To acquire the CDB there could be utmost 3 clock cycle delay. On an average, temporal cost of the proposed technique (visible to the end user) is around 2.4%. Figure 3 compares existing solutions with proposed one, with respect to spatial and temporal overheads.

5 Conclusion and Future Work

In this paper, a novel technique to handle permanent hardware faults in the execution unit of an out-of-order processor is proposed. The proposed technique uses FPGA for reconfiguring and replacing the faulty unit. With increasing number of commercial hybrid processors where ASIC and FPGA are fabricated on the same chip, this paper utilizes the FPGA for handling ASIC hardware faults. This helps in huge optimization of spatial overhead. The granularity is maintained at unit level. Hence a fine balance is achieved between time overhead, spatial overhead and percentage of units covered for faults. In future, with redundant reconfiguration of the faulty unit, the time overhead can be further reduced, thus providing a better cost–benefit ratio.

Acknowledgements We thank VIT University for providing Xilinx tools, Cadence tools and FPGA Kintex Board.

References

1. Maxfield, Clive: Design Warriors Guide to FPGA. Elsevier Publications, USA (2011)
2. Maniatakos, Michail: Instruction-level impact analysis of low-level faults in a modern microprocessor controller. IEEE Trans. Comput. **59**, 1260–1273 (2010)

3. Li, Y., Kim, Y.M., Mintarno, E., Gardner, D., Mitra, S.: Overcoming early-life failure and aging challenges for robust system design. In: IEEE Design and Test of Computers, Special Issue on Design for Reliability and Robustness, p. 1 (2013)
4. Oliveira, R.S., Semiao, J., Teixeira, I.C., Santos, M.B., Teixeira, J.P.: On-line BIST for performance failure prediction under aging effects in automotive safety critical applications. In: Test Workshop (LATW), pp. 64–71 (2011)
5. Zhang, Y., Li, H., Li, X.: Software-based self-testing of processors using expanded instructions. In: IEEE Asian Test Symposium, pp. 415–420 (2010)
6. Cook, J.J., Zilles, C.: A characterization of instruction-level error derating and its implications for error detection. In: IEEE Conference on Dependable Systems and Networks, pp. 482–491 (2008)
7. Li, M.L., Ramachandran, P., Sahoo, S.K., Adve, S.V., Zhou, Y.: Trace-based microarchitecture-level diagnosis of permanent hardware faults. In: Proceedings of the International Conference on Dependable Systems and Networks (DSN), pp. 22–31 (2008)
8. Gherman, V., Massas, J., Evain, S., Chevobbe, S., Bonhomme, Y.: Error prediction based on concurrent self-test and reduced slack time. In: Design, Automation & Test in Europe Conference & Exhibition (DATE), pp. 1–6 (2011)
9. Muller, S., Scholzel, M., Vierhaus, H.: Hierarchical self-repair in Heterogenous multicore systems by means of a software-based reconfiguration. In: ARCS Workshops (ARCS), pp. 1–7 (2012)
10. Imran, N.: Self-Configuring TMR scheme utilizing discrepancy resolution. In: International Conference on Reconfigurable Computing and FPGAs, pp. 398–403 (2011)
11. Jacobs, A., George, A.D., Cieslewski, G.: Reconfigurable Fault Tolerance: a Framework for Environmentally Adaptive Fault Mitigation in Space, NSF Center for High-Performance Reconfigurable Computing Conference, pp. 199–204 (2010)
12. Agarwal, M., Balakrishnan, V., Bhuyan, A., Kyunglok, Kim., Paul, B.C.: Optimized Circuit Failure Prediction for Aging: Practicality and Promise International Test Conference, Santa Clara, CA, pp. 1–10 (2008)
13. Banu, M.F., Poornima, N.: BIST using genetic algorithm for error detection and correction. In: 2012 International Conference on Advances in Engineering Science and Management (ICAESM), pp. 588–592 (2012)
14. Harini, S., Pattabiraman, V.: Hybrid multi-core Asic architecture with FPGA for self-repairing of hard faults. Int. J. Pure Appl. Math. 101(8), 905–914 (2015)
15. Hauser, J.R.: Augmenting a microprocessor with reconfigurable hardware. In: Proceedings of the IEEE Symposium on Field-Programmable Custom Computing Machines, pp. 24–33 (1997)
16. Zuchowski, P.S., Reynolds, C.B., Grupp, R.J., Davis, S.G., Cremen, B.: Bill Troxel, a hybrid ASIC and FPGA architecture. In: IEEE Conference on CAD, pp. 187–194 (2002)

Research on the Improvement of Civil Unmanned Aerial Vehicles Flight Control System

Kunlin Yu

Abstract The flight control system is the UAV core component, the civil UAV flight control system is studied in this paper. In view of the current ArduPiloMega open-source flight control system has the disadvantages of measurement is not accurate in low altitude height and cannot realize automatic obstacle avoidance technology, the ArduPiloMega hardware and software system design is improved in this paper, the ultrasonic ranging sensors of a temperature compensation and automatic obstacle avoidance function is installed, the driver is modified, through the software to achieve the switching of barometric altimeter and ultrasonic sensor. To overcome the defect of measurement in low altitude stage is not accurate and cannot automatically avoid obstacles, make the small unmanned aerial vehicles take-off and landing more safe and stable. And the flight test was carried out to verify the improved flight control system, the results show that the theory and method are correct.

Keywords Flight control system · Hardware design · Software design
The ultrasonic sensors

1 Introduction

The flight control system is the UAV core component, the quality of flight control system directly determines the flight performance of aircraft, Due to the function of APM open-source system is powerful, ground station support, strong extensibility and favored by users, use more widely, but in the use of the process there are some problems: APM with barometric altimeter of plate mechanical integration can only give a roughly height in the process of UAV flight altitude at takeoff and landing stage or low altitude flight stage, the barometric altimeter sensitivity is not enough to realize high accuracy measurement at this time, and cannot automatically avoid obstacles

K. Yu (✉)
Department of Aeronautical mechanical and electrical equipment maintenance, Changsha
Aeronautical Vocational and Technical College, Changsha 410124, Hunan, China
e-mail: ykl6990701@163.com

© Springer Nature Singapore Pte Ltd. 2019
S. K. Bhatia et al. (eds.), *Advances in Computer Communication and Computational
Sciences*, Advances in Intelligent Systems and Computing 759,
https://doi.org/10.1007/978-981-13-0341-8_34

[1]. So I have to optimize and ameliorate the flight control system's hardware and software.

2 The System Hardware Optimize

2.1 System Function

Unmanned aerial vehicle flight control system is mainly to complete two big functions. One is the stability and control the attitude of unmanned aerial vehicle, the other is to realize unmanned aerial vehicle tracking control and mission planning. At the same time in order to facilitate the ground control station for the unmanned aerial vehicle control and communications, increase the remote and telemetry.

2.2 System Components

The flight control system generally consists of subsystems of flight control computer, sensors, and servo actuation [2].

The sensor system collects the UAV's attitude and other information, and put it into the computer system [3]. Computer System is to process the sensor data and plan the control instruction. The servo actuator receives control instructions to control the rudder surface to change or maintain the status of the UAV. Unmanned aerial vehicle flight control system is the control center of the whole unmanned aerial vehicle, relies on the control center to complete the function of high accuracy autonomous navigation, autonomous flight control, task management, etc. The flight control computer as the core, and with various sensors and control actuators constitute a closed-loop control system [4], the improved APM flight control system structure is shown in Fig. 1.

2.3 System Working Principle

In APM flight control system, uses a two-stage PID control method, the first level is navigation level, the second level is control level, the navigation grade PID control is through the algorithm given the aircraft needed pitching angle, throttle and roll angle, and then to control level control calculating. The task of control level is based on the needed pitch angle, throttle, roll angle, combined with the aircraft's current attitude to calculate the appropriate steering rudder control, so that the aircraft to maintain a predetermined pitch angle, roll angle, and direction angle. Finally, the control level of the steering gear is converted into a specific PWM signal output to the actuator [5].

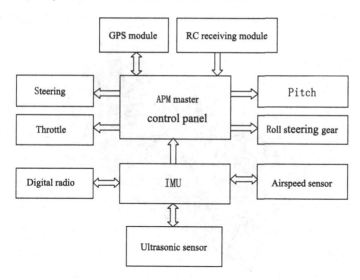

Fig. 1 The improved APM flight control system structure

2.4 Add the Ultrasonic Sensors

Due to mechanical integration barometric altimeter in the ArduPiloMega flight control system board can give a roughly height in the process of unmanned aerial vehicle flight, but in the takeoff and landing or low altitude flight stage, Due to low altitude, pressure range is very small, the barometric altimeter sensitivity is not enough to realize high accuracy measurement to height at this time, therefore I added ultrasonic ranging sensor KS101B module with temperature compensation and with automatic obstacle avoidance function in ArduPiloMega flight control system, with relatively high accuracy ultrasonic sensor KS101B in the takeoff and landing stage and low altitude flight stage instead of barometric altimeter effect, use this to compensate for the inadequacy of the barometric altimeter, at the same time to achieve a smooth takeoff and landing. The ultrasonic ranging sensor module KS101B is shown in Fig. 2.

The ultrasonic ranging sensor module uses the I2C bus interface, wiring is simple, only needs to connect power VCC, GND, data SDA and clock SCL, a total of four lines, In the free port of ArduPilotMega I2C welding four pin with dupont line connected to the ultrasonic module KS103: GND port connect to GND port, VCC port connect to VCC port, SCL port connect to SCL port, SDA ports connect to SDA port, The connection of KS103 and APM as shown in Fig. 3.

Distance measuring range is 1 cm–6.5 m, as far as is 6.5 m, recently is 1 cm, and almost has no blind area. 4.5 m high ranging precision is 1 mm, repeated measurement data is stable and the performance is reliable with temperature compensation function.

Fig. 2 Ultrasonic ranging sensor module KS101B

Fig. 3 The connection of KS103 and APM

Fig. 4 The flight control
system program flow

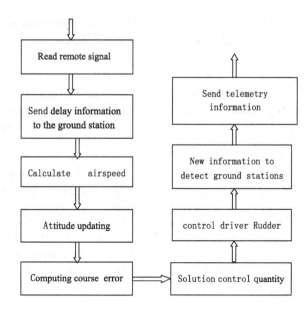

3   The System Software Optimize

3.1   Program Flow

Flight control system first read remote control information, then send delay information to the station and calculate airspeed error and heading error, according to the error size calculate the control quantity to control the drive rudder and finally test station new message and send remote sensing information to station [6]. The specific program flow chart is shown in Fig. 4 [7].

3.2   The Addition of Ultrasonic Sensor Driver

Add the I2C interface KS101B driver under AP_RangerFinder driver classification. Mainly include the newly added basic class file RageFinder_I2C.h, RageFinder_I2C.cpp, the newly added implementation class AP_RangeFinder_KS103.h, AP_RangeFinder_KS103.cpp [8–10].

This design mainly is unified interface methods, convenient compatible with different types I2C interface ultrasonic ranging sensor. Specific-type ultrasonic ranging sensor implementation class inherits from the base class. Because the base class containing virtual functions, so the actual use must build the derived class object.

3.3 The Landing Stage Program Optimization Design Under the Effect of Ultrasonic Altimeter

In the key-independent landing part, when the distance is less than 5 m between the plane and the ground, ultrasonic distance sensor begins to activate, when updating a flight control height value, read the ultrasonic ranging height value and save in sonar ALT variables used as low height reference, and begin to accurately measure the relative height of the aircraft in real time.

Effect of ultrasonic module is less than 5: When flight control executes MAV_LAND command, monitor the ultrasonic ranging sensor measurements. According to the error of the ultrasonic ranging value correct barometric altimeter, and according to the revised altitude control aircraft landing attitude, in the process of landing, once the plane from landing point horizontal distance is less than 2 s voyage or less than 3 m, APM will close the throttle and maintain the current course. When there is no power sliding if ultrasonic ranging height is less than 0.8 m, then control the plane Angle of elevation is 0°, realize the flattening and fall to the ground [11].

4   Flight Experiment Analysis

After the plane landed, aircraft flight control board is connected to the ground station, export flight log, use the data analysis tools with the ground station to analyze the various parameters of the aircraft. Verify the flight control system each function is normal.

4.1   Tremor Curve

Through the analysis of the flight control system flight log, you can see after adding an ultrasonic sensor, the aircraft overall vibration decreased obviously. Before adding an ultrasonic sensor, the aircraft's vibration frequency particularly fast, and the amplitudes are relatively large, in the whole process of aircraft flight, vibration from beginning to end have continued; And after added ultrasonic sensors can clearly see from the table, the plane overall vibration significantly decreases, and the vibration frequency and amplitude are reduce exponentially, especially in the process of take-off and landing stage and fixed high flight jitter coefficient is almost zero, So it can be concluded that adding ultrasonic sensor has the expected effect, reduce the flight process effectively improves the flying qualities of the aircraft (Figs. 5 and 6).

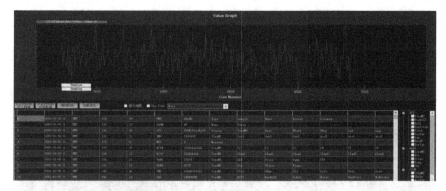

Fig. 5 Before adding an ultrasonic sensor

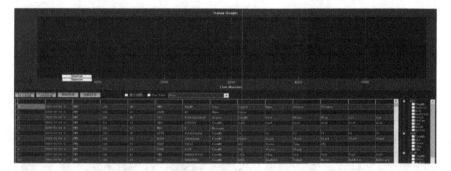

Fig. 6 After adding an ultrasonic sensor

4.2 IMU Overall Noise Level

The Fig. 7 shows the overall noise of the IMU (inertial measurement module) before the ultrasonic sensor is added. It can be seen that the noise intensity is very large and the frequency is very dense. The noise of the IMU module will adversely affect the overall attitude stability of the aircraft [12].

The Fig. 8 shows the overall noise of the IMU (inertial measurement module) after the ultrasonic sensor is added. It can be seen that the noise intensity is significantly reduced as compared with before, and the frequency is also significantly reduced. Only in take-off and landing stage, there is obvious noise. It can be seen that after the ultrasonic sensor is installed, the noise of the IMU module is obviously reduced, and the ultrasonic module plays an obvious role.

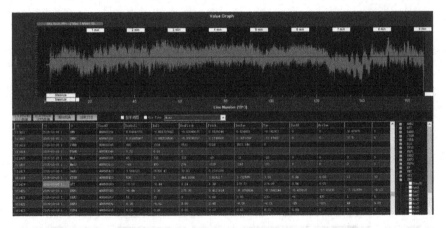

Fig. 7 Before an adding ultrasonic sensor

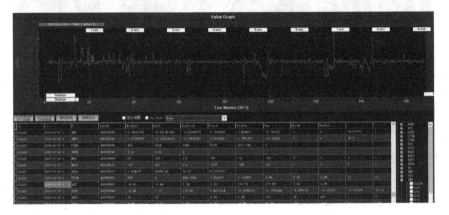

Fig. 8 After adding ultrasonic sensors

4.3   Jitter Coefficient of Takeoff and Landing Stage

Through the analysis the flight log of flight control system, can be seen after adding ultrasonic sensors, the whole aircraft's vibration decreased greatly. Before adding an ultrasonic sensor, the aircraft's vibration frequency is particularly fast, and the amplitude are big, in the whole process of aircraft flight, vibration from beginning to end have continued; and after adding an ultrasonic sensor can clearly see from the figure, the aircraft's overall vibration significantly decreases, and the vibration frequency and amplitude are exponentially reduced, especially in the process of takeoff and landing stage and fixed high flight jitter coefficient is almost zero, so it can conclude that an ultrasonic sensor is added to achieve the desired effect and effectively improve the flight quality of the aircraft (Figs. 9 and 10).

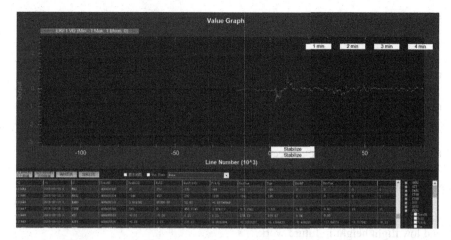

Fig. 9 Jitter curve of takeoff process

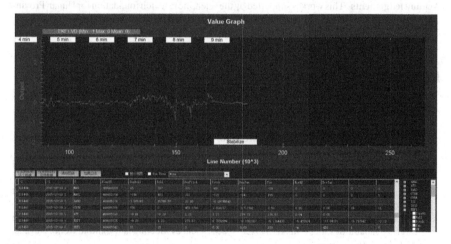

Fig. 10 Jitter curve of landing process

By comparing several groups of curve can be concluded that after optimization of the APM flight control system flutter coefficient and noise significantly is reduced in a low automatic cruise, take-off and automatic landing stage of the system. Thus proved that the optimization is effective and successful, through the ultrasonic probe to make up for error caused by sensor accuracy of original APM flight control system in low altitude and take off and land stage, make flying stable and reliability more higher.

5 Conclusion

The ArduPiloMega hardware and software system design is improved in this paper, the ultrasonic ranging sensors of a temperature compensation and automatic obstacle avoidance function is installed, the driver is modified, through the software to achieve the switching of barometric altimeter and ultrasonic sensor. To overcome the defect of measurement in low altitude stage is not accurate, make the small unmanned aerial vehicles take-off and landing more safe and stable, the test results show that the theory and method are correct.

The APM system optimization in this paper is mainly aimed at low altitude flight stage (the flight height is less than 6 m) and take-off and landing stage, barometric altimeter is used to measure flight altitude when flight height is over 10 m. I will study automatic obstacle avoidance technology of unmanned aerial vehicle in high and low flight height in future work.

Acknowledgements This work is supported by the scientific research foundation for Hunan Provincial Department of Education of China (Project numbers is 16C0011).

References

1. Ciesluk, J., Gosiewki, Z.: The image brightness control system dedicated for the autonomous unmanned aerial vehicle. J. KONBiN 32(1), 71–82 (2014)
2. Liu, P., Meng, Z., Wu, Z.: A robust flight control system design for a small-scale UAV helicopter. Trans. Jpn Soc. Aeronaut. Space Sci. 56(2), 96–103 (2013)
3. Yedavalli, K.R., Belapurkar, K.R.: Application of wireless sensor networks to aircraft control and health management systems. J. Control Theory Appl., pp. 28–33, Jan 2011
4. Chen, Y., Pei, H., Chen, Y.: The hardware design and implementation of a small UAV flight control system. In: Computer Engineering and design, pp. 2159–2162, Oct 2010
5. Moschetta, J.-M., Bataille, B., Thipyopas, C., et al.: On fixed-wing micro-air vehicles with hovering capabilities. In: 46th AIAA Aerospace Sciences Meeting and Exhibit, Reno, Nevada, pp. 1–13 (2008)
6. Shima, T., Rasmussen, S.: UAV Cooperative Decision and Control challenges and Practical Approaches, pp. 140–145 (2009)
7. Tang, X., Tao, G., Joshi, S.M.: Adaptive actuator failure compensation nonlinear MIMO systems with an aircraft control application. Automatica, 1869–1883 (2007)
8. Qingqi, L.U., Dandan, X.: Embedded Linux system transplant, Foreign Electronic Measurement Technology, pp. 78–81, Dec 2014
9. Lin, T., Yanan, Z., Xinfeng, Y.: Research and implementation for embedded Linux system migration. Microcomput. Appl., 12–14 (2016)
10. Han, Y., Lin, Z., Zhou, X.: Transplantation of UAV flight control program on Raspberry Pi. J. Zhejiang Univ. Sci. Technol., pp. 433–438 (2016)
11. Lungu, R., Lungu, M., Grigorie, L.T.: Automatic control of aircraft in longitudinal plane during landing. IEEE Trans. Aerosp. Electron. Syst. 49(2), 1338–1350 (2013)
12. Peiliang, Xu, Shi, Chuang, Fang, Rongxin, et al.: High-rate precise point positioning (PPP) to measure seismic wave motions: an experimental comparison of GPS PPP with inertial measurement units. J. Geodesy 87(4), 361–372 (2013)

Smart Water Dispenser for Companion Animals

Yonghwan Lee, Hwaju Cho and Sungyoung Kim

Abstract As the number of companion animals is increasing with several reasons, related problems such as caring and feeding are also increasing. The main problem is who and how can care the animals when their owners leave them and are staying outside. The animals should be able to drink fresh water steadily and consistently during the absence of their owner in the house. This problem can be addressed with the advanced IoT devices based on ICT technology. In this paper, we develop a smart water dispenser system that supplies the oxygen-rich fresh water and controls water supply from remote site. This system also monitors and alarms the level of the water stored and the states of the water dispenser by using the smartphone at remote site. We show the prototype of this dispenser, some conceptual diagrams of the key components in the dispenser and the implementation of the dedicated smartphone app.

Keywords Water dispenser · IoT · ICT · Companion animals · Wi-Fi MCU module · Water-level sensors · Smartphone app

Y. Lee
School of Electronics Engineering, Kumoh National Institute of Technology, Gumi, Gyeongbuk 39177, South Korea
e-mail: yhlee@kumoh.ac.kr

H. Cho
ITLog Co, Gumi, Gyeongbuk 39177, South Korea
e-mail: Hwaju.Cho@itlog.co.kr

S. Kim (✉)
Department of Computer Engineering, Kumoh National Institute of Technology, Gumi, Gyeongbuk 39177, South Korea
e-mail: sykim@kumoh.ac.kr

© Springer Nature Singapore Pte Ltd. 2019
S. K. Bhatia et al. (eds.), *Advances in Computer Communication and Computational Sciences*, Advances in Intelligent Systems and Computing 759,
https://doi.org/10.1007/978-981-13-0341-8_35

1 Introduction

According to the several statistics about pet ownership and demographics, the percentage of worldwide families feeding companion animals is increasing steadily and rapidly [1, 2]. It is caused from some reason that the notable decrease in birth rate, the remarkable growth in single-person household and so on. People get to know that feeding companion animals can give them as much pleasure as raising their children. In recent years, the term companion animal has become more common than the term pet. The reason for this is that animals are considered much like as a companion, not just a target of rearing.

There are, however, some problems in living with the companion animals indoors. The main problem is who and how can we care the animals when their owners leave them and are staying outside. The animals should eat, drink, and play even when their owners are not with them. Some problem can be addressed with some advanced IoT devices based on ICT technology. For example, automatic feeding systems can feed the animals regularly [3, 4], and housebreaking system based on IoT can train the animals where they should defecate or urinate [5–8].

Water is very important to the companion animals. Water is essential for the animals to survive [9, 10]. Water also helps in controlling their temperature. If they do not get enough water, they will have an abnormality in the kidney, urethra, and bladder. In addition, the metabolic function and the excretion may be lowered, and the abnormalities of stones and dermatitis may occur. In general, the recommended amount of daily water intake of a dog can be calculated as Eq. 1. Where, R is recommended amount of water in milliliter and w is the weight of the dog in kilogram.

$$R = (70 \times w)^{0.75} \times 1.6 \tag{1}$$

A conventional water dispenser has a structure that stores water in a bottle of water and replenishes as much as a pet drinks water. Water that has stayed at water dispenser for the long time can lose its freshness and sometimes impurities can be formed. In this case, dust, debris, and impurities can be accumulated on the water tank, and so the freshness of the water will deteriorate. Eventually, some animals deny to drink such a staled water and can suffer from the lack of the water in their body. To keep fresh water in the water tank, water should be spun by the mechanical appliances.

A fusion of ICT and pet industry has emerged as a blue ocean [11, 12]. The market of the pet industry has been growing steadily. As we mentioned before, the fresh water is essential for caring the companion animals. In this paper, we introduce a new and smart water dispenser system that provides the fresh water to the companion animals in safe and intelligent manner. This system is a smart system because it can control the supply of the water and monitor the working states of the dispenser by using smartphone app connecting through Wi-Fi [13] at remote site. The water dispenser system continuously circulates the water insides the tank to keep the water fresh. This continuous circulation allows air to flow through the water and can provide

Table 1 Percent of the household owningthe companion animals and the market size of related pet industries (from 2012 Samsung Economics Research Institute)

Country	Household (%)	Market size[a]
Korea	17.9	$0.79
Japan	27	$14.03
U.S.	62	$53.33
U.K.	47	$2.05
German	33.3	$3.47

[a]Billions of dollars

oxygen-rich fresh water. It is also required to detect that the system is fallen down by a external impact. The system should minimize the spillage of water even when it falls onto the floor due to the external impact or when drinking.

2 Market Size of Pet Industries

According to the U.S companion animal ownership statistics [14], 62% of total households in U.S. owns companion animals. More specifically, 36.5% of U.S. households breeds puppies, and 30.4% for cats including overlaps. Average number owned per household is 1.6 for puppy and 2.1 for cats. Table 1 shows another statistics of the percentage of the households owning animals and the market size of the related pet industries in 5 countries. You can understand that pet industries have a large market.

Figure 1 shows the trend of expenditures in industries of companion animals. As you can see, the expenditures are increasing rapidly. In developed countries such as the United States and Japan, various services such as beauty, fashion, hotel, obscenity, and insurance for companion animals have developed in advance of Korea, and they are gradually segmented and specialized. Especially, due to the effects of low fertility and aging, single-person household is the main consumer of pet products and has a tendency of "individualism" oriented consumption, and "petting" of companion animals has a tendency to prefer well-being and luxury.

3 Smart Water Dispenser

The proposed smart water dispenser is designed to meet the following conditions. The dispenser continuously circulates the water inside the dispenser by using the water pump to supply the oxygen-rich fresh water. The dispenser also provides functions to check and control the water supply controller and the operation of the dispenser based on smartphone app and Wi-Fi connection. In addition, this dispenser can minimize

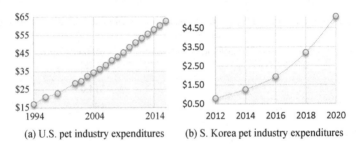

(a) U.S. pet industry expenditures   (b) S. Korea pet industry expenditures

Fig. 1 Total pet industry expenditures in U.S. and South Korea

the water spillage when the dispenser falls onto the floor due to external impact
or when animals are drinking water. We design and develop some functions of the
dispenser as follows:

- Development of circulating water supply system
- Design and development of the water spillage minimization mechanism when the
  dispenser falls down onto floor
- Development of the WI-FI communication module to control the water pump
- Implantation of the smartphone application to control and monitor the water supply
  controller
- Detection of low-level water by the water-level sensor (for smartphone notifica-
  tion).

Figure 2 show the conceptual diagram of the proposed smart water dispenser. As
you can see in the figure, this dispenser consists of several parts such as the guard
preventing water spillage, the equipment filtering water, the valve preventing water
from flowing backward, the sensor detecting the water level and the fallen-down and
the pump-supplying water.

The conceptual diagram of the circulating water supply system in the dispenser
is illustrated in Fig. 3. The water supplier is designed to satisfy some conditions as
follows.

- Easy to disassemble and clean each part
- Minimizing the water spillage even when falling down during operation
- Small and low noise underwater water pump
- Circulating water system using a small water pump to keep the water oxygen-rich
  and fresh by making air passing through the water.

Water pump can be controlled and monitored by the water supply controller in
the dispenser, which is controlled by the smartphone app via Wi-Fi at remote site.
The controller consists of Wi-Fi MCU module and sensor to detect the status of the
dispenser when it is fallen down. The controller has several functions as follows.
Figure 4 shows the block diagram for connecting between the remote smartphone
and the Wi-Fi module in the dispenser.

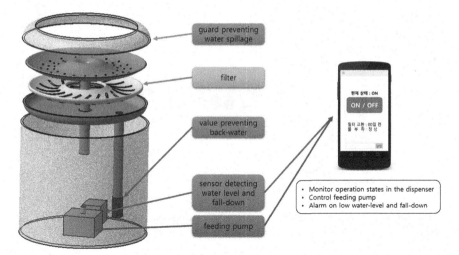

Fig. 2 Conceptual diagram of the proposed smart water dispenser

Fig. 3 Conceptual diagram
of the circulating water
supply system

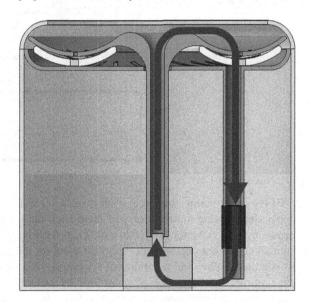

- Controlling the operation of the water pump using a smartphone at remote site
- Monitor/alarm the low water level (no water) and the states of the water dispenser using the smartphone at remote site
- Construction of the relay server for Wi-Fi communication: It is impossible to connect to a router at home from outside (smartphone) using P2P method, so the relay server and the related software should be constructed to be accessed regardless of the location of a user.

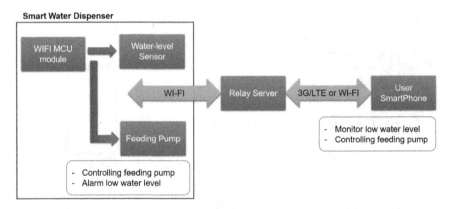

Fig. 4 Block diagram for connecting between remote smartphone and Wi-Fi module in the dispenser

Fig. 5 Concept to detect the level of the water by sensor

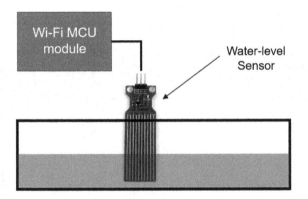

If the water pump runs even when there is no water, the life of the product can be shorten causing malfunction and the animal cannot drink water. Therefore, by detecting the low water level (no water) and sending the alarm, the owners of the animals can be notified the water shortage situation and can also stop the water pump so that the stability and reliability of the product can be improved. Figure 5 shows the concept of the detection of water level by using the sensor. Figure 6 shows the prototype of the non-contact capacitance-type water-level sensing circuit and Fig. 7 shows the user interface of the smartphone app.

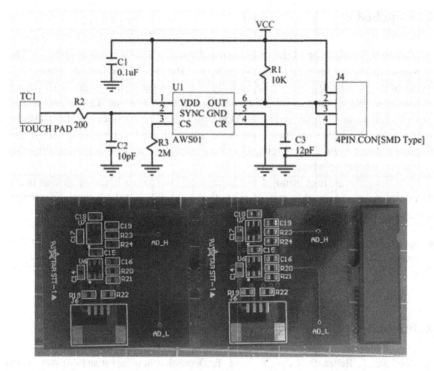

Fig. 6 Prototype of non-contact capacitance-type water-level sensing circuit

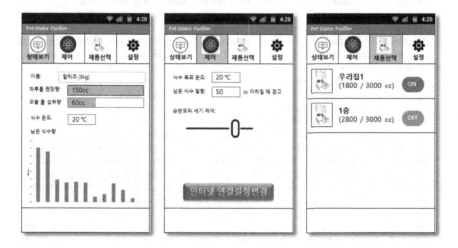

Fig. 7 User interface of smartphone app for controlling and monitoring the smart dispenser

4 Conclusion

In this paper, we developed the smart water dispenser for companion animals. The dispenser can be useful even when the owners of the animals go outside and there is no one at house. The dispenser continuously circulates the water inside the dispenser by using the water pump to supply the oxygen-rich fresh water. The dispenser also provides functions to check and control the water supply controller and the operation of the dispenser based on the smartphone app and Wi-Fi connection. Also, this dispenser can minimize the water spillage when the dispenser falls onto the floor due to external impact or when drinking.

Reflecting the current trend of the increasing number of households that breed the companion animals, this system can help the animals and their owners for happy and convenient life.

Acknowledgements Following are results of a study on the "Leaders in INdustry-university Cooperation +" Project, supported by the Ministry of Education and National Research Foundation of Korea (2017-D-G040-010105).

References

1. Enmarker, I., Hellzn, O., Ekker, K., Berg, T.: Depression in older cat and dog owners: the Nord-Trondelag health study. Aging Mental Health **19**(4), 347–352 (2015)
2. Howell, T.J., Mornement, K., Bennett, P.C.: Pet dog management practices among a representative sample of owners in Victoria, Australia. J. Vet. Behav. Clin. Appl. Res. **12**, 4–12 (2016)
3. Lancheros, C., Diana, J., Suarez, D. R., Hernandez, L.: Solid dosing system for feeding dogs with remote communication module. In: 10th Iberian Conference on Information Systems and Technologies, pp. 1–5 (2015)
4. Karyono, K., Nugroho, I.: Smart dog feeder design using wireless communication, MQTT and Android client. In: International Conference on Computer, Control, Informatics and its Applications, pp. 191–196 (2016)
5. Park, W.H., Cheong, Y.G.: IoT smart bell notification system: design and implementation. In: 19th International Conference on Advanced Communication, pp. 298–300 (2017)
6. Teh, P.L., Ling, H.C., Cheong, S.N.: NFC smartphone based access control system using information hiding. In: IEEE Conference on Open Systems, pp. 13–17 (2013)
7. Kenny, C.H., Chai, C.T., Tan, K.L., YeaDat, C., Alvey, H.C., Tan, R.W.: Development of home monitoring system with integration of safety and security modules. In: 2012 IEEE Conference on Sustainable Utilization and Development in Engineering and Technology, pp. 277–282 (2012)
8. Dhanisha, G., Seles, J.M., Brumancia, E.: Android interface based GCM home security system using object motion detection. In: International Conference on Communications and Signal Processing, pp. 1928–1931 (2015)
9. Lin, X.: Troubleshooting and maintenance of cooling water circulator. J. Chin. Electron Microsc. Soc. **24**(4), 394 (2005)
10. Markides, C.N., Gupta, A.: Experimental investigation of a thermally powered central heating circulator: pumping characteristics. Appl. Energy **110**, 132–146 (2013)

11. Memon, M.H., Kumar, W., Memon, A., Chowdhry, B.S., Aamir, M., Kumar, P.: Internet of Things (IoT) enabled smart animal farm. In: International Conference on Computing for Sustainable Global Development, pp. 2067–2072 (2016)
12. Chao, H., Pin, S., Yueh, C.H.: IoT-based physiological and environmental monitoring system in animal shelter. In: Seventh International Conference on Ubiquitous and Future Networks, pp. 317–322 (2015)
13. Zhuang, Y., Li, Y., Lan, H., Syed, Z., El-Sheimy, N.: Smartphone-based WiFi access point localisation and propagation parameter estimation using crowdsourcing. Electron. Lett. **51**(17), 1380–1382 (2015)
14. Weng, H.Y., Hart, L.A.: Impact of the economic recession on companion animal relinquishment, adoption, and euthanasia: a Chicago animal shelter's experience. J. Appl. Anim. Welfare Sci. **15**(1), 80–90 (2012)

An Automated System for Assisting and Monitoring Plant Growth

Satien Janpla, Kunyanuth Kularbphettong and Supanat Chuandcham

Abstract The purposes of the research are to design and develop the automation control plant monitoring system and evaluate the effect of using this prototype. This system is able to control significant environmental factors that affect plant growth including temperature, humidity, light and water. To facilitate this process, the Arduino board was adapted to program the various sensors used to control temperature and humidity for cultivation. The applications of this prototype include the preparation of the plant nutrient solution, to design the automated system to control the opening—the leading solution for automated control system. The result of the experimental project demonstrated that the system works well in the automated mode.

Keywords Automation control · Monitoring · Arduino

1 Introduction

In the twenty-first century, Agriculture is experiencing a Paradigm Shift and entering the era of "Agricultural Version 4.0", which is considered a major transformation of the agricultural sector. It is revolutionary as farmers are now required to possess knowledge of farming, technology, processing, and marketing strategy. Farmers are now in an era of agricultural technology, with intelligent agricultural farming or otherwise known as smart farming which aims to produce food to feed the world's population in the future. Farmers focus on farming or cultivation with a high accu-

S. Janpla · K. Kularbphettong (✉) · S. Chuandcham
Computer Science Program, Suan Sunandha Rajabhat University,
Bangkok 10300, Thailand
e-mail: kunyanuth.ku@ssru.ac.th

S. Janpla
e-mail: satien.ja@ssru.ac.th

S. Chuandcham
e-mail: supanat.ch@ssru.ac.th

© Springer Nature Singapore Pte Ltd. 2019
S. K. Bhatia et al. (eds.), *Advances in Computer Communication and Computational Sciences*, Advances in Intelligent Systems and Computing 759,
https://doi.org/10.1007/978-981-13-0341-8_36

racy being more environmentally friendly while increasing agricultural production and protect the environment. Thailand's agriculture is highly competitive, diversified and specialized and its exports are very successful internationally and Agricultural production is a whole accounts for an estimated 9–10.5% of Thai GDP [1, 2]. Thai government is attempting to enhance agricultural productivity by launching campaign Thailand 4.0 transforms the traditional economy into an innovation-driven economy and educates farmers and improves the living standard of farmers [3]. Thailand 4.0 should be changed to make a major shift from traditional agriculture to modern agriculture by focusing on Smart Farming.

Therefore, this research aims to develop IoT Solutions for Agriculture automation systems to provide the convenient ways of control the growth of plants. The remainder of this paper is organized as follows. Section 2 presents the system overview of this project. Section 3 we describe the experimental setup based on the purposed model and Sect. 4 shows the results of this research. Finally, the conclusion and future research are presented in Sect. 5.

2 Related Works

In the past, there are limited technologies for farmers to correlate production techniques and crop yields with land variability. However, agriculture is now changing the way farmers and agribusinesses view the land from which they reap their profits. Smart Farm or Smart Agriculture is the new management system for agricultural business by combining agricultural science and engineering electronics and computer technology to enhance the quality of farmer's life. The Internet of Things (IoT) is an internetworking system linked with computing devices, mechanical and digital machines, objects, animals, or people that are provided with unique identifiers that enable these objects to collect and exchange data [4, 5]. The Internet of Things has opened up extremely productive ways to cultivate agricultural products with the use of cheap, easy-to-install sensors, and an abundance of insightful data they offer [6].

According to Jain et al. [7], a remote monitoring system investigates temperature, moisture, and water level in the paddy crop field and analyzed data to a central server. A smart system provided and monitored an animal farm remotely by using microcontrollers, related sensors, and an IP Camera through the Internet [8]. Shahzadi et al. [9] presented an expert system based on the Internet of Things (IoT) for Cotton crop to minimize the losses due to diseases and insects/pests. Also, Mir Sajjad Hussain Talpur and et al. discussed the use of IoT in animal product supply chain by integrated RFID and network database technology and it can effectively improve the supply chain situation [10]. Using IOT for Agricultural Environment Monitoring, the prototype collected the field environment information to investigate and real-time collect information to monitor the field [11].

Fig. 1 System overview

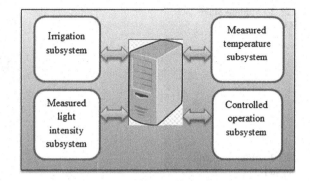

3 System Overview

The system overview of this research consists of four subsystems: irrigation subsystem, measured light intensity subsystem, measured temperature subsystem and controlled operation of light water pump and cooling fan subsystem as shown in Fig. 1. Every subsystem component is integrated with different sensors and devices with wireless communication modules used to interconnect between server and computing devices. Users can control the board through the web application and database is an interactive function used to manage the microcontroller. All subsystems are used to measure indicated levels when sensor works the value is less than or equal to the value set by user. The subsystem stopped only when the value from the sensor is greater than the controlled values. Also, the server sends and receives information from the user end through internet connectivity. Users can select to operate the system in two modes; automatic and manual. In automatic mode, system controls the installed devices automatically whereas, in manual mode, the system controls the operation through the web application.

4 The Experimental Setup

This section discusses the equipment and experimental procedures used to collect the data presented within this paper. The equipment was presented first, followed by the experimental procedures. The experiment performed in a greenhouse setting to evaluate the automate system to minimize outdoor variables and the lab equipment was consists of the following.

Fig. 2   Arduino Mage 2560
R3 [12]

Fig. 3   Arduino ethernet
shield [13]

4.1   Arduino Mage 2560 R3

The Arduino Mega 2560 is a large microcontroller board in the Arduino family and it has more features with ATmega2560. It is used to develop more complex projects and users can connect it to a computer with a USB cable or power [12] (Fig. 2).

4.2   Arduino Ethernet Shield

Arduino Ethernet Shield is used to connect your Arduino to the Internet and it can control and track devices such as take a look on a mobile phone to check the value of installed sensor, or turn off electrical appliances over the Internet (Fig. 3).

4.3   Soil Moisture Sensor

Soil moisture sensor is the device to measure the volumetric water content in soil and it can be connected to a microcontroller using an analog moisture input. To measure soil moisture value, PCB card is used to test and an LM393 ICP amplifier is to measure the pressure difference between the pressure pressures measured in the soil (Fig. 4).

If the voltage measured from the ground piece is higher, it will cause the logic to drop logic 1 to pin D0 and if the ground piece is low, logic 0 is released to pin D0.

Fig. 4 Soil moisture sensor
[13]

Fig. 5 Temperature and
humidity sensor [13]

Pin A0 is directly connected to circuit used moisture in the soil and gives the voltage from 0 to 5 V. However, if the soil is very sluggish, the pressure released will be less.

4.4 Temperature and Humidity Sensor

This project was used DHT22 sensor to measure temperature and humidity and it uses a capacitive humidity sensor and a thermistor to measure the surrounding air, and spits out a digital signal on the data pin [13] (Fig. 5).

4.5 Light-Dependent Resistor

Light dependent resistor sensor is one type of resistor whose resistance varies depending on the amount of light falling on its surface. When the light falls on the resistor, then the resistance changes. These resistors are often used in many circuits where it is required to sense the presence of light (Fig. 6).

Fig. 6 Light-dependent
resistor [13]

Fig. 7 High-pressure water
pump and other equipment
[13]

Table 1 The measure levels
of LDR

Condition	Indicated value (lx)
No light	0
Light	1–5
Gray	6–9
Dark	10

4.6 High-Pressure Water Pump and Other Equipment

High-Pressure Water Pump can pump water for using in agriculture with a 12 V battery and water pump will automatically stops working when high thermal protection
system alerts and after cooling for 5–10 min, the pump will work normally. Also,
relay interface board is able to control various appliances and other equipment with
large current. It can be controlled directly by microcontroller (Fig. 7).

The sensors measure the intensity of light or LDR and analog signal from pin to
pin is based on the incident light to the sensor. The LDR has shown as Table 1 to
indicate the measure levels.

The sensors measure the moisture of soil were collected by plugging the sensors
with PCB board and Table 2 was described the percentage value (Fig. 8).

The proposed system included three sensors to measure soil moisture, temperature and humidity and light dependent resistor and the sensors are installed in the
greenhouse and all signals were sent to the Arduino Mega 2560 R3 for processing

Table 2 The measure levels of the soil's moisture

Condition	Indicated value (%)
Dry soil	0–32
Humid soil	33–74
In water (soil soggy)	75–100

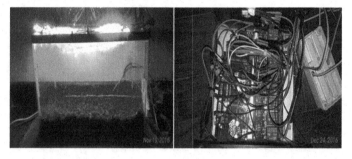

Fig. 8 The equipment

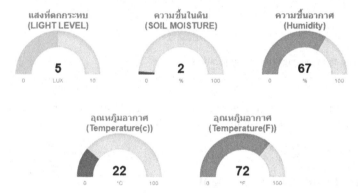

Fig. 9 The example page of web application

and control output devices to work together. The pump will be executed when soil moisture sensor checked the value in the soil lower value as determined.

The Arduino sends signal to spray moist to increase the humidity and Arduino will instruct the fan to operate when the temperature is lower and LEDs will work when the brightness gauge is low. Moreover, all sensor data is sent to the web server so that the user can view the data via Web and the operation data of the input and output devices are stored in the database. The experimental performance of automation control plant monitoring system was shown in Figs. 9 and 10 and the sampling this research was 40 sunflower seedlings.

The experimental design was random and the experimental set is allowed on the acquisition of land by setting temperature 30–35°, soil moisture 50–60%, and the intensity of light at 2–5. The system records the values obtained from the sensors

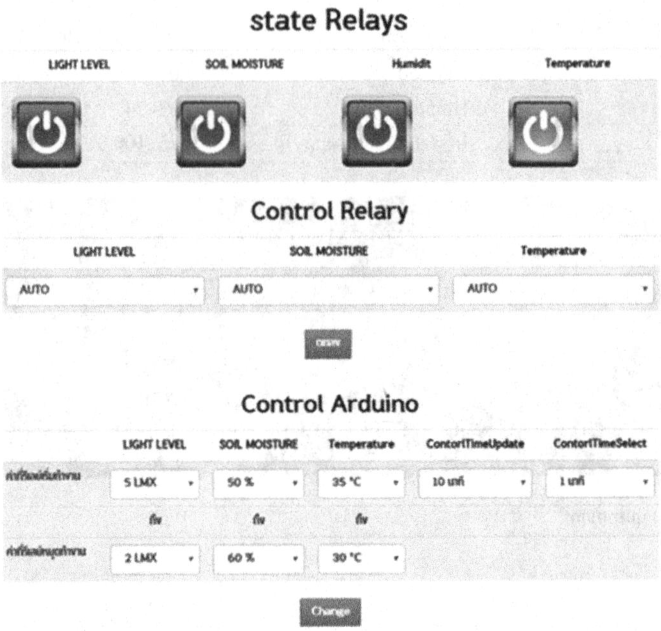

Fig. 10 The example page of web application

every 10 min. Moreover, the application retrieves information from the database every 1 min and plants grown in normal conditions.

5 The Results of Experiment

To evaluate the experimental performance, this research was compared the results between the greenhouse set and outdoor set. The result was found that the amount plant grown in greenhouse with harvest was 25% and the growth plants were 10%. Germination from seed was rate at 12% and the seeds do not grow at 53%. While the sapling of sunflower was grown in normal conditions, the number of plants grown was at 10% and the growth plants were 2%. Germination from seed was rate at 15% and the seeds do not grow at 73% as presented in Figs. 11 and 12.

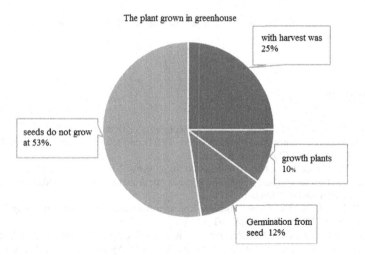

Fig. 11 The result of the plant grown in greenhouse

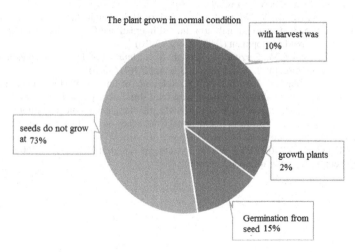

Fig. 12 The result of the plant grown in normal condition

6 Conclusion

In the automated control plant monitoring system sensor was successful in assisting plant growth and the automated system components were connected automatically with Arduino and wireless communication among various subsystems. Plants grown in greenhouses had a survival rate of 47.5%, and another had a survival rate of 27.5%. This result reflects that the application enhances the cultivation. Also, all observations and experimental tests were presented that this project is a complete solution to cover with all activities to cultivate plants.

Acknowledgements The authors gratefully acknowledge the financial subsidy provided by Suan Sunandha Rajabhat University.

References

1. Agriculture in Thailand. https://en.wikipedia.org/wiki/Agriculture_in_Thailand
2. Macroeconomic Strategy and Planning Office. Thai Economic Performance in Q1 and Outlook for 2017. NESDB ECONOMIC REPORT
3. Royal Thai Embassy., What is Thailand 4.0? http://thaiembdc.org/thailand-4-0-2/
4. Definition Internet of Things (IoT), IoT Agenda. http://internetofthingsagenda.techtarget.com/definition/Internet-of-Things-IoT
5. CERP-IoT., Internet of Things Strategic Research Roadmap (2009). http://www.grifsproject.eu/data/File/CERP-IoT%20SRA_IoT_v11.pdf
6. Raja Lakshmi, P., Devi Mahalakshmi, S.: IOT based crop field monitoring and irrigation automation. In: International Conference on International System and Control at Nov 2016
7. Jain, R., Kulkarni, S., Shaikh, A., Sood, A.: Automatic irrigation system for agriculture field using wireless sensor network (WSN) (2016)
8. Memon, M.H., Kumar, W., Memon, A., Chowdhry, B.S., Aamir, M., Kumar, P.: Internet of Things (IoT) enabled smart animal farm. In: 3rd International Conference on Computing for Sustainable Global Development (INDIACom) (2016)
9. Shahzadi, R., Tausif, M., Ferzund, J., Suryani, M.A.: Internet of Things based expert system for smart agriculture. Int. J. Adv. Comput. Sci. Appl. **7**(9) (2016)
10. Talpur, M.S.H., Shaikh, M.H., Talpur, H.S.: Relevance of Internet of Things in animal stocks chain management in Pakistan's perspectives. Int. J. Inf. Educ. Technol. **2**(1) (2012)
11. Cheng-Jun, Z.: Research and implementation of agricultural environment monitoring based on Internet of Things. In: 2014 Fifth International Conference on Intelligent Systems Design and Engineering Applications, pp. 15–16 (2014)
12. ARDUINO MEGA 2560 REV3. https://store.arduino.cc/usa/arduino-mega-2560-rev3
13. Equipment. http://www.arduinoberry.com/

Efficient Rub-Impact Fault Diagnosis Scheme Based on Hybrid Feature Extraction and SVM

Alexander Prosvirin, Jaeyoung Kim and Jong-Myon Kim

Abstract Rub-impact faults of rotor blades are commonly observed in rotating machines, which may cause considerable damage to the equipment. The complex nature of rubbing faults makes it difficult to utilize conventional feature extraction techniques for their diagnosis. In this paper, a new method is proposed for the diagnosis of rub-impact faults of different intensities, using a one-against-all multiclass support vector machines (OAA-MCSVM) classifier that is trained on hybrid features extracted from the intrinsic mode functions (IMFs) of the vibration signal. The raw vibration signal is decomposed into IMFs using empirical mode decomposition (EMD). For each IMF, features such as its total energy and its energy as a fraction of the original vibration signal are calculated directly; whereas its frequency-domain features are extracted by first calculating its frequency spectrum using the fast Fourier transform (FFT). The proposed approach is tested on rubbing fault data obtained using an experimental testbed. The results demonstrate that this approach is effective in differentiating no-rubbing condition and rubbing occurring with various intensities.

Keywords Empirical mode decomposition · Fault diagnosis · Feature extraction
Rub-impact · Rubbing · Support vector machines

A. Prosvirin · J. Kim · J.-M. Kim (✉)
School of Electrical, Electronics and Computer Engineering,
University of Ulsan, Ulsan 44610, South Korea
e-mail: jongmyon.kim@gmail.com

A. Prosvirin
e-mail: a.prosvirin@hotmail.com

J. Kim
e-mail: kjy7097@gmail.com

1 Introduction

Rotating machines such as turbines are widely used for power generation and usually operate under severe operating conditions characterized by high temperatures and high rotational speeds. A small clearance is maintained between the turbine blades and its stator to reduce air reluctance. Rubbing occurs when the rotor blades interact with the stator, either due to expansion because of increasing temperature or due to faults such as misalignment of the rotor and self-excited vibrations [1]. If not detected at an early stage, it can cause severe damage to the rotating machine and significantly increase the maintenance costs. Thus, detecting the onset of rubbing and determining the intensity of the rub-impact fault is an important problem in fault diagnosis of rotating machines.

Rubbing faults are known as highly nonlinear and non-stationary faults [2]. Hence, traditional signal processing techniques, which are developed for linear and stationary signals, are not effective in diagnosing these faults.

For an analysis of complex nonlinear faults, time-frequency-domain signal analysis techniques, such as Wavelet decomposition [3, 4] and Hilbert-Huang transform (HHT) with empirical mode decomposition (EMD) [5–7] have been widely used by researchers to extract the discriminative information from the rubbing fault signals. Yanjun et al. [3] proposed wavelet packet eigenvalue extraction and SVM for rub-impact fault diagnosis. This approach classifies various rubbing faults with high accuracy but does not provide a comparison with the normal state of the system. In [4], four-level discrete wavelet decomposition with a variety of wavelet functions and pre-filtering algorithms was utilized for feature extraction, whereas a radial basis function neural network was used for classification. Authors demonstrated the importance of choosing the appropriate filters and mother wavelet functions for the analysis of vibration signal. From this study, it can be seen that choosing of proper mother wavelet function and filtering technique is not an easy task for the analysis of a complex nonlinear signal, such as rubbing fault signal. Another approach, an adaptive data-driven decomposition of the original vibration signal into intrinsic mode functions (IMFs) using EMD, allows a precise observation of transients in fault signal. However, authors in [5, 6] mainly focused on observation and representation of rubbing faults in various graphical forms of frequency spectrum without attempting to perform classification of fault conditions. Yibo et al. [7] proposed an extraction of maximum singular values from each of IMFs to create feature vectors and their further utilization for the classification of different rub-impact fault models using support vector machines (SVM). This approach differentiates between various rubbing faults; however, it does not consider distinguishing the normal condition when no fault is presented in a signal and ignores variations in rubbing intensities.

In this paper, a new efficient approach is proposed for the diagnosis of rub-impact faults of various intensities using a one-against-all multiclass support vector machines (OAA-MCSVM) classifier that is trained on hybrid features extracted from the rubbing fault signal, i.e., the raw vibration signal, through EMD and FFT. These hybrid features include the total energy of each IMF and its energy as a fraction of the total

energy of the raw vibration signal, which are calculated directly in time domain and
frequency-domain features of each IMF, which are obtained by first computing the
frequency spectrum of this component using FFT. Existing work indicates that the
frequency spectrum of the rubbing fault signal alone does not provide clear information about sub-synchronous and high-frequency components of the rubbing fault
signal due to its complex nonlinear nature [6]. Therefore, in this study, EMD is used
to pre-process the rubbing fault signal by decomposing it into a finite number of
well-behaved oscillating components, i.e., IMFs. The FFT is then applied to each
of these extracted components to obtain valuable information about each rubbing
condition. Similar approaches have been successfully utilized for fault diagnosis of
rolling-element bearings in [8]. Moreover, energy features in time [9] and frequency
[10] domain with signal decomposition techniques have also been effectively used
to resolve problems of fault diagnosis. The hybrid features calculated for all the
extracted components of a rubbing fault signal are pooled into a single feature vector. Such feature vectors constructed for many rubbing fault signals with various
intensities are used to train OAA-MCSVM classifier. The trained classifier is then
used to differentiate unknown rubbing fault signals.

The remainder of this paper is organized as follows. Section 2 explains the proposed framework for intelligent rub-impact fault diagnosis. Section 3 presents the
experimental setup and describes the rubbing fault data used in this study. Section 4
presents and discusses the experimental results, and Sect. 5 concludes this paper.

2 The Proposed Method for Rub-Impact Fault Diagnosis

The proposed method for the diagnosis of rub-impact faults of bladed rotor is illustrated in Fig. 1. It involves the extraction of hybrid features from the rubbing fault
signals and then using those features to train OAA-MCSVM classifier. The trained
OAA-MCSVM is then used to differentiate unknown signals for different rubbing
faults and the normal condition, where there is no rubbing between the rotor blades
and the stator.

2.1 Hybrid Feature Extraction

Empirical mode decomposition (EMD) is an adaptive signal decomposition technique which is based on the assumption that any signal consists of a finite series
of IMFs representing different intrinsic modes of the signal oscillations. The main
advantage of EMD is that the decomposition of original signal is fully data-driven,
and no prior knowledge and constraints are required, which are typical of other techniques for signal decomposition. EMD decomposes the original signal using a sifting
process. To be considered as IMF, the function should satisfy the two conditions: (i)
the number of extrema and zero-crossing points is equal or differ by at most one, and

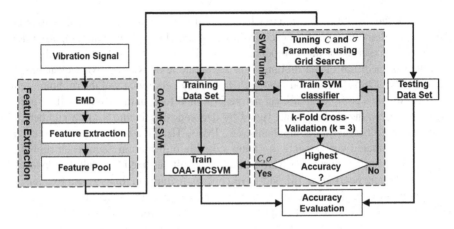

Fig. 1 The proposed algorithm for intelligent rub-impact fault diagnosis

(ii) the mean of the envelopes defined by local maxima and minima is equal to zero at any point. The details on EMD and sifting process can be found in [11].

Extracted IMFs contain frequency components of the decomposed signal in decreasing order, where the first extracted IMF is carrying the highest oscillation content, whereas the last IMF contains the lowest one. Note that the number of extracted components depends on the power of frequency content presented in the raw vibration signal. Thus, for the signals with higher frequency contents, a larger number of IMFs will be extracted. It is usually considered that the first few IMFs contain the most explicit information about the properties of the original signal. Thus, in this study, features are extracted from only the first eight IMFs.

When rub-impact fault occurs in a system, significant changes in amplitudes of the fundamental frequency and its harmonics can be observed. High-frequency components of the signal are also affected with impulses, produced by this fault. Thus, with the increase of the rubbing fault intensity, the energy of the vibration signal changes as well. The changes of magnitudes for various frequency components can be clearly seen by calculating the energies of extracted IMFs and their energy ratios, which provide sufficient insight into the fault type. However, for more accurate fault diagnosis, it is useful to explore the frequency spectra of extracted IMFs to examine these processes in frequency domain. To obtain frequency-domain features, FFT is applied to each of IMFs. Although FFT is not effective in capturing the transient changes when directly applied to non-stationary signals, the frequency spectra of extracted IMFs demonstrate that various frequency components, i.e., the fundamental frequency, its harmonics, and sub-synchronous frequencies, can be clearly observed after signal decomposition using EMD. Each obtained IMF contains specific oscillating component of the original signal and can be treated as a frequency band. From frequency spectra of extracted IMFs presented in Fig. 2, it can be observed that IMF 7 mostly consists of the fundamental frequency 1X (equal to 40 Hz) and its 2X harmonic. Higher order harmonics are observed in frequency spectrum of IMF 6.

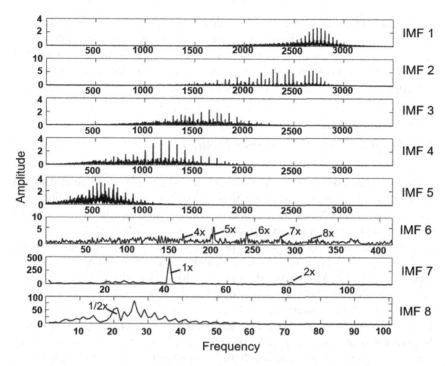

Fig. 2 Frequency spectra of extracted IMF from the rub-impact fault signal

Table 1 Time and frequency-domain statistical feature parameters extracted from IMFs

Time and frequency-domain statistical features			
Energy $\left(f_1^i\right)$	$\sum_{n=1}^{N} C_i^2(n)$	Energy ratio $\left(f_2^i\right)$	$f_1^i / \left(\sum_{i=1}^{L} f_1^i\right)$
Freq. center $\left(f_3^i\right)$	$\frac{1}{N}\sum_{n=1}^{N} fr_i(n)$	RMS freq. $\left(f_4^i\right)$	$\sqrt{\frac{1}{N}\sum_{n=1}^{N} fr_i^2(n)}$
Root var. freq. $\left(f_5^i\right)$	$\sqrt{\frac{1}{N}\sum_{n=1}^{N}\left(fr_i(n) - f_3^i\right)^2}$	Freq. peak $\left(f_6^i\right)$	$\max(fr_i)$
Freq. spectrum energy $\left(f_7^i\right)$	$\sum_{n=1}^{N} fr_i^2(n)$	Freq. spectrum energy proportion $\left(f_8^i\right)$	$f_7^i / \left(\sum_{i=1}^{L} f_7^i\right)$

i—is the number of IMF component; C_i—is a particular IMF component; fr_i—spectral component of C_i, and L—is a total number of extracted components

IMF 8 contains the 1/2X and other harmonics in the range of sub-synchronous frequencies. Some high-frequency components, including resonance frequencies, can be observed in IMFs 1, 2, 3, 4, and 5 as well. The presence of sub-synchronous frequencies and fluctuations of high frequencies, including the resonance frequency, are usually described as essential features of a rub-impact fault. Hence, frequency-domain features are useful in differentiating various rubbing conditions. All the extracted features are given in Table 1.

The resultant feature pool consists of a total of $N_C \times N_S \times N_{channels} \times N_{IMF} \times N_F$ features. Here, N_c is the number of bladed rotor conditions simulated in this study, N_S is the number of samples for each condition, $N_{channels}$ is the total number of channels used by vibration sensors, N_{IMF} is the number of IMFs used for feature extraction and N_F is the number of extracted features.

2.2 Support Vector Machines Classifier

SVM is a binary classification algorithm [12] and is considered as one of the most accurate and robust classification methods. In contrast to other classification techniques, SVM is not affected by the curse of dimensionality as it is insensitive to the number of dimensions of the input feature vector.

SVM solves the binary classification problem by finding the largest margin hyperplane that separates the two classes in the feature space of the training vectors. New data samples are labeled based on the sign of the hyperplane function, i.e., a positive sign indicates that the new instance of data belongs to the current class, whereas a negative sign indicates otherwise. Let $(x_1, y_1) \ldots (x_i, y_i)$ be the training dataset with N samples, where $x_i \in \mathbb{R}^D$ is the D-dimensional feature vector, $y_i = ([+1, -1])$ are the class labels and $i = 1, 2, \ldots, N$. The generalized problem of finding the optimal hyperplane can be expressed in dual form by applying Lagrangian multipliers [13]:

$$\max W(\alpha) = \sum_{i=1}^{N} \alpha_i - \frac{1}{2} \sum_{i,j}^{N} a_i a_j y_i y_j \varphi(x_i)^T \varphi(x_j)$$

$$subject\ to \sum_{i=1}^{N} a_i y_i = 0,\ 0 \le a_i \le C,\ i = 1, 2, \ldots, N, \tag{1}$$

where a_i and a_j are Lagrange multipliers, x_i and x_j are two input training vectors, C is a penalty coefficient, which determines the tolerance to misclassification errors and has to be tuned to reflect the noise in the data. The dot product of $\phi(x_i)$ and $\phi(x_j)$ can be replaced with kernel function $K(x_i, x_j)$ by Mercer's theorem [12]. As a kernel function, Gaussian radial basis function (RBF) for linearly non-separable data is utilized in this paper.

As a binary classifier, SVM cannot be employed directly to solve multiclass classification problems. To overcome this constraint, the one-against-all (OAA) scheme is implemented, where each class is trained individually against all the other classes.

The parameters C of objective function and σ of RBF kernel should be fine-tuned, as C affects the generalization performance of classifier, whereas σ has an impact on the decision boundary. In this paper, the optimal values for these parameters are selected using grid search algorithm with k-fold cross validation as was shown in Fig. 1. The values for the parameters C and σ were chosen from the predetermined ranges, i.e., $C \in \{2^{-5}, 2^{-3}, \ldots, 2^{15}\}$ and $\sigma \in \{2^{-2}, 2^{-1}, \ldots, 2^{7}\}$, respectively.

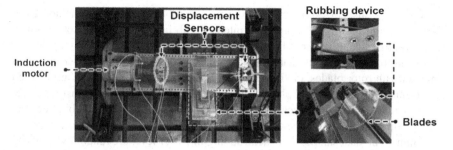

Fig. 3 Experimental testbed for rub-impact fault simulation

After selection process, the best combination of C and σ is then used for training the OAA-MCSVM classifier.

3 Experimental Testbed and the Rubbing Fault Data

3.1 Experimental Testbed

The experimental testbed used to collect the rubbing fault data by simulating rub-impact faults of different intensities is shown in Fig. 3. Two vibration sensors are installed at the opposite ends of the shaft. The displacement values of the rotor are recorded, when there is rubbing between the rotor and the stator, and when no rubbing has been observed. Each sensor records the displacements in both the horizontal and vertical directions using different channels. Thus, a total of four channels are utilized by the two vibration sensors.

The experiment is performed at a constant rotational speed of 2400 revolutions per minute (RPM), and the signal is sampled at a rate of 65.536 kHz. The local rub-impact fault is simulated by creating the interaction between rubbing device and a rotor with 12 blades.

The experiment is repeated five times to record five data sets, where each data set consists of seven groups of signals. Each group of signals corresponds to a different rubbing intensity including the normal state when no rubbing is observed. The total duration of each recorded signal is 180 s; however, each signal is divided into 180 slices of 1 s each for feature extraction.

3.2 Training and Testing Data Configuration

The classification performance of the proposed fault diagnosis method is validated over five data sets, with 1260 feature vectors each. Each data set consists of seven groups of signals, corresponding to seven various rubbing intensities, including no-rubbing condition, slight rubbing, and intensive rub-impact faults. For convenience, each group of signals is named based on the movement distance of the rubbing device. Thus, normal condition is marked as "7 mm" when the rubbing device does not interact with rotor blades, the intensity levels of slight rub-impact fault are marked as "6 mm","5 mm","4 mm", "3.5 mm," and "3 mm", where the intensity of rubbing increases with the decrease of the movement distance of the rubbing device. The group of samples, corresponding to the intensive rub-impact fault is marked as '2.5 mm'. The data samples for each rubbing intensity level are randomly split into three folds, where two folds are used for training, and one fold is used for testing. More specifically, each training set consists of 840 feature vectors, whereas the testing set has 420 feature vectors. The training set contains 120 randomly selected feature vectors for every rub-impact fault intensity level, whereas the testing set comprises of the remaining 60 feature vectors for each intensity level.

4   Results and Evaluation

To evaluate the efficacy of the proposed approach in classifying rubbing faults of rotor blades with various intensity levels, it is compared to other feature extraction methods which use a different set of features. Classification for the comparison is done by means of OAA-MCSVM classifier. The first two feature extraction techniques are represented by frequency-domain feature set (i.e., frequency center, frequency RMS and root variance frequency) extracted using FFT, and combined feature pool which is based on the extraction of time and frequency-domain statistical feature parameters, such as peak value, RMS, kurtosis, crest factor, clearance factor, impulse factor, shape factor, entropy, skewness, square mean root, 5th and 6th normalized moments, mean value, peak-to-peak ratio, kurtosis factor, frequency center, frequency RMS, and root variance frequency. These two feature extraction methods are referred to as FFT, and TFD, respectively. Another two specific approaches extract features by combining EMD and maximum singular values (SV) [7], and Wavelet Packet (WP) and SV [3] (these approaches are referred to as EMD+SV and WP+SV, respectively). Comparing these techniques to the proposed one in this study, experiments are carried out on all the five data sets collected at a fixed rotational speed of 2400 RPM. To perform all the operations on data samples (i.e., signal decomposition, feature extraction, and fault classification routines), MATLAB software is used.

The classification accuracy for each group of signal instances is computed through true positive rate (TPR) evaluation index, whereas the final classification accuracy is computed as follows: sum of numbers of true positive samples for each class in testing data set is divided by the total number of samples, represented in this particular testing data set. The experimental results are given in Table 2.

The results demonstrate that the proposed rub-impact fault diagnosis method outperforms the conventional and field-specific fault diagnosis techniques which are

Table 2 Classification performance comparison

Data set	Method	TPR (%)							Acc. (%)
		2.5 mm	3 mm	3.5 mm	4 mm	5 mm	6 mm	7 mm	
1	Proposed	98.3	98.3	100	98.3	98.3	96.6	96.6	**98.1**
	EMD+SV	98.3	96.6	98.3	96.6	98.3	76.6	81.6	**92.3**
	WP+SV	100	96.6	98.3	50	63.3	98.3	100	**86.6**
	TFD	70	96.6	98.3	95	100	98.3	98.3	**93.8**
	FFT	100	93.3	98.3	63.3	88.3	81.6	85	**87.1**
2	Proposed	100	96.6	96.6	98.3	95	100	96.6	**97.6**
	EMD+SV	98.3	91.6	96.6	91.6	91.6	98.3	98.3	**95.2**
	WP+SV	100	81.6	76.6	81.6	83.3	98.3	100	**88.8**
	TFD	100	80	100	100	100	100	93.3	**96.2**
	FFT	100	48.3	73.3	81.6	91.6	78.3	33.3	**72.3**
3	Proposed	100	100	98.3	100	86.6	98.3	96.6	**97.1**
	EMD+SV	100	100	98.3	100	91.6	91.6	88.3	**95.7**
	WP+SV	100	63.3	65	76.6	96.6	98.3	100	**85.7**
	TFD	100	90	98.3	95	100	98.3	100	**97.3**
	FFT	100	43.3	65	78.3	85	93.3	71.7	**76.6**
4	Proposed	100	98.3	96.6	96.6	96.6	100	93.3	**97.3**
	EMD+SV	100	85	93.3	50	93.3	100	78.3	**85.7**
	WP+SV	100	91.6	83.3	66.6	76.6	100	96.6	**87.8**
	TFD	98.3	100	95	100	93.3	100	91.6	**97**
	FFT	100	100	76.7	73.3	83.3	95	98.3	**89.5**
5	Proposed	100	96.6	98.3	100	98.3	98.3	96.6	**98.3**
	EMD+SV	98.3	90	93.3	86.6	96.6	100	78.3	**92**
	WP+SV	100	96.6	70	68.3	86.6	100	100	**88.8**
	TFD	100	100	100	91.6	98.3	100	98.3	**98.3**
	FFT	100	98.3	83.3	13.3	93.3	98.3	98.3	**83.6**

used for the comparison. The proposed scheme achieves an average classification accuracy of 97.6% over five data sets.

From Table 2, it can be observed that rub-impact fault diagnosis method, which utilizes only frequency-domain features extracted after the FFT of the raw signal, has the lowest average accuracy equal to 81.8%, which is reasonable due to the constraints inherent to this method. Since FFT cannot perform well on nonlinear and non-stationary signals, therefore, the resultant frequency-domain features have a high degree of overlapping in the resultant feature space. This problem degrades the ability of classifier to distinguish between different classes. To improve the classification performance, while using frequency-domain features in cases of non-stationary faults, the signal should be pre-processed or additional statistical features in time domain should be extracted to create a more discriminative feature space. Thus, the combination of time and frequency-domain features results in an average accuracy of 96.5%.

In general, the specific-field feature extraction methods can show better classification performance for specific faults than the general ones. However, the two referenced methods used in this study result in average accuracies of 92.2% and 87.5%, respectively. These results can be explained as follows: the problem resolved in this study differs from those, for which these referenced methods are developed. These feature extraction techniques were proposed for addressing the classification of rub-impact fault types which include such faults as full annular rub-impact, single point rub-impact, and double point rubbing faults. TPRs also demonstrate that these two methods perform well in recognizing the normal and intensive rubbing states, but in case of slight rubbing, the number of features and their quality is not discriminative enough to distinguish various levels of rubbing intensity among the slight rubbing cases.

5 Conclusion

In this paper, a new efficient fault diagnosis method of rub-impact faults with various intensities using OAA-MCSVM and hybrid feature extraction from IMFs of the vibration signal was proposed. The pre-processing performed by EMD allows to extract valuable information from nonlinear fault signals. Thus, highly discriminative features can be extracted from the obtained oscillating components. The experimental results demonstrated that the proposed hybrid features extracted from pre-processed raw vibration signal are effective for diagnosis of rubbing faults of different intensities. The results showed that the proposed approach with hybrid feature pool outperforms conventional and specific feature extraction techniques used for the comparison and achieves an average accuracy of 97.6% over five data sets.

Acknowledgements This work was supported by the Korea Institute of Energy Technology Evaluation and Planning (KETEP) and the Ministry of Trade, Industry & Energy (MOTIE) of the Republic of Korea (Nos. 20162220100050, 20161120100350, 20172510102130). It was also funded in part by The Leading Human Resource Training Program of Regional Neo industry through the National Research Foundation of Korea (NRF) funded by the Ministry of Science, ICT and future Planning (NRF-2016H1D5A1910564), and in part by the Basic Science Research Program through the National Research Foundation of Korea (NRF) funded by the Ministry of Education (2016R1D1A3B03931927).

References

1. Zhang, Y., Wen, B., Leung, A.Y.T.: Reliability analysis for rotor rubbing. J. Vib. Acoust. **124**, 58 (2002)
2. Rubio, E., Jáuregui, J.C.: Time-frequency Analysis for Rotor-rubbing Diagnosis. Citeseer (2011)
3. Lu, Y., Meng, F., Li, Y.: Research on rub impact fault diagnosis method of rotating machinery based on wavelet packet and support vector machine. Presented at the International Conference on Measuring Technology and Mechatronics Automation (2009)
4. Roy, S.D., Shome, S.K., Laha, S.K.: Impact of wavelets and filter on vibration-based mechanical rub detection using neural networks. In: 2014 Annual IEEE India Conference (INDICON), pp. 1–6. IEEE (2014)
5. Zhao, Y., Liu, E., Zhu, J., Zhang, B., Wang, J., Tian, H.: Rub-impact fault diagnosis of rotating machinery based on Hilbert-Huang transform. In: 2015 IEEE International Conference on Mechatronics and Automation (ICMA), pp. 32–36 (2015)
6. Xiang, L., Tang, G., Hu, A.: Analysis of rotor rubbing fault signal based on hilbert-huang transform. Presented at the International Conference on Measuring Technology and Mechatronics Automation (2009)
7. Yibo, L., Fanlong, M., Yanjun, L.: Research on rub impact fault diagnosis method of rotating machinery based on EMD and SVM. In: 2009 International Conference on Mechatronics and Automation, pp. 4806–4810. IEEE (2009)
8. Rai, V.K., Mohanty, A.R.: Bearing fault diagnosis using FFT of intrinsic mode functions in Hilbert-Huang transform. Mech. Syst. Signal Process. **21**, 2607–2615 (2007)
9. He, D., Li, R., Zhu, J.: Plastic bearing fault diagnosis based on a two-step data mining approach. IEEE Trans. Ind. Electron. 3429–3440 (2012)
10. Miao, Q., Wang, D., Pecht, M.: Rolling element bearing fault feature extraction using EMD-based independent component analysis. In: 2011 IEEE Conference on Prognostics and Health Management (PHM), pp. 1–6. IEEE (2011)
11. Sandoval, S., De Leon, P.L.: Theory of the Hilbert Spectrum (2015). arXiv:1504.07554
12. Vapnik, V.N.: An overview of statistical learning theory. IEEE Trans. Neural Netw. **10**, 988–999 (1999)
13. Kang, M., Kim, J., Kim, J.-M., Tan, A.C.C., Kim, E.Y., Choi, B.-K.: Reliable fault diagnosis for low-speed bearings using individually trained support vector machines with Kernel discriminative feature analysis. IEEE Trans. Power Electron. **30**, 2786–2797 (2015)

Diagnosis Approach on Compensation Capacitor Fault of Jointless Track Circuit Based on Simulated Annealing Algorithm

Bao-ge Zhang, Wei-jie Ma and Gao-wu Chang

Abstract Aiming at the disadvantage of low precision, long time consuming and poor practicability of the current test results in the fault detection of jointless track circuit, the fault detection method based on simulated annealing algorithm is proposed. First, the shunt state model on jointless track circuit was established, relationship between the compensation capacitor and the sectional shunt current amplitude envelope was acquired by Two Port Network Theory. The optimal values of the compensation capacitance was obtained using simulated annealing algorithm, Under the optimal values, the curve of shunt current amplitude envelope was plotted. The experimental results show that the detection effect of the capacitance fault detection method based on simulated annealing algorithm is accurate, and detection time is short, and the practicability is strong, compared with the genetic algorithm.

Keywords Component · Jointless track circuit · Compensation capacitor
Simulated annealing algorithm

1 Introduction

According to the principle of jointless track circuit, there are electrical isolation type (resonant), natural attenuation type (superimposed). At present, the resonant ZPW-2000 series has been widely used in railway system in our country. The compensation capacitor and the ballast resistance play important roles for the transmission of the railway signal. In practice, the failure of compensation capacitor will be caused by

B. Zhang (✉) · W. Ma · G. Chang
School of Automation and Electrical Engineering, Lanzhou Jiaotong University, Lanzhou, Lanzhou 730070, Gansu, China
e-mail: zbg4938836@163.com

W. Ma
e-mail: 276497535@qq.com

G. Chang
e-mail: 425158098@qq.com

© Springer Nature Singapore Pte Ltd. 2019
S. K. Bhatia et al. (eds.), *Advances in Computer Communication and Computational Sciences*, Advances in Intelligent Systems and Computing 759,
https://doi.org/10.1007/978-981-13-0341-8_38

bad weather, the vibration, impact of the train, surge voltage caused by unbalanced traction current and some human factors [1], "red tape" phenomenon will endanger, which will Endanger traffic safety [2]. So, establishing a rapid and accurate diagnostic method is necessary [3, 4].

At present, the domestic and foreign scholars have made a lot of research on the fault diagnosis of the compensation capacitor. In the Ref. [5], the partial least squares regression and neural network are used in the fault detection of the compensation capacitor of the jointless track circuit. In the Ref. [6], based on the D-S evidence theory and trend analysis method are proposed for the compensation capacitor failure of the jointless track circuit by Oukhellou et al. In the Ref. [7], Levenberg Marquardt algorithm is used to get the shunt current amplitude envelope signal by piecewise exponential fitting, then instantaneous frequency of signal changes is showed based on the application of generalized S transform, according to the result of frequency variation, the fault position of compensation capacitance can be concluded. In the Ref. [8], The frequency information of locomotive signal is proposed through Adaptive Optimal Kernel Time-Frequency Representation, then the fault location of compensation capacitance is realized through frequency variation.

But there is little research on the fault diagnosis of compensation capacitor based on intelligent algorithm, although the genetic algorithm is proposed to realize the fault diagnosis and detection [9], the method has the shortcomings of low diagnostic accuracy, long time consuming and so on, that is to say, the diagnosis effect is not ideal. Simulated annealing algorithm is a global optimization algorithm, and the probability close to the global optimal value is about 1, Therefore, it has been widely used in various fields [10–12].

In order to improve the performance of the fault diagnosis of the compensation capacitor, a fault detection method based on simulated annealing algorithm is proposed in this paper, the correctness, feasibility, and practicability of the method are verified by the simulation data and the actual data.

2 The Jointless Track Circuit Model

The schematic diagram of the structure of the jointless track circuit is showed in Fig. 1. It is consisted of a transmitter, a transmission cable device, a tuning region, a rail, a parallel compensation capacitor and a receiver [13–15].

According to the transmission direction of electric energy, the equivalent circuit model of the track circuit can be constituted of the four-port network of sending end, the four-port network of the rail line and the four-port network of the receiving end. The shunt state four-port equivalent model of the jointless track circuit is shown in Fig. 2.

$G(x)$ is the total transmission characteristics from transmitter to shunt points, four-port network, that is

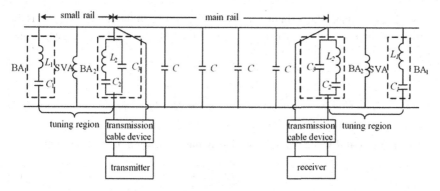

Fig. 1 The structure schematic diagram of the jointless track circuit

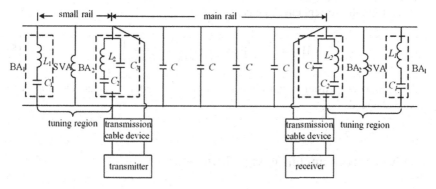

Fig. 2 The shunt state equivalent four-port network diagram

$$\boldsymbol{G}(x) = \begin{bmatrix} g_{11}(x) \ g_{12}(x) \\ g_{21}(x) \ g_{22}(x) \end{bmatrix} = \boldsymbol{N}_\mathrm{p} \times \boldsymbol{N}_\mathrm{b} \times \boldsymbol{T}_{\mathrm{ZF}}(x) \qquad (1)$$

$\boldsymbol{N}_\mathrm{p}$ is the equivalent four-port network for cable equipment of transmitting end.

$\boldsymbol{N}_\mathrm{b}$ is the equivalent four-port network for tuning region of transmitting end.

$\boldsymbol{T}_{\mathrm{ZF}}(x)$ is the equivalent four-port network for rail between transmitter tuning zone and the rail surface branching point.

x is the distance between transmitter and shunt points.

Then, the relationships are described as follows:

$$\begin{cases} \dot{U}_s = g11(x)\dot{U}_{\mathrm{fz}} + g12(x)\dot{I}_{\mathrm{fz}} \\ \dot{I}_s = g21(x)\dot{U}_{\mathrm{fz}} + g22(x)\dot{I}_{\mathrm{fz}} \end{cases} \qquad (2)$$

$\dot{U}_{\text{fz}}$ is the voltage of the shunt point, $\dot{I}_{\text{fz}}$ is the current of the shunt point, $\dot{U}_s$ is the voltage of the transmitter, $\dot{I}_s$ is the current of the transmitter The $g_{11}(x)$ and $g_{12}(x)$ are function of the compensation capacitance.

According to (2), it is easily get

$$\frac{\dot{U}_s}{\dot{I}_{\text{fz}}} = g11(x)\frac{\dot{U}_{\text{fz}}}{\dot{I}_{\text{fz}}} + g12(x) \tag{3}$$

There is the following expressions during the calculation of shunt current:

$$\dot{U}_{\text{fz}}/\dot{I}_{\text{fz}} \approx R_{\text{f}} \tag{4}$$

Here, R_{f} is the equivalent resistance of wheelset at shunt point.

Therefore, the current envelope is:

$$\text{Afz} = \left|\dot{I}_{\text{fz}}(x)\right| = \left|\frac{\dot{U}_s}{g_{11}(x)R_{\text{f}} + g_{12}(x)}\right| \tag{5}$$

When the compensation capacitance value is changed, the current envelope will be also changed.

3 Simulated Annealing Algorithm

Simulated annealing algorithm is referred to as SA algorithm, SA algorithm is a general probabilistic algorithm searching for a solution to a problem in a search space. At the beginning of 1980s, the idea of SA algorithm is proposed by Metropolis which originates from the cooling and annealing process of the solid [16–18]. Three stages are included in the annealing process, First, it is heating process. When the temperature of the solid reaches a certain degree, the solid is dissolved into liquid, and the state of uneven state exists in the solid state, which is the starting point of a balance state for the cooling process. Second, it is isothermal process. The heat exchange is occured between the solid and the surrounding environment, the temperature of the solid is equal to the temperature of the surrounding environment when the heat exchange reaches the state of dynamic equilibrium. Third, it is cooling process. It is the aim to weaken orderly the thermal motion of particles, the energy of the system is decreased, and the crystal structure will be obtained.

Algorithm process is divided into two layers cycle, at any temperature, the new solution is generated by random disturbance, and the value change of the objective function is calculated, which decide whether the new solution will be accepted. Because the initial temperature is set up relatively high in the algorithm, which causes the new solution having energy increased may be accepted at the initial time, so that it can jump out of local minimum solution, then by slowly decreasing the temperature, the global optimal solution will be eventually converged. At low temperature, the function value is already very small, but a worse solution is still not ruled out, therefore, the best feasible solution (historical solution) is also recorded, which is output together with the last one solution accepted before the termination of the algorithm process.

The implementation steps of the SA algorithm are as follows:

1. Initialization: The initial temperature T_0 is set large enough, ordering $T = T_0$, taking arbitrarily initial solution x_0, and calculating value of the objective function $E(x_0)$ with x_0, determining the number of iterations of per T, which is equal to the length of Markov chain.
2. Making T equal to the next value T_i in the cooling schedule.
3. The present solution $x_m(R_m, C_m)$ is updated by random disturbance, which will create a new solution x_n, and the value of the objective function $f(x_n)$ is calculated with x_n, then

$$\nabla E = f(x_n) - E(x_m)$$

4. If $\nabla E < 0$, x_n is accepted as the new present solution, that is $x_m = x_n$, otherwise the value of acceptance probability $\exp(-\nabla E/T_i)$ is calculated, the random number rand with uniform distribution is produced in region (0, 1). If $\exp(-\nabla E/T_i) >$ rand, x_n is also accepted as the new present solution, otherwise the new present solution x_m is retained, T_i is the current temperature.
5. Under T_i, the disturbance and acceptance process are repeated for L times (length of Markov chain), that is, repeating steps (3) and (4) for L times.
6. It is determined whether to meet the termination conditions, if the conditions are right, the present solution x_m is optimum solution, and x_m is output, the program process is end. Otherwise, attenuation function is attenuated with T, then it is returned to step (2).

4 Experimental Simulation of the Fault Diagnosis of the Compensation Capacitor Based on SA

SA method is adopted for the fault diagnosis of compensation capacitor, First, selecting the decision variables and the design objective function are as shown below:

The decision variables: compensation capacitor $C_i(i = 1\sim12)$, here, $C_i \in [0, C_z]$, C_z is the standard value.

The design objective function: The shunt current amplitude envelope is I_{sd} with $i(i = 1\sim13)$, The simulation results I_{fz} of the corresponding segment can be obtained by the formula (4), then the objective function is defined with the difference between I_{sd} and I_{fz}, as

$$E_i = 1 - \text{xcorr}(I_{sd}, I_{fz}) \tag{6}$$

Here, xcorr is cross correlation function.

According to the size of the solution space, the value of the initial temperature is between 100 and 150, and the other parameters are used by default. Main function is called through saengine for iterative searching of SA algorithm, in the loop iteration process, sanewpoint and saupdates are two key functions. The main function of sanewpoint is to generate new solutions and determine whether to accept a new solution, specific implementation methods are as follows:

The new solution generated by perturbation is

The method that new solution can be accepted is: The probability of accepted the new solution is subordinated to the Boltmann probability distribution, the expression is as follows:

$$P = \begin{cases} 1, & \nabla E < 0 \\ \frac{1}{1+\exp\left[\frac{\nabla E}{T_i}\right]}, & \nabla E \geq 0 \end{cases} \tag{7}$$

The main function of saupdates is to update the number of iterations, the optimal solution, the temperature, and the cooling treatment and so on, exponential method is used to realize decreasing temperature function, the formula is as follows:

$$T_i = q^k \times T_0 \tag{8}$$

Here, q is the cooling constant, k is the annealing parameter, generally meaning the current iteration number, T_0 is the initial temperature.

$$x_n = x_m + alpha1 * T_i$$
$$x_n = alpha2 * ub + (1-alpha2) * x_m$$
$$x_n = alpha2 * lb + (1-alpha2) * x_m$$

Here, *alpha1* is the random number generated randn function in the (1, 1), *alpha2* is the random number generated rand function in the (0, 1), *ub* is upper bound for the compensation capacitor value, *lb* is the lower bound of the compensation capacitor value, the method can be used to generate new solution based on the current solution, which can be ensured that the new solutions should be in variable range. among them, the 11 compensation capacitor standard values are 40 μF, one compensation capacitance standard value is 1 μF.

The flow chart of the fault diagnosis method of the jointless track circuit based on SA algorithm is shown in Fig. 3.

Under laboratory conditions, the actual simulation values, optimal values obtained using corresponding SA algorithm, and optimal values obtained based on the genetic algorithm(GA) [9] are shown in Table 1.

Table 1 shows that: the maximum absolute error using SA algorithm for fault diagnosis method is 1.47, the maximum absolute error based on the genetic algorithm

Fig. 3 The total flow chart
of fault diagnosis method

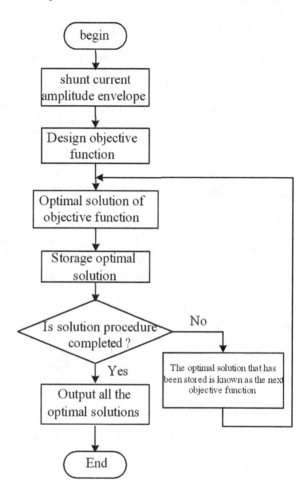

reaches 3.36, so the SA algorithm for fault diagnosis method can obtain more accurate optimization results.

In practice, personnel carry outs fault detection by measuring shunt current amplitude envelope. Therefore, according to Eq. (5), the actual simulation value and the optimal values based on SA algorithm in Table are respectively plotted corresponding shunt current amplitude envelope, as shown in Fig. 4.

It is easy to find in Fig. 4, the shunt current amplitude envelope based on SA optimal values agrees well with the shunt current amplitude envelope with the actual values, two curves almost coincide with each other, which indicates that the SA algorithm has strong practicability.

Table 1 Simulation results and error

Compensating capacitor	Actual simulation values/μF	Optimal values based on GA/μF	Optimal values using SA/μF	Absolute error based on GA/μF	Absolute error using SA/μF
–	–	–	–	–	–
C_1	40	40	40	0	0
C_2	1	1.08	1.44	0.08	0.44
C_3	40	39.58	39.5	0.42	0.5
C_4	40	36.68	39.91	3.32	0.09
C_5	20	18.2	20.08	1.8	0.08
C_6	40	36.64	39.85	3.36	0.15
C_7	40	40	39.98	0	0.02
C_8	40	39.7	39.89	0.3	0.11
C_9	40	39.36	39.59	0.64	0.41
C_{10}	40	39.29	38.53	0.71	1.47
C_{11}	40	40	39.34	0	0.66
C_{12}	40	40	40	0	0

Fig. 4 The shunt current amplitude envelope

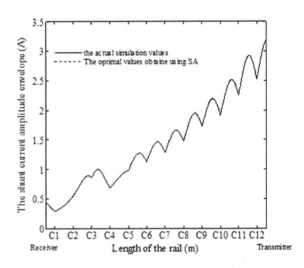

5   Conclusions

First, the four-port network equivalent model of the jointless track circuit shunt state is the established according to the jointless track circuit structure, the expressions of the shunt current amplitude envelope is derived through the using two-port theory. The capacitor fault diagnosis method of the jointless track circuit based on SA algorithm is proposed and researched, setting compensate capacitance as decision variables, setting the shunt current amplitude envelope as the objective function, the optimization calculation are objective function, the optimal values are obtained through calculation. Compared with the actual value, the error of the optimal values is very small. It is proved that the algorithm is correct. Compared with the diagnosis method based on genetic algorithm, the maximum absolute error using SA algorithm for fault diagnosis method is 1.47, the maximum absolute error based on the genetic algorithm reaches 3.36, the SA algorithm for fault diagnosis method can obtain more accurate optimization results. So the SA algorithm for fault diagnosis method is more suitable for the fault diagnosis of the jointless track circuit.

Acknowledgements This work was supported by the Research Foundation of Colleges and Universities of Gansu Province under Grant 213052; National Natural Science Foundation under Grant 51367010.

References

1. Sun, S.P., Zhao, H.B.: Fault detection method on compensation capacitor of jointless track circuit based on phase space reconstruction. J. China Railw. Soc. **34**(10), 79–84 (2012)
2. Ran, Y.: Study on the measuring on-line method of insulation resistance of ballast track circuit, pp. 1–10. Beijing Jiaotong University, Beijing (2011)
3. Zheng, F.-L.: Analysis and discussion on reason and protection to failure of compensation capacitor. Railw. Comput. Appl. **18**(11), 52–53 (2009)
4. Debiolles, A., Oukhellou, L., Denoeux, T., et al: Output coding of spatially dependent subclassifiers in evidential framework, application to the diagnosis of railway/vehicle transmission system. In: 9th International Conference on Information Fusion, pp. 1–6. IEEE Press, New York (2006)
5. Debiolles, A., Oukhellou, L., Aknin, P.: Combined use of partial least squares regression and neural network for diagnosis tasks. In: Proceedings of the 17th International Conference on Pattern Recognition, vol. 4, pp. 573–576. IEEE Press, New York (2004)
6. Oukhellou, L., Debiolles, A., Denoeux, T., et al.: Fault diagnosis in railway track circuits using Dempster-Shafer classifier fusion. Eng. Appl. Artif. Intell. **23**(1), 117–128 (2010)
7. Zhao, L.H., Xu, J.J., Liu, W.N., et al.: Fault detection on compensation capacitor of jointless track circuit based on Levenberg Marquardt algorithm and generalized S-transform. Control Theory Appl. **27**(12), 1612–1622 (2010)
8. Zhao, L.H., Cai, B.G., Qiu, K.M.: Fault diagnosis on compensation capacitor of jointless track circuit based on the HHT and DBWT. J. China Railw. Soc. **33**(3), 49–54 (2011)
9. Zhao, L.H., Ran, Y.K., Mu, J.C.: Fault diagnosis method for compensation capacitor of jointless track circuit based on genetic algorithm. China Railw. Sci. **31** (2010)
10. Hayat, S., Kausar, Z.: Mobile robot path planning for circular shaped obstacles using simulated annealing. In: 2015 International Conference on Control, Automation and Robotics (TCCAR), pp. 69–73. IEEE, Singapore (2015)
11. Lyden, S., Haque, M.E.: A simulated annealing global maximum power point tracking approach for PV modules under partial shading conditions. IEEE Trans. Power Electron. **31**(6), 4171–4181 (2016)
12. Acharya, S., Saha, S., Thadisina, Y.: Multiobjective simulated annealing-based clustering of tissue samples for cancer diagnosis. IEEE Trans. Power Electron. **20**(2), 691–698 (2016)
13. Zhao, L.H., Bi, Y.S., Liu, W.N.: Fault diagnosis system for compensation capacitor of jointless track circuit based on hierarchical immune mechanism. J. China Railw. Soc. **35**(10), 73–81 (2013)
14. Xilin, L.I.: Research on the remote monitoring system of the compensation capacitance of the jointless track circuit, pp. 20–40. Beijing Jiaotong University, Beijing (2015)

15. Liang, N.: The research and application of the central air conditioning control strategy based on Trnsys, pp. 46–56. Institutes of Technology of South China, Guangzhou (2013)
16. Mirhosseini, S.H., Yarmohamadi, H., Kabudian, J.: MiGSA: a new simulated annealing algorithm with mixture distribution as generating function. In: 2014 4th International eConference on Computer and Knowledge Engineering (ICCKE), pp. 455-461. IEEE, Mashhad (2014)
17. Li, J.-Z., Xia, J.-W., Zeng, X.-H., et al.: Survey of multi-objective simulated annealing algorithm and its applications. Comput. Eng. Sci. **35**(8), 77–88 (2013)
18. Liu, Y., Ma, J.Q., He, T., et al.: Hybrid simulated annealing-hill climbing algorithm for fast aberration correction without wavefront sensor. Opt. Precis. Eng. **20**(2), 213–219 (2012)

Author Index

© Springer Nature Singapore Pte Ltd. 2019
S. K. Bhatia et al. (eds.), *Advances in Computer Communication and Computational Sciences*, Advances in Intelligent Systems and Computing 759,
https://doi.org/10.1007/978-981-13-0341-8

427

Printed in the United States
By Bookmasters